Absorption
Chillers
and
Heat Pumps

Keith E. Herold
Reinhard Radermacher
Sanford A. Klein

CRC PRESS

Boca Raton   London   New York   Washington, D.C.

Published in 1996 by
CRC Press
Taylor & Francis Group
6000 Broken Sound Parkway NW, Suite 300
Boca Raton, FL 33487-2742

No claim to original U.S. Government works
Printed in the United States of America on acid-free paper
15 14 13 12 11 10 9 8 7 6

International Standard Book Number-10: 0-8493-9427-9 (Hardcover)
International Standard Book Number-13: 978-0-8493-9427-5 (Hardcover)
Library of Congress catalog number: 95-45325

**The software mentioned in this book is now available for download on the CRC Web site at:
http://www.crcpress.com/e_products/downloads/default.asp**

1006332945

Library of Congress Cataloging-in-Publication Data

Herold, K.E.
    Absorption chillers and heat pumps / Keith E. Herold, Reinhard
    Radermacher, Sanford A. Klein.
        p. cm.
    Includes bibliographical references and index.
    ISBN 0-8393-9427-9 (alk. paper)
    1. Heat exchangers. 2. Heat—Radiation and absorption. 3. Heat pumps.
    I. Radermacher, Reinhard. II. Klein Sanford A., 1950- .
TJ263.H47   1996
621.402'5—dc20   95-45325

Taylor & Francis Group
is the Academic Division of Informa plc.

**The software mentioned in this book is now available for download on the CRC Web site at:
http://www.crcpress.com/e_products/downloads/default.asp**

Visit the Taylor & Francis Web site at
http://www.taylorandfrancis.com

and the CRC Press Web site at
http://www.crcpress.com

Dedication

This book is dedicated to Professor Dr. Georg Alefeld (March 2, 1933 to August 25, 1995). Dr. Alefeld started his career with the investigation of the relaxation effects in lattices and proceeded to working with hydrogen in metals. During the last 15 years of his career he focused on applied thermodynamics, in particular heat conversion systems, absorption heat pumps and heat transformers. He established himself as a leader in this field with a comprehensive research effort including extensive experiments and the development of a basic theory and underlying concept for the design rules of multi-stage absorption systems and cascades renewing the interest in the exploration of advanced cycles.

PREFACE

The advances made in absorption technology during our careers have always motivated and encouraged us to work in this field. A frustrating part of that endeavor has been the lack of a systematic and detailed treatment of the fundamentals of the technology in the English literature. A recent book titled <u>Heat Conversion Systems</u> (Alefeld and Radermacher, 1994) focusses on the advanced aspects. This lack may be due, in part, to the interdisciplinary nature of the technology as evident by the fact that work in the field resides in the Department of Physics at the Technical University of Munich, in the Department of Chemical Engineering at the Royal Institute of Technology in Stockholm and in Departments of Mechanical Engineering in many other locations, including our own. A full understanding of absorption technology involves contributions from all these fields and others. The physics of the technology is rich including corrosion chemistry, physical chemistry, coupled heat and mass transfer, fluid dynamics, materials science, thermodynamics and more. It is safe to say that there are a number of poorly understood aspects of the technology that provide a fertile workspace for innovative minds.

Activity in the field of absorption technology has fluctuated considerably over the years. In particular, the industry in the U.S. is currently expanding activity in the manufacturing, research and development arenas. An indication of the rising interest can be seen by monitoring publications in the field. One such measure is the number of patents issued in the U.S. as indicated in the chart below. It is interesting to note that the number of patents issued stayed relatively constant for several years in the 1980s and then fell off dramatically around 1989. The resurgence in the industry in recent years is reflected very clearly in the patents issued since 1991. The resurgence of absorption technology is a particularly exciting time with many new developments and new engineers and companies entering the field. It is hoped that this book will promote the technology and support the resurgence by making absorption technology more accessible.

The objectives of this text include an introduction to the field and the presentation of sufficient detail about the technology to form a basis for reading the extensive literature found in technical journals. The treatment is designed for engineers or other science-educated

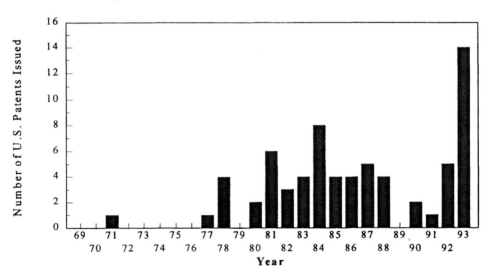

Number of U.S. patents issued per year on the subject of absorption chillers and heat pumps

individuals at the senior undergraduate or first year graduate level. An undergraduate understanding of thermodynamics, heat transfer, and fluid flow is assumed as prerequisite. From this base, the fundamentals peculiar and unique to absorption technology are developed. Particular emphasis is placed on mixture thermodynamics and its practical application to absorption system analysis. Both analytical and graphical methods are employed. Emphasis is placed on using computers to enhance the analytical capability as well as on developing a sound intuitive sense for the technology. Numerous examples are provided throughout the text to guide the reader through the concepts. End of chapter problems are included to provide the instructor with exercises for further reinforcing the concepts.

Software is supplied to supplement the text. A special version of the Engineering Equation Solver, EES, is supplied with the book and numerous examples from the text are provided in EES source code form. EES is a commercial package with powerful equation-solving capabilities. Another major feature of EES is the built-in thermodynamic property routines for many fluids including water/lithium bromide and ammonia/water. This unique combination of equation solving and built-in property routines makes EES an excellent tool for studying and understanding absorption systems. To complement the basic EES capabilities, the disk supplied includes over fifty EES programs tied to the treatment in the text. These include models for single-effect, double-effect and triple-effect cycles; Type II cycles; and GAX cycles. Appendix C provides information on cycle modeling that describes the process involved in setting up a model in EES for a new cycle. Thus, the reader has all the tools needed to perform a wide range of absorption cycle analyses.

ABSIM (an acronym for ABsorption SIMulation) is another powerful code which could be used to solve many of the examples given in this book. ABSIM is a modular computer code for simulation of absorption systems. It was developed with DOE/ORNL sponsorship by Dr. Gershon Grossman. This modular code is based on unit subroutines containing the governing equations for the system's components and on property subroutines containing thermodynamic properties of the working fluids. Eleven absorption fluids are presently available in the code's property database, and twelve units are available to compose practically every absorption cycle of interest. ABSIM may be used for evaluating new cycles and working fluids and to investigate a system's behavior in off-design conditions, to analyze experimental data and to perform preliminary design optimization. A graphical user interface enables the user to draw the cycle diagram on the computer screen, enter the input data interactively, run the program and view the results either in the form of a table or superimposed on the cycle diagram. Special utilities enable the user to plot the results and to produce a PTX diagram of the cycle. A copy of the latest developmental version of ABSIM along with user's manual and several examples can be obtained from the address given in Appendix F.

The book is organized into four major sections. Chapters 1 and 2 provide an introduction to the technology and a discussion of absorption cycle fundamentals. Chapters 3 and 4 cover properties of working fluids and a description of the thermodynamic processes involving mixtures. Chapters 5 to 8 discuss water/lithium bromide technology. Chapters 9 to 12 discuss ammonia/water technology. Appendices are provided covering cycle modeling, concentration measurements, and properties of fluids.

A project of this magnitude involves the contributions of many individuals and organizations. We cannot mention all of the contributions due to space constraints. One group that deserves mention is our students who have enthusiastically provided continuous stimulation through their questions and comments. Extensive assistance was provided by Sanjoy Sanyal, Yun-Ho Hwang, Sunil Mehendale, and Mohit Pande. A significant factor in making this project viable was the support of the U.S. Department of Energy, Oak Ridge National Laboratory and the Oak Ridge Institute for Science and Education through sabbatical support for K. E. Herold during 1994-1995. The University of Maryland also provided sabbatical support during this time. The manuscript was reviewed and improved by Donald K. Miller and by Joseph Murray. The staff at CRC Press patiently guided us through the publication process.

Finally we would like to thank our families for their support of this project. Yanli, Paula, and Jan have endured the effort and contributed through their patient support.

Keith E. Herold
Reinhard Radermacher
Sanford A. Klein
December 1995

CONTENTS

List of Figures

List of Tables

List of Examples

List of EES Files

File names marked with asterisk can be dowloaded from the CRC Press World Wide Web server by accessing the CRC Press home page at http://www.crc.com. Follow the prompts to the section for Absorption Chillers and Heat Pumps. All other files are included on the floppy disk supplied with the book.

Nomenclature

ϵ	Effectiveness	U	Overall heat transfer coefficient,	
η	Efficiency		internal energy	
μ	Chemical potential	V	Volume, velocity	
v	Specific volume	W	Mechanical work rate	
ν	Partial molal volume	x	Mass fraction	
ρ	Density	x_q	Vapor quality	
σ	Surface tension	y	Thermal conductivity	
ψ	Helmholtz free energy, $\psi = u + pv$			

Subscripts

A	Area	A	Component A
AHP	Absorption heat pump	B	Component B
AR	Absorption refrigeration system	c	Condenser
c	Molar concentration	e	Evaporator
CFC	Chlorofluorocarbon refrigerant	d	Desorber
COP	Coefficient of performance	a	Absorber
c_p	Specific heat at constant pressure	h	Hot
C_{sf}	Surface fluid coefficient	h	Hot internal
c_v	Specific heat at constant volume	c	Intermediate internal
D	Diffusion coefficient	e	Low internal
EES	Engineering equation solver	R	Refrigeration
F	Solution circulation ratio	p	At constant pressure
g	Gibbs free energy; $g = h + pv$,	T	At constant temperature
	acceleration of gravity	v	Vapor
h	Enthalpy, heat transfer coefficient	l	Liquid
H	Extensive enthalpy	H	Water
HT	Heat transformer	N	Ammonia
k	Thermal conductivity	in	Inlet
l	Liquid	s	Surface
L	Length	st	Saturation
m	Mass flow rate	fg	Liquid to vapor
M	Molecular weight	lm	Log mean
N	Number of moles	w	Wall
Nu	Nusselt number	va	Vapor phase
p	Partial pressure	la	Liquid phase
p_i	Partial pressure	rec	Rectifier, reflux cooler
P	Pressure	hx	Heat exchanger
q	Heat flux, heat transfer per unit mass	shx	Solution heat exchanger
Q	Heat transfer rate, vapor quality	rev	Reversible
r	Latent heat	eva	Evaporator
R	Universal gas constant, thermal resistance	des	Desorber
S	Entropy	con	Condenser
SHX	Solution heat exchanger	abs	Absorber
T	Temperature	sc	Subcooler

Chapter 1

INTRODUCTION

1.1   Heat Pumps

The term "heat pump" refers to a group of technologies that transfer heat from a low temperature to a high temperature. Such a transfer requires a thermodynamic input in the form of either work or heat. This is made clear in the Clausius statement of the second law of thermodynamics which can be stated as:

> *It is impossible for any system to operate in such a way that the sole result would be an energy transfer by heat from a cooler to a hotter body.*

The thermodynamic implications of the second law for heat pump technology are the factors that complicate the application and understanding of this technology. Even when the second law is not explicitly applied in a heat pump analysis, its requirements are in force implicitly through the properties of the working fluids.

The type of energy input, whether heat or work, changes the details of the technology needed to provide the heat pumping function. Absorption technology, which is the focus of this book, is an example of a heat-driven technology. A schematic diagram of a simple absorption cycle is shown as Figure 1.1. The details of the schematic will be discussed fully starting in Chapter 2. The main point here is to focus on the energy transfers between the cycle and its surroundings. It is possible to transfer heat from a low to a high temperature while supplying only heat as the driving energy. By eliminating the need for a work input, an absorption cycle

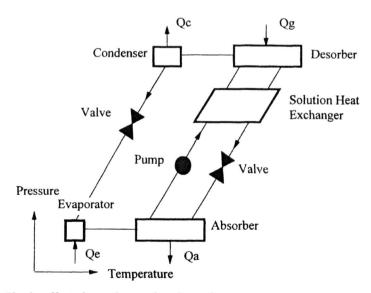

Figure 1.1 Single-effect absorption cycle schematic

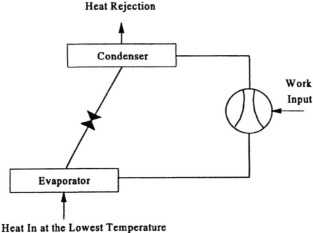

Figure 1.2 Vapor compression heat pump

provides a unique solution for a range of technological problems from solar cooling to steam-driven refrigeration. Many other heat-driven heat pump technologies have been conceived but absorption technology is by far the most widely applied. Other heat-driven technologies that have been demonstrated include adsorption (solid/vapor), Stirling cycle, ejector and magnetic refrigeration. Many other ideas have appeared in the literature.

Heat pumps that require a work input provide solutions to a complementary range of technological problems. The vapor compression heat pump is the most widely used variant of this technology. The vapor compression heat pump is widely used for residential and commercial heating and cooling, food refrigeration and automobile air-conditioning, among other uses. This technology is illustrated in Figure 1.2. A mechanical compressor, typically driven by an electric motor, provides the work input that drives the heat transfer from the low temperature to the high temperature.

The emphasis in this book is on absorption heat pumps that are heat-driven. It should be kept in mind that the basic function, that is, transferring heat from a low temperature to a high temperature, can be implemented in a wide range of technologies of which absorption machines are just one family.

1.2   Heat-Driven Heat Pumps

The simplest heat-driven heat pump is a device that transfers heat at three temperature levels. There are two variations of this arrangement that represent two fundamentally different heat pumping types. The first, which we will call Type I, is illustrated in Figure 1.3. In this type of heat pump, the driving heat is input at the highest temperature level and the product is either 1) refrigeration at the lowest temperature or 2) heating at the intermediate temperature. The second type, Type II, is illustrated in Figure 1.4. In a Type II heat pump, the driving heat is input at the intermediate temperature level, and the product is the heat provided at the highest temperature level. Type II heat pumps, which are also known as heat transformers, temperature boosters or temperature amplifiers, are useful for upgrading the temperature of a waste heat stream to a useful level. A consequence of the second law is that this upgrade in temperature

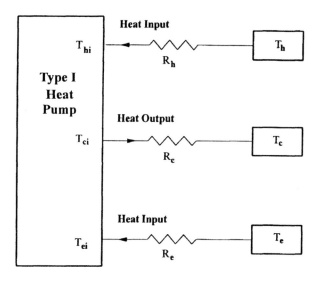

Figure 1.3 Type I heat pump

requires a portion of the input heat to be rejected at a lower temperature.

The blocks with temperatures indicated on Figures 1.3 and 1.4 as T_h, T_c and T_e represent the thermal boundary conditions with which the absorption machine must interact. The highest temperature is at the top and the lowest is at the bottom of both diagrams. Internal to the absorption machine, another set of temperatures is designated with the subscript I. At each temperature level, the heat transfer interaction between the absorption machine and the surroundings must occur through a heat exchanger represented by the sawtooth thermal resistance. These diagrams emphasize the key role of temperature in understanding absorption technology.

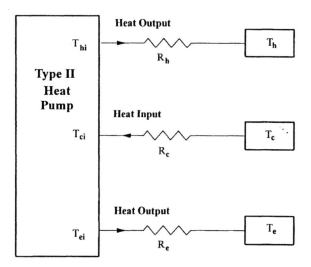

Figure 1.4 Type II heat pump

Type I absorption heat pumps are commonly used in commercial buildings as water chillers for air-conditioning applications. They can be found in thousands of mechanical rooms across the world. Type II heat pumps have been demonstrated in laboratories, including facilities run by the authors, and have been demonstrated in a few industrial installations. The economics of Type II heat pump utilization with current technology and at current energy prices is not favorable. However, a full understanding of heat-driven heat pumping requires an understanding of the Type II heat pump concept.

1.3   Description of Current Absorption Products

Three distinct absorption technologies are currently marketed in the U.S. These are 1) water/lithium bromide chillers, 2) ammonia/water chillers and 3) ammonia/water/hydrogen refrigerators. Within each of the three categories there is some differentiation among products but it is useful to divide the technology in this way to facilitate the discussion.

1.3.1   Water/Lithium Bromide Chillers

Water/lithium bromide technology is discussed in detail in Chapters 5 to 8. Water/lithium bromide is an absorption working fluid which has been used widely since the 1950s when the technology was pioneered by several manufacturers in the U.S. This working fluid utilizes water as the refrigerant and is therefore limited to refrigeration temperatures above 0°C. Absorption machines based on water/lithium bromide are typically configured as water chillers for air-conditioning systems in large buildings. Machines are available in sizes ranging from 10 to 1500 Ton (Note: Ton is a unit of refrigeration capacity, 1 Ton = 12,000 BTU/hr = 3.517 kW). The coefficient of performance (COP) of these machines, defined as the refrigeration capacity divided by the driving heat input, typically varies over the range $0.7 < COP < 1.2$ depending on the particular cycle configuration. These machines have a reputation for consistent, dependable service among mechanical room operators. The main competing technology is vapor compression chillers and the choice between the two depends strongly on economic factors.

1.3.2   Ammonia/Water Chillers

Ammonia/water technology is discussed in detail in Chapters 9 to 11. Ammonia/water is an absorption fluid that has been used since the late 1800s at which time it was used for ice production prior to the introduction of vapor compression technology. This working fluid utilizes ammonia as the refrigerant. Thus, the role of water is distinctly different between ammonia/water and the water/lithium bromide discussed in Section 1.3.1. One advantage of ammonia as refrigerant is that the allowable refrigeration temperature is much lower (the freezing temperature of ammonia is -77.7°C). However, the toxicity of ammonia is a factor that has limited its use to well-ventilated areas. Ammonia/water absorption chillers are commonly sold as air-conditioning components and this use is regulated in some densely populated urban areas. The ability to provide direct gas-fired and air-cooled air-conditioning is the primary selling point of this technology. Machines are available in capacity ranging from 3 to 25 Ton (10 to 90 kW) with coefficient of performance typically around 0.5. These units have a niche market since there are few competing gas-fired technologies suitable for many applications.

1.3.3   Ammonia/Water/Hydrogen Refrigerators

Ammonia/water/hydrogen refrigerators are discussed in detail in Chapter 12. This technology is a direct descendant of that patented by the Swedes von Platen and Munters in

1921. Domestic refrigerators based on ammonia/water/hydrogen have been continuously available since that time. Machines based on this cycle are manufactured by numerous companies throughout the world. In the U.S. the refrigerators are used primarily for the recreational vehicle market where the lack of need for any electric input is the selling point. Internationally, the technology has a significant niche market as hotel room refrigerators. The selling point here is the silent operation.

1.4   Outlook for Absorption Technology

Absorption technology is currently experiencing a resurgence of interest from end users in the U.S. who need a dependable heat-driven heat pumping technology. This resurgence follows a period of approximately 20 years of decline of the absorption industry in the U.S.

This decline was driven by political decisions in the early 1970s to limit usage of natural gas due to expected supply shortfalls. Although the predicted supply shortfalls never materialized, the damage was done to the industry. The three manufacturers of absorption equipment saw their business dry up and production of absorption machines was reduced to less than 10% of earlier production levels. Companies were forced to reassign staff and reallocate resources away from the absorption business center. In the meantime, the Japanese absorption industry boomed due to Japanese domestic demand and product innovation.

The current resurgence is market-driven. Significant market forces which have stimulated absorption demand include peak electric power rate premiums, gas utility rebate programs and the CFC replacement issue. When demand for absorption machines began to rise in the early 1990s, the original manufacturers found themselves in the position of having to license the new technology from Japanese manufacturers. This is the industry status as of the time of this writing. The U.S. absorption manufacturers are selling machines based on Japanese developments made during the U.S. industry decline. However, the U.S. manufacturers are now conducting aggressive product research to regain their lead.

In addition to the product developments pioneered by industry, research on advanced cycles and fluids has been performed at a number of laboratories world-wide. These efforts have resulted in a series of intriguing technology options that are currently being investigated by industry to determine to what extent they are practical. The resurgence in interest in the subject is a dramatic change for those investigators who have labored in this field in the last 20 years and represents an exciting new era in absorption heat pump development.

Chapter 2

ABSORPTION CYCLE FUNDAMENTALS

This chapter explains the concept of absorption systems in terms of idealized energy conversion cycles. It provides a description of the operation of absorption systems within the larger context of what is thermodynamically possible. Conventionally, in this chapter, all temperatures are absolute values.

2.1   Carnot Cycles

The Carnot cycle can be used as an example for an idealized energy conversion cycle. Figure 2.1 shows a Carnot cycle for power generation on a temperature-entropy diagram (T,s-diagram). The process line AB represents the isothermal addition of heat Q_2 at the temperature T_2 to a working fluid. We use here the convention that an arrow pointing to a process line represents energy supplied to the cycle. BC represents the isentropic production of work, CD the isothermal rejection of heat Q_1 at the temperature T_1 and DA the isentropic input of work. When it is assumed that all processes are reversible, then the area enclosed by ABCD represents the net amount of work produced, W, and the area CDEF the amount of thermal energy Q_1 rejected by the cycle assuming E and F are at T = 0 K. For a work-producing process as shown in Figure 2.1 the processes follow a clockwise direction in the T,s-diagram. The sum of the two areas is the amount of heat Q_2 (thus area ABFE) supplied to the cycle as required by the first law.

$$Q_2 \; = \; Q_1 \; + \; W \hspace{4cm} 2.1$$

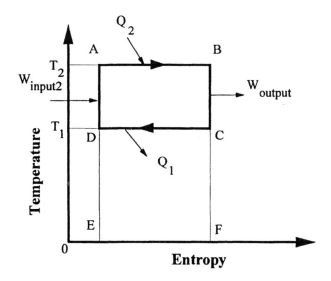

Figure 2.1 The Carnot cycle for power generation on a temperature-entropy diagram

The signs of the quantities in Equation 2.1 are to be selected as indicated in Figure 2.1. All energies are counted as positive in the direction of the arrows.

The efficiency η for power generation is defined as the amount of work produced divided by the amount of heat supplied at the high temperature.

$$\eta = \frac{W}{Q_2} \qquad\qquad 2.2$$

The Second Law of Thermodynamics states for the Carnot cycle of Figure 2.1 that for reversible operation the net entropy production is zero, so that

$$\frac{Q_2}{T_2} - \frac{Q_1}{T_1} = 0 \qquad\qquad 2.3$$

Equation 2.2 can be modified to an expression that contains only temperatures using Equations 2.1 and 2.3 by eliminating W.

$$\eta = \frac{T_2 - T_1}{T_2} \qquad\qquad 2.4$$

The expression in Equation 2.4 is frequently termed the Carnot efficiency factor for power generation.

Figure 2.2 shows a Carnot cycle that is operated as a heat pump cycle. The direction of all processes is reversed as compared to the power generation cycle. Here, the temperatures are chosen such that heat Q_0 is added to the working fluid at T_0 along the process line GH; the fluid is compressed isentropically HI and heat Q_1 is rejected at T_1, IJ; and the fluid expanded isentropically, JG. The areas represent again energy transfers. The net amount of work input required for this cycle is represented by the area GHIJ and the amount of heat absorbed by the area GHKL. The sum of both areas (area IJLK) represents the amount of heat rejected at T_1.

The performance of a heat pump is described by the ratio of the benefit obtained, the amount of heat available at the high temperature, divided by the expenditure, the net work requirement. Since this value is always greater than 1.0 the term "coefficient of performance"

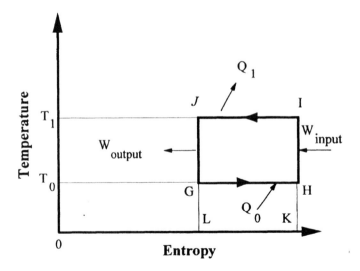

Figure 2.2 The Carnot cycle for heat pumping on a temperature-entropy diagram

(COP) is used customarily.

$$COP = \frac{T_1}{T_1 - T_0}$$

2.5

This equation is derived by using the First and Second Laws for the heat pump process of Figure 2.2 and by following the same procedure as for Equation 2.4. The Carnot cycle of Figure 2.2 can also be used for cooling or refrigeration applications. While the cycle itself does not change, the way we view the application is different. For refrigeration, the heat removed at T_0 is of interest, and the COP is defined as the ratio of the cooling capacity Q_0 over the work input. Subscript R is used to designate that the COP is defined for refrigeration.

$$COP_R = \frac{T_0}{T_1 - T_0}$$

2.6

Equation 2.6 is derived by applying the First and Second Laws as described above for the power generation cycle. For the coefficients of performance of a heat pump cycle and a refrigeration cycle the following relationship can be derived.

$$COP_R + 1 = COP$$

2.7

This expression can be verified by substituting Equations 2.5 and 2.6 into Equation 2.7.

2.2   Absorption Heat Pump, Type I

In order to describe an absorption heat pump with the help of Carnot cycles, now both cycles (the work producing and refrigeration cycle of Figures 2.1 and 2.2) are combined into one device. It is assumed that the amount of work produced by the first cycle, Figure 2.1, is identical to the amount of work required by the second cycle, Figure 2.2. The new device, Figure 2.3, raises the temperature level of heat supplied at T_0 to T_1 by using the thermodynamic availability of the high temperature energy supplied at T_2. The waste heat of the power generation portion of this combined cycle is also rejected at T_1. Thus the total amount of heat rejected at T_1 has two contributions, Q_1' and Q_1''.

This combined cycle represents a heat-pumping device that is driven by the input of heat only. It is the ideal representation of several different heat-pumping concepts. Examples are an engine-driven vapor compression heat pump, a combination of a steam power plant with an electrical vapor compression heat pump or the representation of an absorption heat pump. The coefficient of performance for this latter device is customarily defined as

$$COP_{AHP} = \frac{Q_1' + Q_1''}{Q_2}$$

2.8

for heating applications and as

$$COP_{AR} = \frac{Q_0}{Q_2}$$

2.9

for cooling and refrigeration applications. By applying the First and Second Laws and

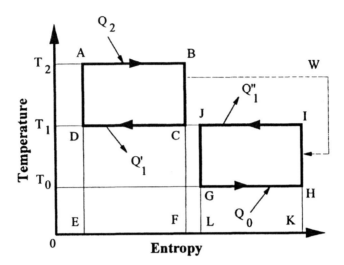

Figure 2.3 Carnot cycles for a combined power-generation/heat-pumping facility such as an absorption heat pump

eliminating either Q_0 or Q_1 (with $Q_1 = Q_1' + Q_1''$), Equations 2.8 and 2.9 can be converted into expressions that depend on temperature only.

$$COP_{AHP} = \frac{T_2 - T_0}{T_2} \frac{T_1}{T_1 - T_0} \qquad\qquad 2.10$$

$$COP_{AR} = \frac{T_2 - T_1}{T_2} \frac{T_0}{T_1 - T_0} \qquad\qquad 2.11$$

As with vapor compression heat pumps and refrigerators, the distinction between a heat pump and a chiller is only a function of the application, not of the operating mode. The analogue to Equation 2.7 is also valid for absorption systems:

$$COP_{AR} + 1 = COP_{AHP} \qquad\qquad 2.12$$

The system described so far, as it pertains to the implementation of absorption heat pumps, is referred to in the literature at times as "Type I Absorption Heat Pump".

2.3   Absorption Heat Transformer, Type II

The combined cycle of Figure 2.3 can be operated in a reversed mode. For this purpose, the directions of all fluid streams and all energy streams are reversed and the combined cycle of Figure 2.4 is obtained. This cycle is referred to as "heat transformer" or as "absorption system, Type II". In this case, the cycle operating between T_1 and T_2 is a heat pump cycle, while the cycle operating between T_1 and T_0 is a power generation cycle (note the circulation directions). Now heat is supplied at T_1 to both cycles. The purpose of this cycle is to use heat at an intermediate temperature level, for example, 80°C waste heat, and to convert it to high temperature heat, for example, 120°C. This conversion is accomplished by using a portion of

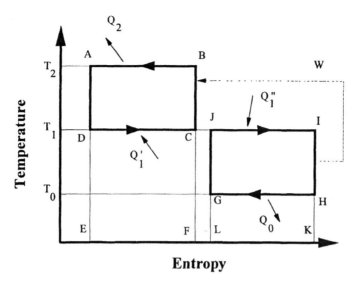

Figure 2.4 Carnot cycles for a combined power generation/heat-pumping facility such as an absorption heat transformer

the waste heat in a power generation cycle that rejects heat at T_0 while it provides the work for the heat pump cycle. The efficiency for a heat transformer cycle is defined as

$$COP_{HT} = \frac{Q_2}{Q_1' + Q_1''} \qquad 2.13$$

COP_{HT} is the amount of useful high temperature heat divided by the total amount of waste heat required. Q_1' and Q_1'' represent the two contributions of waste heat, one to be pumped to T_2 and the second to drive the power plant to do so. Using the respective equations for the First and Second Laws for the heat transformer and eliminating the heat Q_0, the following term for the efficiency is obtained.

$$COP_{HT} = \frac{T_1 - T_0}{T_1} \frac{T_2}{T_2 - T_0} \qquad 2.14$$

This heat transformer is at times referred to as "Type II absorption heat pump".

2.4   Absorption Heat Pump as Combination of Rankine Cycles

Figure 2.5a shows the components of a Rankine heat pump cycle while Figure 2.5b shows a basic Rankine cycle that produces work. By studying the direction of the working fluid streams, it is apparent that the two cycles could potentially be combined into one. For example, when the following conditions are fulfilled, the compressor and turbine can be eliminated:

1) If the stream leaving the boiler of the power generation cycle had the same high pressure level and flow rate as the stream entering the condenser of the heat pump cycle, and

2) If the stream leaving the heat pump evaporator had the same low pressure level and flow rate as the stream entering the power cycle condenser,

then the compressor and turbine can be eliminated. The result is an absorption heat pump.

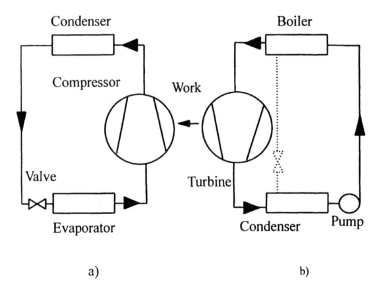

a)   b)

Figure 2.5 Illustration of the combination of two Rankine cycles (a, heat pump; b, power generation) to a combined system

Thus, an absorption heat pump cycle, is a heat-driven refrigeration cycle that can be thought of as a combination of a power cycle and a work-driven heat pump all in one machine.

In order to fulfill condition 1 and 2 and to have different temperature levels for the evaporation process in the boiler of the power cycle and the condenser of the heat pump cycle (they are both at the same pressure), the fluid in the boiler must evaporate at a temperature that is higher than that of the condenser of the heat pump cycle. This effect can be achieved when the vapor in the boiler is evaporating out of a mixture with a fluid of a much higher boiling point. This fluid is usually termed the absorbent or solvent. Only with such a mixture of absorbent and refrigerant in the power generation portion is the absorption heat pump process possible. This requirement explained in more detail in Chapter 3. The absorbent remaining in the boiler must be circulated in its own loop. For this purpose, a dashed line with a pressure-reducing valve is shown in Figure 2.5b. This derivation makes it plausible that the absorption cycle can be viewed as a combination of two Rankine cycles.

2.5   Reversible Analysis with Variable Temperatures

The reversible analysis discussed in Sections 2.1 to 2.3 is based on the assumption that all heat transfers occur at a fixed temperature. This assumption makes the analysis simple, and the resulting models provide insight into the thermodynamics of absorption cycles. However, the heat transfer processes actually occur over a range of temperatures in all the components. The working fluid experiences temperature changes as it passes through the components due to a number of factors including pressure changes, mass fraction changes, superheating, and subcooling. From the perspective of reversible analysis, it is desirable to determine an average temperature which can represent the variable temperature heat transfer processes. The appropriate temperature is the entropic average temperature.

Consider an arbitrary heat transfer process occurring over a range of temperatures. A simple

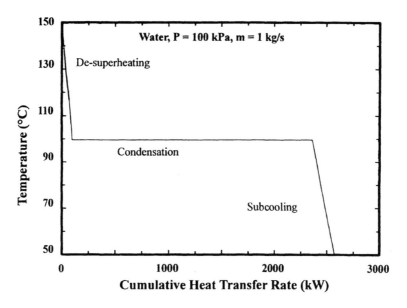

Figure 2.6 Temperature versus cumulative heat transfer rate for condensation of steam at 100 kPa from 150 to 50°C

example is the cooling of water flowing in a pipe. As energy is transferred out of the system, the temperature decreases. An energy balance on the system allows determination of the total energy transfer as a function of the flow rate and the end state temperatures. A reversible analysis, of the type done in Sections 2.1 to 2.3 requires a second law analysis which, in turn, requires knowledge of the temperature at which the heat transfer occurs. Since the temperature of the system is a continuously changing variable, it is necessary to view the process on a differential basis.

Consider the arbitrary heat transfer process shown on temperature vs. cumulative heat transfer rate coordinates in Figure 2.6. The cumulative heat transfer rate is the integral of the heat flux over the heat exchanger area with the zero set arbitrarily at one end of the device. The total entropy flowing with the heat transfer can be written as

$$S = \int \frac{dQ}{T}$$
2.15

where the heat transfer rate dQ is transferred at the temperature T. Note that the temperature corresponds to the temperature of the system boundary selected by the analyst. An average temperature, called the entropic average, can be defined such that the entropy transfer would be the same if the entire heat transfer occurred at the average temperature. The definition is

$$T_{sa} = \frac{Q}{S} = \frac{Q}{\int \frac{dQ}{T}}$$
2.16

The entropic average temperature, as defined in Equation 2.16, reflects the thermodynamic "quality" of a variable temperature heat transfer. Thus, the entropic average temperature is the appropriate temperature to use whenever a variable temperature process is to be analyzed thermodynamically as a constant temperature process. This formulation retains the simplicity of the constant temperature model but makes it applicable to varying temperature.

2.6   Irreversibilities in Absorption Cycle Processes

The reversible cycle analyses discussed in Sections 2.1 to 2.3 represent idealized performance limits for absorption cycles. In real cycles, performance falls below the idealized limits due to irreversible processes such as those listed in Table 2.1. Thermodynamic irreversibilities occur whenever transport occurs. This includes transport of momentum, mass or heat. These transport processes are always accompanied by a finite difference in the driving potential (such as the temperature difference for heat transfer). The irreversibility of a given

Table 2.1 Irreversible effects in absorption technology

Phenomena	Example
Viscous friction	Vapor flow, liquid flow
Thermal mixing	Liquid inlet to desorber
Mass mixing	Liquid inlet to desorber
Heat transfer	Every component
Unrestrained expansion	Expansion valve

process is related to both the quantity of transport and to the potential difference over which it is transported. An introduction to the analysis of irreversibilities in thermal systems can be found in Bejan (1982).

Irreversibility is defined as the entropy generation multiplied by a reference temperature. Although this is helpful, in the sense that it allows a specific definition for this concept, it also introduces some confusion since the choice is arbitrary and the interpretation depends on the temperature used. An alternative that avoids the need for an arbitrary reference temperature is to simply concentrate on the entropy generation rate and to avoid introducing the concept of irreversibility. This method may be preferable for heat pump analysis because it avoids the confusion associated with the arbitrary reference temperature. This method has been applied to absorption heat pump analysis by Alefeld (1987). By analyzing the state at the inlet and outlet of each component, and, with several assumptions about the processes occurring in the components, it is possible to calculate the entropy generation in each component. By comparing the entropy generation in the various components, the analyst can develop an understanding of which components are causing performance degradation and where to invest design efforts to improve performance.

The details of such analyses yield some very informative general results. For a well-designed absorption machine, the major irreversibilities are those associated with the heat transfer processes. Those processes include the solution heat exchanger, which is an internal heat transfer, and the external heat transfers. In the case of the external heat transfer processes, the process on the solution side of the machine is a coupled heat and mass transfer process. The irreversibilities associated with the coupled processes are lumped together in this discussion

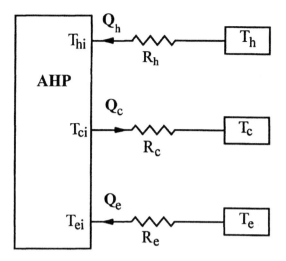

Figure 2.7 Zero-order model schematic

and assigned to the heat transfer. In a typical machine, the other transport processes occurring within the machine are also irreversible, but fluxes and potential differences are low enough so that they are insignificant compared to the heat transfer processes. This result is important because it points toward potential performance improvements. It also points toward the type of models needed to predict the performance of absorption machines. As long as a model predicts the major irreversibilities, it should also predict performance.

2.7   Zero-Order Absorption Cycle Model

The ability to predict the performance of a particular absorption cycle, utilizing a particular absorption working pair, is a fundamental skill for the engineer who needs to determine the applicability of an absorption cycle to a particular situation. The experienced heat pump engineer has rules of thumb and intuition on which to base a decision. A newcomer to the field must have modeling tools to enable the careful evaluation of a given design. In this book, a series of tools are presented which allow absorption cycle performance to be predicted with varying levels of accuracy.

The factors that determine the performance of a thermodynamic cycle in a given application are the irreversibilities, which are losses of thermodynamic availability or exergy. The major irreversibilities in a typical absorption cycle application are associated with heat transfer between the cycle and the surroundings. The simplest single-effect absorption cycle requires heat transfer interactions with the surroundings (two of which occur typically at the same temperature level). Each of these heat transfers have an associated irreversibility or loss. The summation of losses, plus the losses internal to the cycle, determines the approach of the cycle performance to the reversible limit introduced in Chapter 1. The zero-order model is based on the concept of modeling only those processes that contribute the largest irreversibilities, ignoring all other losses. Thus, the internal losses are ignored in the zero-order model. The method results in a model that is easy to solve and to understand and produces excellent predictions of performance trends but one that provides only a rough approximation to absolute performance prediction. A detailed description of the zero-order model can be found in Herold and Radermacher 1990.

The zero-order model is based on the schematic in Figure 2.7. The block marked AHP represents the internal workings of the absorption cycle. The internal workings are modeled as thermodynamically reversible. The zero-order model emphasizes the heat exchanger losses between the AR and the surroundings represented in the figure by thermal resistances denoted R. There are three thermal resistances shown, corresponding to the three temperature levels. When interpreting the model, one must realize that the thermal conductance at the intermediate temperature level represents the sum of the condenser and the absorber heat transfer conductances. The zero-order model consists of the following equations. There is one heat transfer equation for each of the three thermal resistances shown in Figure 2.7.

$$Q_h = \frac{T_h - T_{hi}}{R_h} \qquad\qquad 2.17$$

$$Q_c = \frac{T_{ci} - T_c}{R_c} \qquad\qquad 2.18$$

$$Q_e = \frac{T_e - T_{ei}}{R_e} \qquad\qquad 2.19$$

An overall energy balance on the system requires that the sum of the three heat transfers must be zero. Note that the model assumes steady state and that the pump work and environmental heat losses are neglected.

$$Q_h + Q_e = Q_c \qquad\qquad 2.20$$

The performance of the AR is assumed to be reversible and can be represented as (Equation 2.11)

$$COP = \frac{Q_e}{Q_h} = \frac{T_{ei}}{T_{hi}}\left(\frac{T_{hi} - T_{ci}}{T_{ci} - T_{ei}}\right) \qquad\qquad 2.21$$

The system represented by Equations 2.17 to 2.21 includes five equations in the six unknowns including the three unknown Q's and the three unknown internal temperatures. An additional equation is needed to close the system. One such equation that can be deduced from the working fluid characteristics is

$$T_{hi} - T_{ci} = T_{ci} - T_{ei} \qquad\qquad 2.22$$

Assuming that the condenser and the absorber operate at the same temperature, Equation 2.22 says that the temperature difference between the evaporator and the condenser is the same as between the absorber and desorber. This characteristic is a reasonable approximation for a single-effect machine. For other cycles, Equation 2.22 must be replaced accordingly.

The six equations in six unknowns represented by Equations 2.16 to 2.22 can be solved iteratively. The result is a very simple absorption cycle model that predicts performance trends remarkably well considering the effort required to set up and execute the model. The model does an excellent job of clarifying and defining the physics underlying the shape of the COP curve for an absorption machine. A typical single-effect performance curve is plotted in Figure

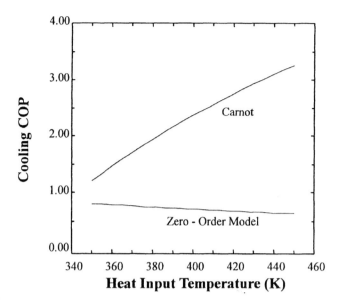

Figure 2.8 Cooling COP for a single-effect absorption machine

2.8 which shows that the COP is relatively insensitive to changes in the heat input temperature. This result seems odd when compared to the Carnot performance, also plotted in Figure 2.8, which was obtained by plotting Equation 2.11. Based on the Carnot model, cycle performance should increase significantly as the heat input temperature increases. Experience shows that the COP curve is flat. This characteristic results from the fact that the irreversibilities in the external heat transfer processes increase significantly as the heat input temperature increases. The irreversibilities increase because both the fluxes and the potential differences increase. As in a real absorption machine, any change in the temperature inputs causes changes in the operating conditions throughout the cycle. These changes are accurately reflected in the result of the zero-order model and thus the shape of the COP curve is predicted accurately

Another quite useful aspect of the zero-order model is in determining the minimum firing temperature for a particular cycle design. The temperatures associated with the zero-order model are illustrated in Figure 2.9, which shows the temperatures internal to the absorption machine as a connected set which are linked by the relationship in Equation 2.22. The external temperatures are also shown. It is useful to define the temperature lift as the difference between the heat rejection temperature, T_c, and the refrigeration temperature, T_e, as

$$\Delta T_{lift} = T_c - T_e \qquad\qquad 2.23$$

It is noted that, in general, the temperature differences between the machine and its surroundings, which represent the heat transfer potential difference, will not be equal at the three levels. These differences will depend on the size of the heat exchangers at each of the three temperature levels. Although these differences are not necessarily equal, it is sometimes useful at the design stage to think in terms of a design temperature difference which would be equal at all three levels. This perspective is taken here, and the heat transfer temperature difference is referred to as ΔT_{ht} where the subscript stands for heat transfer. It is assumed also that the heat rejection temperature and the refrigeration temperature are known for a given design and that the minimum heat input temperature is to be determined. By inspection of Figure 2.9, the following relationship is derived

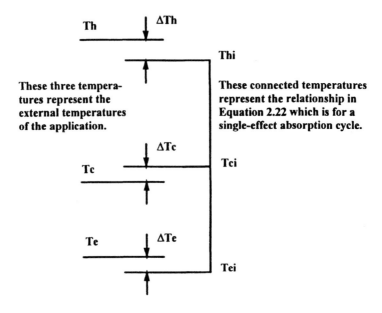

Figure 2.9 Temperatures associated with zero-order model

$$T_h = (\Delta T_{lift} + 2\,\Delta T_{ht}) + 2\,\Delta T_{ht} + T_c$$
$$= \Delta T_{lift} + 4\,\Delta T_{ht} + T_c$$

2.24

This relationship is used in the following example to determine the minimum heat input temperature.

Example 2.1   Determine Minimum Heat Input Temperature

Consider a single-effect absorption cycle. Determine the minimum heat input temperature assuming the rejection temperature is 35°C and the refrigeration temperature is 8°C.

Solution: Direct application of Equation 2.24 requires a specification of the heat transfer driving potential difference. Assuming values of 0, 5 and 10°C yields T_h values of 62, 82 and 102°C.

Observations: The magnitude of the heat transfer driving potential difference has a major impact on the required heat input temperature.

2.8   Absorption Cycle Design Optimization

A discussion of the subject of design optimization belongs in this chapter on fundamentals because all of the subsequent design concepts must ultimately be assembled into an overall design. Thus, it is fundamental to consider the meaning of optimum design as it applies to absorption technology. It is tempting to consider maximization of COP as a design philosophy since this would result in the most efficient energy utilization and the least environmental damage due to machine operation. However, this is not a particularly practical approach because it ignores first cost considerations. Another reason why COP is not an appropriate objective to maximize is that the machine capacity is usually inversely related to COP such that the maximum in COP occurs at zero capacity

A more practical approach is to consider the life cycle cost of a particular machine in a given application. In this approach two designs are compared by weighing the first costs, associated with building the machine, together with the operating costs to determine a single overall cost for each design. The design with the lowest cost is considered the best. Modifications to this method can take opportunity costs and environmental costs into account.

An alternative method of design, which works well for technology which has high first costs like absorption technology, is to choose a design that maximizes capacity for a given first cost. This approach is related to the power cycle capacity maximization efforts that have appeared in the literature. The approach has been applied to refrigeration cycles by Klein (1992). The concept is to allocate the heat exchanger area, based on a fixed total area (approximately fixed cost), so as to maximize heat pump capacity. The zero-order model presented in Section 2.7 provides a particularly simple model on which to base such an optimization.

Example 2.2   Absorption Cycle Optimization

Using the zero-order model determine the optimum distribution of heat exchanger surface area at the three temperature levels. Assume that the U value for all the heat exchangers is identical and equal to 0.5 kW/m^2-K. Assume that the total area available is 10 m^2. Further assume that the heat pump operates between the three temperatures: 200, 50, and -20°C. File: ex2_2.ees.

Solution: Using the zero-order model, search for the heat exchanger area distribution that maximizes heat transfer. Do this by iteratively solving the model for a wide range of different cases until the best one is found. For the conditions given, the maximum capacity was found to be 17.49 kW at a cooling COP of 0.53. The area distribution at the maximum is 2.705 m^2 at the high temperature, 2.286 m^2 at the low temperature and 5.009 m^2 at the   rejection temperature.

Observations: The maximum capacity results at a COP below the maximum possible COP. It is left as an exercise to use COP as the objective function with the same constraints. It is quite possible to demonstrate COP values well above 0.6 for the given conditions. However, the capacity for the high COP cases is significantly less. The maximum COP occurs when the capacity is driven to zero. Since zero capacity is of no value, the objective of maximizing COP comes into question.

HOMEWORK PROBLEMS

2.1   The reversible coefficient of performance (COP) of an absorption refrigeration cycle is a function of the three temperatures of the external heat transfer interactions as given in Equation 2.1.

    a)   Determine the sensitivity of the reversible COP to each of the three temperatures by evaluating the partial derivative with respect to each temperature.

    b)   Consider a baseline set of operating temperatures of 90, 30, and 5°C. Plot the derivatives determined in a) by holding two of the baseline temperatures constant and varying the third around the baseline value (three plots required).

    c)   Based on the results in a) and b), propose a design solution to improve the COP.

2.2   Repeat Problem #1 for an absorption heat pump. Compare the results of Problem #2 to those of Problem #1.

2.3   Repeat Problem #2 for a Type II absorption heat pump.

2.4   Compute the entropic average temperature of the heat transfer process from a stream of liquid water as it is cooled from 50 to 30°C. Assume the process is reversible.

2.5   Compute the entropic average temperature of the heat transfer process as a stream of steam is cooled and condensed. The process is isobaric at 1 bar. The starting temperature is 150°C and the ending temperature is 50°C. Assume the process is reversible.

2.6   Use the zero-order model to predict the impact on coefficient of performance and capacity of varying the size of the evaporator heat exchanger. Start from the baseline configuration obtained in Example 2.2 and plot COP and capacity vs. heat exchanger UA.

2.7   Derive an expression similar to Equation 2.24 for a double effect absorption cycle.

2.8   Repeat Example 2.2 for a more realistic set of U values in the desorber, absorber, condenser and evaporator of 1.0, 0.25, 0.5 and 0.5 kW/m^2-K, respectively.

2.9   It has been found from experience that the cooling COP of an absorption cycle is a strong function of the ratio between the high and low temperatures imposed on the cycle. Demonstrate this by running the zero-order model for a range of values for this ratio. In particular, include several cases that have the same ratio but which have different absolute temperatures.

2.10   Using the structure of Example 2.2, maximize the cooling COP under the same constraints and interpret the results in comparison to Example 2.2.

2.11   Using the result of Problem 2.7, determine the minimum heat input temperature for a double effect cycle using the data from Example 2.1.

Chapter 3

PROPERTIES OF WORKING FLUIDS

The performance and efficiency of reversible cycles are independent of the properties of the working fluids in use. However, the performance and efficiency of a real machine are determined to a large degree by the properties of the working fluids. Both the first cost and the operating cost of an absorption machine are strongly dependent on the working fluid properties. This chapter describes the properties of ammonia/water and water/lithium bromide mixtures, which are conventional absorption working fluids. Typical diagrams of fluid properties are discussed and the determination of the properties necessary for cycle calculations are explained in detail.

Many other working fluids have been considered for absorption machines. These include water/sodium hydroxide, water/sulfuric acid, ammonia/sodium thiocyanate and hundreds of other mixtures. The literature contains numerous studies on properties of alternative working fluids. Some of these fluids are for specialized applications, such as high temperature, while others are proposed as possessing improved properties over the conventional fluids. The reason why none of these alternative fluids have gained a market foothold is that the combination of properties exhibited by the conventional fluids is hard to compete against. In general, proposed alternatives may address one weakness of the conventional fluids while contributing several additional weaknesses of their own.

The properties desirable in an absorption working fluid have been listed by other authors. A summary of these properties is given in Table 3.1. The desirable properties are sometimes mutually exclusive. An example is the need for 1) a high affinity between the absorbent and the refrigerant and 2) a low heat of mixing (needed for Type I cycles). It is apparently not possible to find a fluid mixture that meets all of the criteria. Thus, one must consider the compromises presented by existing fluids.

The key tradeoffs that are available are illustrated in Table 3.1, where the conventional fluids are ranked according to the various desirable properties. Both conventional fluids utilize refrigerants that have high latent heat. This is advantageous in cycle design because it minimizes the refrigerant flow rate. However, neither fluid exhibits ideal vapor pressure characteristics. Ammonia pressures are inconveniently high while water pressures are inconveniently low. Lithium bromide is advantageous as an absorbent because it is essentially non-volatile, resulting in cycle designs that avoid the need for a rectifier. Water is advantageous as an absorbent because it does not crystallize (in the property ranges of interest). Numerous other properties are important in addition to the ones highlighted here. The preferred working fluid for a given application depends on the overall mix of properties. A key property is the freezing point of water which restricts the use of water/lithium bromide (and other systems utilizing water as refrigerant) to temperatures above 0°C. The restrictive nature of the requirements has resulted in very little success for fluids other than the conventional pairs. This situation may be changing now as high-temperature applications become more attractive.

The chapter is divided into two major sections. Section 3.1 covers an analytical treatment of thermodynamic properties, and Section 3.2 covers a graphical treatment of thermodynamic properties.

Table 3.1 Absorption working fluid properties

Property	Ammonia/Water	Water/Lithium Bromide
Refrigerant		
High latent heat	Good	Excellent
Moderate vapor pressure	Too high	Too low
Low freezing temperature	Excellent	Limited application
Low viscosity	Good	Good
Absorbent		
Low vapor pressure	Poor	Excellent
Low viscosity	Good	Good
Mixture		
No solid phase	Excellent	Limited application
Low toxicity	Poor	Good
High affinity between refrigerant and absorbent	Good	Good

3.1   Analytical Treatment of Thermodynamic Properties

Thermodynamic property relations for mixtures is a subject that has been well-known in chemical thermodynamics for many years going back to Gibbs (1876). More recent general treatments of chemical thermodynamics include Lewis and Randall (1961) and Rowlinson and Swinton (1982). In recent years, mixtures have become commonly used in energy conversion cycles and, hence, mechanical engineers have become interested in the subject. The subject, as it applies to energy conversion cycles, is treated partially in a wide range of textbooks, but no single source exists which provides a treatment directed at engineers who understand pure fluid thermodynamics but who do not have a background in mixtures. This chapter is designed to fill that void. The objective is to review the subject of mixture thermodynamic property relations and to highlight state-of-the-art approaches to dealing with mixtures.

The emphasis is on thermodynamic properties and, specifically, on the relationships between the thermodynamic properties of a mixture. These relationships allow the practitioner to derive thermodynamic properties from measured experimental data, to check the consistency of data and to apply the laws of thermodynamics to an engineering system.

3.1.1   Property Relations for Systems of Fixed Composition

The properties of a mixture are related through their definitions and through the concepts of classical thermodynamics, including the First and Second Laws. This discussion is focused on a general binary mixture so as to avoid the complexity associated with additional mixture components. Once the binary mixture analysis is understood, it can be extended to a mixture with an arbitrary number of components without excessive effort.

The internal energy of a system, or more exactly changes in the internal energy, are defined by the First Law, which can be written for a closed system as

$$du = \delta q - \delta w \tag{3.1}$$

where the heat is assumed positive into the system, and the work is a assumed positive out of the system. To obtain information about the equilibrium properties of a system, one can consider a reversible process. For such a process the definition of entropy changes can be written as

$$dq = Tds \tag{3.2}$$

The work can be written as

$$dw = pdv \tag{3.3}$$

Equation 3.3 assumes that volume change work is the only relevant work mode. Substituting Equations 3.2 and 3.3 into Equation 3.1 yields a fundamental property relation:

$$du = Tds - pdv \tag{3.4}$$

Based on the definitions of the enthalpy, Gibbs free energy and Helmholtz free energy, three other forms of Equation 3.4 are obtained as

$$dh = Tds + vdp \tag{3.5}$$

$$dg = vdp - sdT \tag{3.6}$$

$$d\psi = -pdv - sdT \tag{3.7}$$

Equations 3.4 to 3.7 provide simple relationships between property changes for a closed system. Although a quasi-equilibrium process was used in the derivation of Equation 3.4, the final result is a relationship between the properties of a closed system which can be interpreted independent of any process. Equations 3.4 to 3.7 are frequently used to determine changes in a property based on known changes in other properties. These equations apply to mixture systems and multi-phase systems, as well as to single-phase pure fluids, as long as a system of fixed composition is under consideration. In all cases, the system is assumed to be in equilibrium, and the properties represent the average property for the entire system.

An additional series of useful property relations can be obtained by assuming functional forms, as indicated by the functions in Equations 3.4 to 3.7. For example, from Equation 3.6, the following functional form is implied

$$g = g(T,p) \tag{3.8}$$

From calculus, the differential of g can be written as

$$dg = \left(\frac{\partial g}{\partial T}\right)_p dT + \left(\frac{\partial g}{\partial p}\right)_T dp \tag{3.9}$$

Equating Equations 3.6 and 3.9 yields

$$v = \left(\frac{\partial g}{\partial p}\right)_T \tag{3.10}$$

and

$$-s = \left(\frac{\partial g}{\partial T}\right)_p \tag{3.11}$$

Since the properties of interest possess exact differentials, the mixed derivative of g, obtained by differentiating both of Equations 3.10 and 3.11, can be equated to yield

$$\left(\frac{\partial v}{\partial T}\right)_p = -\left(\frac{\partial s}{\partial p}\right)_T \tag{3.12}$$

This relation is one of the so-called Maxwell relations. An analysis similar to the one that resulted in Equations 3.10 to 3.12 can be applied to the other differentials in Equations 3.4, 3.5, and 3.7. The result yields nine more differential equations representing relationships between thermodynamic properties. Many other such relations can be developed. As an example, note that by interpreting Equation 3.5 with respect to changes in temperature at constant pressure, an expression for the specific heat, c_p, in terms of the entropy is obtained

$$c_p \equiv \left(\frac{\partial h}{\partial T}\right)_p = T\left(\frac{\partial s}{\partial T}\right)_p \tag{3.13}$$

Equation 3.13 is useful if one is interested in the differential of entropy in terms of the independent variables pressure and temperature. Starting from s = s(T,p)

$$ds = \left(\frac{\partial s}{\partial T}\right)_p dT + \left(\frac{\partial s}{\partial p}\right)_T dp$$
$$= \frac{c_p}{T} dT - \left(\frac{\partial v}{\partial T}\right)_p dp \tag{3.14}$$

where Equations 3.12 and 3.13 have been utilized in the second line. Substituting Equation 3.14 into Equation 3.5 yields a very useful expression for changes in the enthalpy in terms of easily measurable properties.

$$dh = c_p dT + \left[v - T\left(\frac{\partial v}{\partial T}\right)_p\right]dp \tag{3.15}$$

Equation 3.15 contains the same information as contained in Equation 3.5, but Equation 3.15 is often more useful because the independent variables are those that are typically controlled and measured in experiments. Many insights can be obtained from examination of Equation 3.15. For example, for liquids away from the critical point, the entire term in the brackets can often be neglected, resulting in a simplified approximation for changes in enthalpy (dh = c_p dT). The validity of such approximations can be routinely evaluated by referring to the exact expression in Equation 3.15.

3.1.2   Open System Property Relations

In general, mixture thermodynamics implies changes in composition of the system. Such changes can result from chemical reactions or mass flow into or out from the system. In this treatment chemical reactions are assumed to be unimportant. Thus, the focus in this section is on open systems where transfer of mass between the system and the surroundings is allowed. The composition of a system can be specified in many ways and in this treatment mass fraction is used. For a binary mixture, a functional form for the Gibbs free energy can be assumed as follows

$$g = g(T,p,x_1) \qquad 3.16$$

where x_1 is the mass fraction of component 1. By definition, mass fraction changes imply that one is considering an open system. The differential of the function in Equation 3.16 can be written as

$$dg = \left(\frac{\partial g}{\partial T}\right)_{p,x_1} dT + \left(\frac{\partial g}{\partial p}\right)_{T,x_1} dp + \left(\frac{\partial g}{\partial x_1}\right)_{T,p} dx_1 \qquad 3.17$$

The first two derivatives on the right-hand side of Equation 3.17 are identical to those in Equations 3.10 and 3.11 with the exception that the constant composition specification is explicit in Equation 3.17. Thus, dg can be written as

$$dg = -sdT + vdp + \left(\frac{\partial g}{\partial x_1}\right)_{T,p} dx_1 \qquad 3.18$$

The remaining partial derivative in Equation 3.18 is related to the chemical potentials, μ_i, of the components in the mixture defined as the partial molal Gibbs free energy

$$\mu_i = \left(\frac{\partial G}{\partial N_i}\right)_{T,p,Nj} \qquad 3.19$$

where N_i is the number of moles of component I and the subscript Nj implies that the number of moles of all other components is held constant.

Because it is more convenient to work in terms of the intensive Gibbs free energy and the mass fraction, the following expressions (which can be derived directly from the definition) will be used as the defining expressions for the chemical potentials in a binary mixture.

$$\mu_1 = g + (1 - x_1)\left(\frac{\partial g}{\partial x_1}\right)_{T,p} \qquad 3.20$$

$$\mu_2 = g - x_1\left(\frac{\partial g}{\partial x_1}\right)_{T,p} \qquad 3.21$$

In this form, it is convenient to interpret the chemical potential on a mass basis instead of a molar basis and it is convenient to refer to these as the partial mass Gibbs free energies.

Example 3.1   Relationship Between Partial Molal and Partial Mass Properties

---

Consider subcooled liquid ammonia/water at 20°C, 1000 kPa and x = 0.5. Determine 1) the partial mass Gibbs free energy for each of the components, 2) the corresponding partial molal Gibbs free energy, 3) the mole fraction as a function of mass fraction, and 4) the molar enthalpy of the mixture state. File: ex3_1.ees.

Solution: 1) Using Equations 3.20 and 3.21 and the properties of ammonia/water, the following values are obtained. First, the mixture Gibbs free energy is obtained from the mixture enthalpy and entropy as

$$h \;\; = \;\; -149.82 \;\; J/g$$

$$s \;\; = \;\; 0.14114 \;\; J/g\text{-}K$$

$$g \;\; = \;\; h \;\; - \;\; Ts \;\; = \;\; -191.20 \;\; J/g$$

Then the partial derivative is approximated numerically by using a small increment in x of Δx = 0.0001. Based on this, the partial derivative is

$$\left(\frac{\partial g}{\partial x} \right)_{T,p} \;\; = \;\; 41.147 \;\; J/g$$

Finally, the chemical potentials are calculated directly from Equations 3.20 and 3.21 as

$$\mu_1 \;\; = \;\; -170.62 \;\; J/g$$

$$\mu_2 \;\; = \;\; -211.77 \;\; J/g$$

Since the mass fraction used is the mass fraction of ammonia, the subscripts are interpreted as 1 for ammonia and 2 for water.

2) The relationship between partial mass properties and partial molal properties is a simple unit conversion. Partial properties are expressed on the basis of one unit of the material. Thus, they can be expressed as above on a unit mass basis, or they can be expressed on a molar basis as follows by introducing the molecular weight of the species (M_1 = 17.031 (NH_3) and M_2 = 18.015 (H_2O)).

$$\overline{\mu_1} \;\; = \;\; \mu_1 M_1 \;\; = \;\; -2905.8 \;\; J/mol$$

$$\overline{\mu_2} \;\; = \;\; \mu_2 M_2 \;\; = \;\; -3815.0 \;\; J/mol$$

where the overbar denotes molar quantities.

3) The mole fraction is directly related to the mass fraction as shown next. First, define the mole fraction of component 1 as

$$x_{N,1} \;\; = \;\; \frac{N_1}{N_1 \;\; + \;\; N_2}$$

where N represents the number of moles of a component. Now, since the number of moles is related to the mass through the molecular weight as m = NM, we have

$$x_{N,1} = \frac{m_1/M_1}{m_1/M_1 + m_2/M_2}$$

$$= \frac{x_1}{x_1 + \dfrac{M_1}{M_2}(1 - x_1)} = 0.514$$

For ammonia/water, the mass and mole fractions have similar values due to the fact that the molecular weights of the two constituents are close in value.

4) The conversion to a molar basis of the mixture enthalpy, determined on a unit mass basis as a part of the solution of part 1) above, is more complicated than the conversion performed in part 2) because mixture properties are expressed on the basis of a unit of mixture. Thus, the mixture enthalpy, given as h = -149.82 J/g, is interpreted as the enthalpy per gram of mixture. The conversion to a molar basis requires the effective molecular weight of the mixture, M_{eff}, which can be calculated as

$$M_{eff} = x_{N,1} M_1 + (1 - x_{N,1}) M_2 = 17.509 \; g/mol$$

where the units are interpreted as grams of mixture per mole of mixture. Once the effective molecular weight is determined, the conversion is straightforward as

$$\overline{h} = h M_{eff} = -2623.2 \; J/mol$$

Observations: The entire treatment in this text is done on a mass basis. However, some developments in the literature are found on a molar basis. The connection between these two approaches is one-to-one, but it can be confusing. This example is meant to illustrate the main types of manipulations that are of interest.

The relationship between the chemical potential and the mixture properties can be found by equating g in Equations 3.20 and 3.21, solving for the derivative and substituting into Equation 3.18 yields

$$dg = -sdT + vdp + (\mu_1 - \mu_2) dx_1 \qquad 3.22$$

Equation 3.22 represents a generalized version of Equation 3.6 applicable to binary mixtures. The last term involving the chemical potentials represents the energy changes associated with changes in composition.

The chemical potential of a component in a mixture is a measure of the contribution to the Gibbs free energy associated with that component. It plays a similar role in defining the enthalpy and internal energy associated with each component. For example, the enthalpy can be written as

$$dh = Tds + vdp + (\mu_1 - \mu_2)dx_1 \qquad 3.23$$

The chemical potential is also a key variable in defining and understanding phase equilibrium in mixtures. Thus, the chemical potential is seen as a key variable which is as important as the temperature (thermal potential) or pressure (mechanical potential) in understanding mixture thermodynamics. Completely apart from any consideration of chemical reactions, the chemical potential accounts for the energy and entropy of interaction between molecules as they mix.

Because of its fundamental role, some additional discussion of the chemical potential is appropriate to gain some insight into its nature.

Based on the definition of μ as the partial mass Gibbs free energy, g can be written as

$$g = x_1 \mu_1 + (1 - x_1) \mu_2 \qquad\qquad 3.24$$

Thus, the mixture Gibbs free energy is the mass fraction weighted average of the partial mass Gibbs free energies of the components. Differentiating this expression, assuming constant temperature and pressure, yields

$$dg = x_1 d\mu_1 + \mu_1 dx_1 + (1 - x_1) d\mu_2 + \mu_2 d(1 - x_1) \qquad\qquad 3.25$$

but from Equation 3.22, under these conditions, dg is

$$dg = (\mu_1 - \mu_2) dx_1 \qquad\qquad 3.26$$

Then, by equating Equations 3.25 and 3.26 one obtains what is often called the Gibbs-Duhem equation

$$0 = x_1 d\mu_1 + (1 - x_1) d\mu_2 \qquad\qquad 3.27$$

This equation provides a relationship between changes in the chemical potentials of the components in a binary mixture at constant temperature and pressure. The utility of Equation 3.27 is in understanding the shape of the chemical potentials as a function of composition. It is also occasionally useful for calculating the chemical potential of one component when the chemical potential of the other component is known.

The chemical potential of a pure component is equal to the Gibbs free energy of that component as can be deduced from Equation 3.20 by setting x_1 to 1. Thus, the chemical potential of a pure component has a well-defined, finite value. The chemical potential of a component goes to negative infinity as the mass fraction of that component goes to zero. This can be deduced from Equation 3.27 by considering the approach to the pure component 2 state (consider x_1 --> 0). It is required that μ_2 is finite at $x_1 = 0$, so one expects that $d\mu_2$ will be finite so as to allow finite values of μ_2 for the mixture states. However, as x_1 goes to zero, the only way Equation 3.27 can be satisfied is for $d\mu_1$ to get large. In fact, μ_1 must go to negative infinity at $x_1 = 0$ since, as is discussed later in the section on phase equilibrium, mass always transfers in the direction of decreasing chemical potential. Thus, the absence of any molecules of component 1 represents a discontinuity in the chemical potential.

The shape of the chemical potentials for an ideal mixture are plotted in Figure 3.1 to provide the reader with an idea of the nature of the chemical potential. For this plot, the molecular weights of ammonia and water have been used to provide a mass fraction scale. The chemical potential concept is key to understanding mixture thermodynamics, mixture phase equilibrium and mass transfer in multi-phase systems. The introduction provided here provides a clear method of relating the chemical potential to other thermodynamic variables. The applications to phase equilibrium and mass transfer are discussed in Section 3.1.7.

3.1.3   Equations of State

The usual meaning of the term equation of state is a relationship between pressure, volume and temperature of a substance. In the mixture context, this concept must be generalized to include composition as a variable. It is convenient to generalize it further to cover any equation

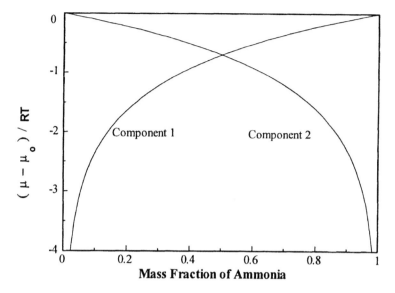

Figure 3.1 Chemical potential for an ideal mixture

containing complete thermodynamic information about the substance. This implies that all other thermodynamic properties can be calculated from the equation of state. The most useful of such equations allow thermodynamic properties to be calculated explicitly through the use of the property relations discussed in Sections 3.1.1 and 3.1.2. Such property calculations are simplified if the equation of state is constructed as a fundamental function (Moran and Shapiro, 1992) as shown next.

The concept of a fundamental function is useful in doing the bookkeeping associated with property relations and in forming equations of state. Based on Equation 3.6, for example, the Gibbs free energy is defined as the fundamental function associated with the independent variables temperature and pressure. Similarly, if one were to choose the temperature and volume as independent variables, the Helmholtz free energy would be the fundamental function. This distinction has no physical significance but is made for convenience in keeping the analysis organized. Once the independent variables are chosen for a given analysis, the remainder of the analysis can frequently be simplified if it is done in terms of the fundamental function. If an equation of state is formed as an expression relating the fundamental function to the so-called natural variables associated with that function (for example, $g = g(T,p)$), then all other thermodynamic properties can be derived from the fundamental function with a minimum of effort, starting with Equations 3.10 and 3.11. This approach has been used for pure fluids (Harr et al., 1984; Harr and Gallagher, 1978) as well as for mixtures (Ziegler and Trepp, 1984; Herold, 1985; Ibrahim and Klein, 1993). For the mixture applications an additional independent variable is needed to define the mixture composition.

Once the fundamental function is known as a function of the independent variables, other properties can be computed from it. As a summary, the following relations hold (based on the choice of temperature, pressure and mass fraction as independent variables and Gibbs free energy as the fundamental function).

$$v = \left(\frac{\partial g}{\partial p}\right)_{T,x_1} \tag{3.28}$$

$$u = g - T\left(\frac{\partial g}{\partial T}\right)_{p,x_1} - p\left(\frac{\partial g}{\partial p}\right)_{T,x_1} \tag{3.29}$$

$$h = g - T\left(\frac{\partial g}{\partial T}\right)_{p,x_1} \tag{3.30}$$

$$-s = \left(\frac{\partial g}{\partial T}\right)_{p,x_1} \tag{3.31}$$

$$c_p = -T\left(\frac{\partial^2 g}{\partial T^2}\right)_{p,x_1} \tag{3.32}$$

The chemical potential can also be computed from the Gibbs free energy according to Equations 3.20 and 3.21. All of these property expressions have in common that they do not require integration. This is one key advantage of the fundamental function approach since integration tends to be more cumbersome to implement when dealing with property relations.

As an example of an equation of state constructed using the fundamental function approach, consider a fluid that is incompressible and has a constant specific heat. The equation of state for such a fluid has the form

$$g(T,p) = h_o - T_o s_o + v[p - p_o] - s_o[T - T_o] + c_p\left[T - T_o - T\ln\left(\frac{T}{T_o}\right)\right] \tag{3.33}$$

This fundamental function reproduces all of the thermodynamic behavior of such a fluid in a single function. It is left to the reader to perform the operations represented by Equations 3.28 to 3.32 and to verify the expected behavior. In fact, the form of Equation 3.33 was determined by integrating from the assumed characteristics (i.e., constant volume and specific heat). This type of reverse engineering of the form of the equation of state is quite useful in choosing a functional form to fit experimental data. Of course, in most cases of interest the model will be more complicated than that given in Equation 3.33. The reason for including Equation 3.33 is as a simple but concrete example of the fundamental function concept. Numerous more complicated examples appear in the literature cited. Harr and Gallagher (1978) give a very useful discussion of the method including details about overfitting, underfitting and multi-property fitting, all of which are topics beyond the scope of the present treatment.

Thermodynamic Consistency

One of the advantages of the equation of state approach is that it can lead to superior thermodynamic consistency between different properties for a particular fluid (or substance). If all properties are derived from a single function, then the only inconsistencies should be those resulting from finite precision arithmetic. In the past, thermodynamic data were often calculated and tabulated for future use. Due to the coarse tabulations, it is quite difficult to obtain accurate estimates of related properties from such tabulations. Both differentiation and integration of such data lead to significant errors which tend to propagate and multiply as one performs the calculation. If one then attempts to calculate the original property from derived properties, it frequently does not match the original data. This is thermodynamic inconsistency. The problem of thermodynamic inconsistency arises frequently when entropy data for a given fluid are computed from enthalpy data. Inconsistencies between the entropy and enthalpy data sets, then, tend to obscure the entropy analysis which often involves small differences between large numbers and thus requires highly accurate data.

By utilizing an analytical form for the fundamental function, derivatives can be expressed exactly, and all properties computed from the function are inherently consistent. This inherent consistency is quite useful in that it ensures that the properties satisfy the laws of thermodynamics. However, it must be realized that just because the properties resulting from a given equation of state are self-consistent does not imply that they are correct. It is still up to the analyst to formulate a meaningful equation of state. It should be noted that the equation of state approach is not necessary to achieve thermodynamic consistency. Careful work using other methods can also minimize error. However, the advantage of the equation of state approach is that the consistency comes automatically and the analyst can focus effort on formulating appropriate models.

Example 3.2   Thermodynamic Consistency

Determine the thermodynamic consistency of the NBS equation of state for water (Haar et al., 1984) around the subcooled liquid state defined by the temperature of 20°C and the pressure of 100 kPa. File: ex3_2.ees.

Solution: There are many consistency tests that could be performed if more than one property is available. In this case, the properties predicted by the equation are:

h	=	83.927 J/g	c_p	=	4.183 J/g-K
u	=	83.827 J/g	s	=	0.29618 J/g-K
v	=	0.001002 m³/kg			

Based on these data, a first check is to see whether the enthalpy and internal energy are consistent according to their definition. The definition of enthalpy is h = u + pv and the calculation yields exact agreement through eight digits.

A second check is to see whether the enthalpy and the specific heat values are consistent. To do this, the derivative of the enthalpy with respect to temperature must be estimated from the equation of state. This was done with a temperature increment of 0.001 K and the result is

$$\left(\frac{\partial h}{\partial T}\right)_P = 4.183 \ J/g{-}K$$

These data were found to be consistent up to five digits of accuracy.

A third check can be done on the entropy. To facilitate this, the Gibbs free energy is needed. This can be calculated from the definition as g = h - Ts. The derivative of the Gibbs free energy is related to the entropy according to Equation 3.11. The derivative was evaluated numerically here with a temperature increment of 0.001 K. The result is

$$\left(\frac{\partial g}{\partial T}\right)_P = -0.29619 \ J/g\text{-}K$$

This matches the entropy out to four digits with a 10% error in the fifth digit.

Observation: The NBS equation of state for water is a highly consistent formulation. The method used here involving numerical derivatives with a temperature increment of only 0.001 K would be expected to introduce some inconsistency. Thus, the consistency tests must be performed very carefully to get meaningful results. If such tests as these were applied to typical data tabulations, much lower consistency would be expected.

3.1.4   Mixture Volume

When two miscible species are mixed, the resulting volume is not generally equal to the mass weighted average of the volumes of the pure components. Mixtures where such a simple relationship does hold are termed ideal mixtures. When the molecules of a mixture do not interact significantly, that is, when there are not significant attractive or repulsive forces between the dissimilar molecules, then the mixture can approach ideal behavior. A low pressure gas mixture is an example where such conditions can occur. However, for most liquid mixtures, the molecules interact significantly and real mixture effects must be considered.

As an example, consider the volume versus mass fraction plot in Figure 3.2. The data for the plot are for liquid ammonia/water and were generated from a Gibbs free energy equation of state (Ibrahim and Klein, 1993). The heavy line represents mixture volume at a pressure of 10 bar and a temperature of 20°C (subcooled liquid states). The pure component volumes are found at the limits of mass fraction. An ideal mixture would be represented by the straight line connecting the endpoints. The fact that the actual volume falls below the ideal mixture line indicates that the mixture exhibits a reduction in volume when mixing occurs. This reduction is quantified here in two ways: 1) using partial volume and 2) using the concept of the volume change of mixing. These perspectives are equivalent and they both lead to some useful techniques for analyzing volume changes.

First, consider the volume changes in terms of the partial molal volume. The partial molal volume is defined as

$$v_i = \left(\frac{\partial V}{\partial N_i}\right)_{T,p,Nj} \tag{3.34}$$

As was done for the partial Gibbs free energy in Equations 3.19 to 3.21, it is convenient to express the partial volumes in terms of the mixture specific volume and the mass fraction as

$$v_1 = v + (1-x_1)\left(\frac{\partial v}{\partial x_1}\right)_{T,p} \tag{3.35}$$

$$v_2 = v - x_1\left(\frac{\partial v}{\partial x_1}\right)_{T,p} \tag{3.36}$$

Since it is convenient to interpret these volumes in mass units, a better term might be to call these partial mass volumes. Equations 3.35 and 3.36 are represented graphically on Figure 3.2

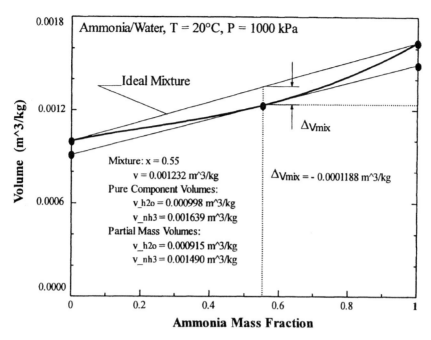

Figure 3.2 Volume versus mass fraction

by the tangent line to the mixture volume curve. The intersections of the tangent line with the pure component axes are the partial mass volumes. Example numerical values are given based on a mass fraction of x = 0.55 in Figure 3.2.

The partial mass volumes represent the contribution that each mixture component makes to the total volume. Thus, one can add up the mass fraction weighted sum of the partial mass volumes to obtain the total volume as follows

$$v = x_1 v_1 + (1-x_1)v_2 \qquad\qquad 3.37$$

In the example in Figure 3.2, the partial mass volumes at the point $x_1 = 0.55$ are both less then the corresponding pure component volumes. This is consistent with the volume reduction that occurs when ammonia and water are mixed. From this perspective, a portion of that reduction is attributed to each component according to the definition in Equation 3.34.

An alternative perspective is to formulate the real mixture characteristics as an ideal mixture with a term added to account for the non-ideal behavior. This term is called the volume change of mixing or the excess volume. From this perspective, the mixture volume is expressed as

$$v = x_1 v^o_1 + (1-x_1)v^o_2 + \Delta v_{mix} \qquad\qquad 3.38$$

The volume change of mixing is shown on Figure 3.2. The negative value indicates the volume reduction that this mixture experiences.

The volume characteristics of aqueous lithium bromide are somewhat different than that of ammonia/water. When in solution, the salt ionizes and occupies more volume than when in the solid phase. Thus, when solid lithium bromide is added to pure water, the volume of the mixture is greater than the volume of the individual components. This implies that the volume change of mixing, as defined in Equation 3.38, is positive. However, the magnitude of the volume changes is quite small. This change can be quantified by computing the partial volumes. At a salt mass fraction of 0.6 the computed volumes are $v_{H2O} = 1.018$ cm^3/g, v_{LiBr} = 0.3072 cm^3/g. Values are computed numerically from an empirical volume function based on the data of Uemura and Hasaba (1964). These can be compared against the pure component volumes $v_{H2O,Pure} = 1.002$ cm^3/g and $v_{LiBr,Pure} = 0.2887$ cm^3/g (Handbook of Chemistry and Physics, 1978). The fact that the partial values are greater than the pure component volumes is consistent with the volume expansion upon mixing.

3.1.5   Mixture Energy Properties

The energy properties referred to here include the enthalpy and internal energy. The entire analysis applies to both properties. To make for a streamlined treatment, the discussion is in terms of enthalpy. Many of the concepts and details introduced in the previous sections discussion of mixture volume also apply to the enthalpy. For completeness, these concepts are repeated in terms of the enthalpy to allow the unique aspects of the energy properties to be efficiently introduced.

When two miscible species are mixed, the resulting enthalpy is not generally equal to the mass weighted average of the enthalpies of the pure components. As an example, consider the enthalpy versus mass fraction plot in Figure 3.3. The heavy line represents mixture enthalpy of liquid ammonia/water (Ibrahim and Klein, 1993) for a pressure of 10 bar and a temperature of 20°C (subcooled liquid states). The fact that the actual enthalpy falls below the ideal mixture line indicates that the mixture exhibits a reduction in enthalpy when mixing occurs. This reduction is quantified here in two ways: 1) using partial mass enthalpy and 2) using the concept of the enthalpy change of mixing.

First, consider the enthalpy changes in terms of the partial molal enthalpy defined as

$$h_i = \left(\frac{\partial H}{\partial N_i} \right)_{T,p,N_j} \qquad 3.39$$

As was done for the partial Gibbs free energy in Equations 3.19 to 3.21 and the volume in Equations 3.34 to 3.36, it is convenient to express the partial enthalpies in terms of the mixture specific enthalpy and the mass fraction as

$$h_1 = h + (1 - x_1) \left(\frac{\partial h}{\partial x_1} \right)_{T,p} \qquad 3.40$$

$$h_2 = h - x_1 \left(\frac{\partial h}{\partial x_1} \right)_{T,p} \qquad 3.41$$

Equations 3.40 and 3.41 are represented graphically on Figure 3.3 by the tangent line to the mixture enthalpy curve. The intersections of the tangent line with the pure component axes are the partial mass enthalpies. Numerical example values are given on Figure 3.3 for an arbitrary mass fraction of x = 0.55.

The partial mass enthalpies represent the contribution that each component makes to the mixture enthalpy. Thus, one can add up the mass fraction weighted sum of the partial mass enthalpies to obtain the total enthalpy as follows

$$h = x_1 h_1 + (1 - x_1) h_2 \qquad 3.42$$

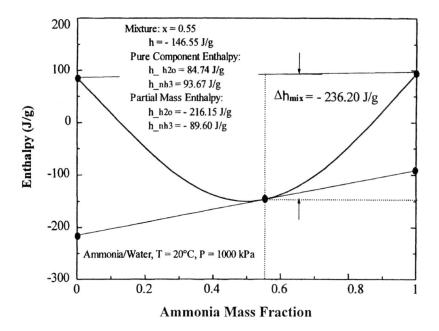

Figure 3.3 Enthalpy versus mass fraction

In the example in Figure 3.3, the partial mass enthalpies at the point $x_1 = 0.55$ are both less than the corresponding pure component enthalpies. This observation is consistent with the exothermic energy release that occurs when ammonia and water are mixed. From this perspective, a portion of that reduction is attributed to each component according to the definition in Equation 3.42. It should be noted that the fact that the ideal mixture line is almost parallel to the abscissa is a coincidence for this particular mixture and the chosen reference states.

An alternative perspective is to formulate the real mixture characteristics as an ideal mixture with a term added to account for the non-ideal behavior. This term is called the enthalpy change of mixing or the excess enthalpy. From this perspective, the mixture enthalpy is expressed as

$$h = x_1 h^o_1 + (1 - x_1) h^o_2 + \Delta h_{mix} \qquad\qquad 3.43$$

The enthalpy change of mixing is shown on Figure 3.3. The negative value indicates the enthalpy reduction that this mixture experiences.

The enthalpy of aqueous lithium bromide follows a similar pattern as water/ammonia, with both mixtures exhibiting a large exothermic heat release upon mixing (i.e., a negative heat of mixing). For such mixtures the partial mass enthalpy of each of the components is expected to be less than the pure component enthalpy. At a mass fraction of 0.6 LiBr and a temperature of 20°C, the partial mass enthalpies are $h_{H2O} = -278.0$ J/g and $h_{LiBr} = 313.9$ J/g. The pure water enthalpy at 20°C is $h_{H2O,Pure} = 83.93$ J/g. The corresponding pure LiBr state is a solid, crystalline salt. To make a meaningful comparison, the reference state enthalpy for the salt must be chosen consistent with the solution enthalpy reference state (i.e., h = 0 at T = 0°C and x = 0.5 mass fraction LiBr). An estimation of the consistent enthalpy for the solid salt is possible based on a solid specific heat of 0.598 J/g-K (Foote) and a simple mixing experiment. The experiment involves mixing equal masses of anhydrous LiBr and pure water at room temperature and recording the temperature of the resulting mixture. The result is that the enthalpy of the solid is 322 J/g at 20°C. These values are all consistent with the exothermic mixing associated with aqueous lithium bromide.

Specific Heat

A property related to the enthalpy is the constant pressure-specific heat defined as

$$c_p = \left(\frac{\partial h}{\partial T} \right)_{p, x_1} \qquad \qquad \textbf{3.44}$$

The relationship of the mixture-specific heat to the specific heats of the components can be found by applying the definition in Equation 3.44 to either Equations 3.42 or 3.43. Application to Equation 3.42 results in

$$c_p = x_1 \left(\frac{\partial h_1}{\partial T} \right)_{p, x_1} + (1 - x_1) \left(\frac{\partial h_2}{\partial T} \right)_{p, x_1} \qquad \qquad \textbf{3.45}$$

$$= x_1 c_{p,1} + (1 - x_1) c_{p,2}$$

The terms in Equation 3.45 are partial mass specific heats of the two mixture components. From this perspective, the specific heat of the mixture is viewed as having contributions from each of the two mixture components. Alternatively, by applying the definition in Equation 3.44 to Equation 3.43 one obtains

$$c_p = x_1 \left(\frac{\partial h^o_1}{\partial T} \right)_{p, x_1} + (1 - x_1) \left(\frac{\partial h^o_2}{\partial T} \right)_{p, x_1} + \left(\frac{\partial \Delta h_{mix}}{\partial T} \right)_{p, x_1} \qquad \qquad \textbf{3.46}$$

$$= x_1 c^o_{p,1} + (1 - x_1) c^o_{p,2} + \Delta c_{p, mix}$$

From this perspective, the mixture specific heat is viewed as consisting of the ideal mixture specific heat plus a correction term to account for real mixture effects. Both of the perspectives represented in Equations 3.45 and 3.46 are equivalent and both are useful.

Reference States

One aspect of the energy properties that is not relevant to the volumetric properties is the need for a reference state. Energy properties are defined only as differences, based on the First Law. Absolute energy values have no intrinsic meaning. However, because it is convenient in utilizing enthalpy values in analyzing engineering systems, common practice is to arbitrarily define a zero energy or enthalpy and to express all values relative to this reference. It should be noted that this applies to pure fluids as well as to mixtures. It is often possible to ignore the reference state when dealing with a pure fluid as long as a single data source is used for all calculations. However, if two or more data sources are used with different reference states, it is necessary to convert all the data to the same reference state before use. This problem is more frequently encountered in dealing with mixtures because each mixture component contributes its own reference state and the possibility of encountering a reference state mismatch increases with the number of choices available.

To enable tabulation or specification of enthalpy values for a pure substance, a reference state must be chosen. For example, the reference state commonly chosen for water is the liquid phase at triple point conditions. The internal energy of this state is arbitrarily assigned the value zero. Note that any value could be assigned but that zero is a convenient choice. Since the enthalpy is defined in terms of the internal energy, it is not possible to choose a reference value for both. The pv term in the definition of the enthalpy contributes a small term to the enthalpy of liquid water at the reference state. As an alternative, one could choose to specify the enthalpy and to compute the internal energy.

It is usually possible to deduce the reference state from examining a tabulation of energy data. For a pure fluid the enthalpy in terms of the reference state enthalpy can be expressed as

$$h(T,p;T_o,p_o) = h_{ref}(T_o,p_o) + \Delta h(T,p;T_o,p_o) \qquad 3.47$$

The functional notation is used in Equation 3.47 to emphasize the role of the reference state. This result gives a simple recipe for reference state corrections for a pure fluid. In general, to force two data sets to the same reference state, a constant must be added to all the values in one of the sets so that h_{ref} is the same for both sets. The procedure is slightly more complicated for a mixture because two reference states are involved.

The two pure component enthalpy values appearing in Equation 3.43 each can be written as in Equation 3.47. With this substitution, Equation 3.43 becomes

$$
\begin{aligned}
h(T,p,x_1; T_{o,1},p_{o,1}, T_{o,2},p_{o,2}) = \; & x_1\left[h_{ref,1}(T_{o,1},p_{o,1}) + \Delta h_1(T,p;T_{o,1},p_{o,1})\right] + \\
& + (1-x_1)\left[h_{ref,2}(T_{o,2},p_{o,2}) + \Delta h_2(T,p;T_{o,2},p_{o,2})\right] + \qquad 3.48 \\
& + \Delta h_{mix}(T,p,x_1)
\end{aligned}
$$

For a binary mixture, two reference states affect the mixture enthalpy as can be seen in Equation 3.48. The reference state enthalpies contribute according to the relative amount of each component in the mixture. A similar result is obtained if the mixture enthalpy is viewed as in Equation 3.42. In that case, the values of each of the partial mass enthalpies depend on the reference state enthalpy for that component and the reference state enthalpies appear in the mixture enthalpy expression in identical form to that in Equation 3.48.

The implication of Equation 3.48 is that the mixture enthalpy depends on the reference state enthalpies according to

$$
\begin{aligned}
h_{ref} &= x_1 h_{ref,1} + (1-x_1)h_{ref,2} \\
&= h_{ref,2} + x_1\left(h_{ref,1} - h_{ref,2}\right)
\end{aligned}
\qquad 3.49
$$

Equation 3.49 implies that to compare two enthalpy data sets which use different reference states it is necessary to add a linear function of mass fraction to one of the data sets so that the enthalpy values of both data sets match at two distinct states. The linear function has two parameters that can be computed by matching the two states. One restriction is that the two states must be at different mass fractions. Once the reference state correction has been applied, then any remaining differences between two data sets are substantive.

For the mixture water/lithium bromide, the nature of pure lithium bromide is such that it exists as a solid at room temperature and pressure. However, when mixed with sufficient water it goes into aqueous solution. This behavior complicates the analysis of mixture properties because one does not usually wish to include the solid- to liquid-phase transition in the analysis. For the enthalpy reference state, common practice is to choose a solution state as the reference state in place of the pure lithium bromide state. One common choice is saturated liquid at x = 0.5, T = 0°C. This highlights the fact that the enthalpy reference states chosen for a mixture are actually mixture states. The analyst must choose two different mixture states for binary systems and assign enthalpy values to those states to define the mixture enthalpy. However, with water/lithium bromide as an exception, in practice it is often convenient to choose the pure fluid states (i.e., x = 0 and x = 1.0) to maintain consistency with pure fluid data tabulations and pure fluid equations of state. It should be noted that the two reference states chosen must be at different mass fraction.

3.1.6.   Mixture Entropy

When two components mix, there is an entropy increase even if the molecules of the components do not interact (i.e., an ideal mixture). This entropy change of mixing is associated with the fact that two substances can mix spontaneously (e.g., by diffusion), but a mixture cannot separate into pure components spontaneously. The entropy change of mixing for an ideal mixture is used as a model against which to measure the characteristics of a real mixture. Thus, the entropy changes in an ideal mixture are discussed first.

An ideal gas mixture is one example of an ideal mixture. The entropy changes in an ideal gas can be derived directly from Equation 3.14 as

$$ s = s_o + c_p \ln\frac{T}{T_o} - R\ln\frac{p}{p_o} \qquad 3.50 $$

From an ideal gas mixture perspective, Equation 3.50 applies to each component in the mixture with the pressure interpreted as the partial pressure, p_i. The mixture entropy is then

$$
\begin{aligned}
s &= \frac{f(x_1)}{M_{eff}}\left(\overline{s^\circ_{o,1}} + \overline{c^\circ_{p,1}}\ln\frac{T}{T_{o,1}} - R\ln\frac{p_1}{p_{o,1}}\right) + \frac{(1-f(x_1))}{M_{eff}}\left(\overline{s^\circ_{o,2}} + \overline{c^\circ_{p,2}}\ln\frac{T}{T_{o,2}} - R\ln\frac{p_2}{p_{o,2}}\right) \\
&= \frac{f(x_1)}{M_{eff}}\left(\overline{s^\circ_{o,1}} + \overline{c^\circ_{p,1}}\ln\frac{T}{T_{o,1}} - R\ln\frac{p}{p_{o,1}}\right) + \frac{(1-f(x_1))}{M_{eff}}\left(\overline{s^\circ_{o,2}} + \overline{c^\circ_{p,2}}\ln\frac{T}{T_{o,2}} - R\ln\frac{p}{p_{o,2}}\right) \\
&\quad + \frac{\overline{\Delta s_{mix,ideal}}}{M_{eff}} \qquad\qquad 3.51 \\
&= \frac{1}{M_{eff}}\left[f(x_1)\overline{s^\circ_1} + (1-f(x_1))\overline{s^\circ_2} + \overline{\Delta s_{mix,ideal}}\right] \\
&= x_1 s^\circ_1 + (1-x_1)s^\circ_2 + \Delta s_{mix,ideal}
\end{aligned}
$$

Where the entropy change of mixing for the ideal binary mixture is

$$ \Delta s_{mix,\,ideal} = -\frac{R}{M_{eff}}\left[f(x_1)\ln(f(x_1)) + (1-f(x_1))\ln(1-f(x_1))\right] \qquad 3.52 $$

and the composition function is

$$ f(x_1) = \frac{x_1}{x_1 + \dfrac{M_1}{M_2}(1 - x_1)} \qquad\qquad 3.53 $$

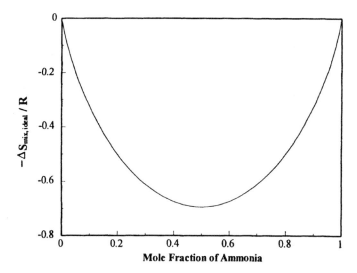

Figure 3.4 Entropy of mixing for an ideal mixture

The function f(x) is actually just the mole fraction which can be expressed in terms of the mass fraction, as is done here, to maintain a consistency in use of mass fraction throughout the development. Although derived for an ideal gas mixture, the expression in Equation 3.52 applies to any ideal mixture. A plot of the entropy change of mixing is provided as Figure 3.4 where the molecular weights of ammonia and water have been used to define the mass function coordinates. The characteristic shape shows a maximum entropy of mixing at a mass faction of 0.486 falling off to zero at both pure component limits. For the ideal mixture, the entropy change of mixing is independent of temperature and pressure.

It is sometimes useful to express the entropy in terms of the partial mass entropies. The partial mass entropy of component 1 in the mixture can be derived from Equation 3.51 as

$$s_1 = s + (1-x_1)\left(\frac{\partial s}{\partial x_1}\right)_{T,p}$$

$$= s^o_1 - \frac{R}{M_{eff}} \ln(f(x_1))$$

3.54

and the partial mass entropy of component 2 is

$$s_2 = s - x_1\left(\frac{\partial s}{\partial x_1}\right)_{T,p}$$

$$= s^o_2 - \frac{R}{M_{eff}} \ln(1 - f(x_1))$$

3.55

The mixture entropy can also be expressed as the sum of the mass fraction weighted partial mass entropies as

$$s = x_1 s_1 + (1-x_1)s_2$$

3.56

It is left to the reader to substitute Equations 3.54 and 3.55 into Equation 3.56 and demonstrate that this approach is equivalent to that in Equation 3.51.

For real mixtures the mixture entropy can be represented as

$$s = x_1 s^o_1 + (1 - x_1)s^o_2 + \Delta s_{mix,\,ideal} + \Delta s_{mix,\,real} \qquad \qquad 3.57$$

or equivalently as in Equation 3.56. For a real mixture, the partial mass entropies in Equation 3.56 will have additional terms beyond those of the ideal mixture given in Equations 3.54 and 3.55.

Although the Third Law of thermodynamics defines absolute entropies, it is common in applications where chemical reactions are not significant to use relative entropy values. This approach involves selecting reference states and assigning reference state entropy values analogous to the method used for enthalpy. This approach works as long as the subsequent analysis using the entropy values involves only differences in entropy. Since the analysis is closely identical to that for the enthalpy, no additional discussion of entropy reference states is needed.

3.1.7   Phase Equilibrium

Phase equilibrium between liquid and vapor plays a key role in the understanding of absorption technology. A general requirement for phase equilibrium is that the thermodynamic potentials must be equal between the phases. The potentials of interest are thermal (temperature), mechanical (pressure), and mass transfer (chemical potential). If the two phases are represented by a single prime and a double prime, phase equilibrium requires that

$$T' = T'' \qquad \qquad 3.58$$

$$p' = p'' \qquad \qquad 3.59$$

$$\mu'_i = \mu''_i \qquad i = 1, \cdots, n \qquad \qquad 3.60$$

where n is the number of mixture components.

The chemical potential of component I, denoted μ_i, was defined previously in Equation 3.19. This property has been called the escaping potential (Lewis and Randall, 1961) since an imbalance in chemical potential between two phases drives a transfer of that component from the phase with the higher potential toward the phase with the lower chemical potential. The transfer tends to equalize the potentials between the two phases until, at equilibrium, there is no net transfer between the phases.

The composition of the equilibrium phases is, in general, not the same. This characteristic is a key aspect of mixture thermodynamics and is an important property in absorption cycle design. As an example, consider the vapor-liquid equilibrium in ammonia-water as plotted in Figure 3.5. Such phase diagrams are sometimes called bubble point or dew point diagrams and are used to visualize certain mixture phase change processes such as cooling of superheated vapor at constant pressure as shown by the line labeled A to B on Figure 3.5. The intersection of line A to B with the dew line indicates the temperature at which the first drop of liquid will form. The composition of that liquid can be read from the diagram as well. .

The energy transfers associated with two-phase processes in mixture systems can be calculated if mixture properties are known. The mass fraction variable must be accounted for and it contributes some complication. The type of analysis required is outlined in the following discussion of the heat of vaporization.

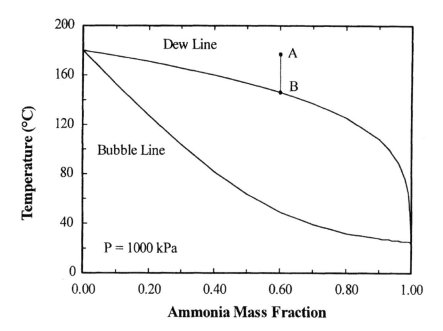

Figure 3.5 NH$_3$/H$_2$O bubble point

Heat of Vaporization

The concept of heat of vaporization requires some additional specification when dealing with mixtures. In particular, it is necessary to fully define the vaporization process under consideration. A constant pressure vaporization requires a different heat than a constant temperature vaporization. There are an infinite number of such processes one could use to define the heat of vaporization of a mixture making it necessary to specify the process being considered whenever the concept is utilized for a mixture. For a pure fluid it is also necessary to specify the process but it is generally implied by a specification of either the temperature or the pressure and both remain constant during the vaporization. Since this is not generally true for a mixture, the specification of the vaporization process becomes necessary for clarity.

Furthermore one can define (Bosnjakovic, 1965) a total heat of vaporization, for complete evaporation of a liquid sample, or a differential heat of vaporization, for evaporation of an infinitesimal amount of mass from a finite sample of liquid. These two quantities are related as is discussed next for the constant pressure process.

Consider the differential vaporization process illustrated in Figure 3.6. A balance on the ammonia mass yields

$$\frac{dm}{m} = \frac{dx}{x - x_v} \qquad\qquad 3.61$$

An energy balance on the same control volume yields

$$dQ = mdh + (h_v - h)dm \qquad\qquad 3.62$$

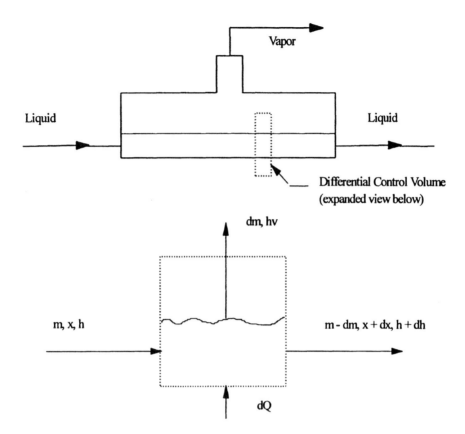

Figure 3.6 Differential vaporization process

Normalizing the heat based on the mass flow rate of the vapor yields

$$q_p = h_v - h - (x_v - x)\left(\frac{\partial h}{\partial x}\right)_p \qquad\qquad 3.63$$

where Equation 3.61 has been used to eliminate the mass flow rates in terms of the mass fraction. Also, it should be noted that the constant pressure specification given in Equation 3.63 is just one of the processes for which one can define such a heat of vaporization. The constant-pressure differential heat of vaporization is denoted q_p.

The energy required for complete vaporization of a binary mixture depends on the conditions of the process. The process implied in Figure 3.6 allows the vapor to escape from the system. If the vapor is kept with the liquid (i.e., in equilibrium with the liquid), some additional energy is needed to heat the vapor as the equilibrium temperature changes. For the complete evaporation process, or for a finite partial evaporation process, the heat requirement can be determined by a simple energy balance. For a complete evaporation process, the result is

$$q_c = h_v - h_l \qquad\qquad 3.64$$

where h_v and h_l are the vapor and liquid enthalpies, respectively, at the entrance and exit of the process. The heat required, q_c, is slightly greater than that obtained from integrating Equation

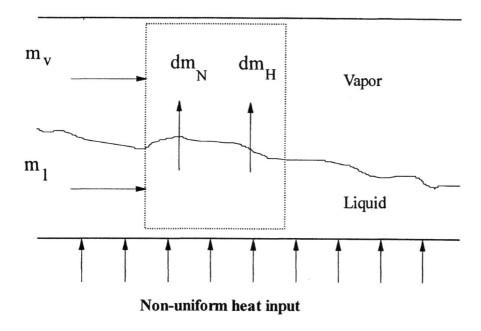

Figure 3.7 Total evaporation process

3.63 over the same process because of the heat required to raise the temperature of the vapor.

In the differential vaporization process illustrated in Figure 3.6, the vapor is ducted away from the surface as it is generated. The vapor leaving the liquid surface is assumed to be at the equilibrium concentration. Furthermore, the liquid is assumed to be well mixed at every location so that the bulk liquid concentration equals the surface concentration. All of these idealizations are not strictly accurate in a real evaporator but they allow some insight into the evaporation process. In particular, the heat of vaporization defined in Equation 3.63 is a property of the fluid which is related to the actual heat required to evaporate the mixture in a real device.

Another evaporation process of interest is total evaporation in a tube with heat applied. This process involves vapor and liquid flowing co-current as shown in Figure 3.7. As a first approximation, assume that the liquid and vapor are in equilibrium at every point along the tube and that the process is isobaric. One difference between this process and that illustrated in Figure 3.6 is that the vapor must be heated, along with the liquid, to maintain the equilibrium temperature. Another difference comes from mass balance considerations. For total evaporation, the exiting vapor composition must match the entering liquid composition. Also the initial vapor bubbles that form in the device have much more of the volatile component. Thus, as the temperature rises through the device, the transfer of mass between the liquid and vapor cannot occur at the equilibrium concentration. What happens instead is that as the temperature rises and the equilibrium concentration shifts accordingly, the chemical potentials act to selectively extract the components in the composition mix needed to reach the end state.

An ammonia mass balance for a differential change in the mass fraction of the liquid gives

$$dm_N = \frac{m_l + \dfrac{x_l}{x_v}m_v\dfrac{\partial x_v}{\partial x_l}}{\dfrac{x_l}{x_v}(1 - x_v) + x_l - 1} dx_l \qquad 3.65$$

Similarly for the mass transfer of the water component

$$dm_H = \frac{1}{x_v}\left[(1 - x_v)dm_N - m_v\frac{\partial x_v}{\partial x_l}dx_l\right] \qquad 3.66$$

The model is completed by enforcing overall mass balance considerations on the liquid and vapor which amounts to

$$m_v + m_l = m_{in} \qquad 3.67$$

where m_{in} is the inlet mass flow rate. Finally, by definition,

$$dm_v = dm_N + dm_H \qquad 3.68$$

The mass balance details in Equations 3.65 to 3.68 are illustrated for a particular case in the following example.

Example 3.3 Total Evaporation of Ammonia/Water Mixture

Consider a system consisting of a heated pipe with a mixture of ammonia/water flowing isobarically at a pressure of 10 bar. Analyze the case where the inlet condition is saturated liquid with a mass fraction of ammonia of 0.5. Assuming the vapor and liquid are in equilibrium at every location within the pipe, trace the mass transfer process as the vapor quality goes from 0 to 1.0. In addition, determine the heat transfer rate needed to drive the evaporation process. File: ex3_3.ees.

Solution: The inventory of mass can be determined by integrating Equations 3.65 to 3.68 over the full range of concentrations. At 10 bar, the saturated states are

Saturated Liquid: T = 63.975°C   Saturated Vapor: T = 153.606°C
                      $x_l = 0.5$ $x_v = 0.5$
                      $x_v = 0.991$ $x_l = 0.098$

Thus, if we consider the liquid mass fraction as the independent variable, the integration proceeds from 0.5 to 0.098. The derivative in Equations 3.65 and 3.66 was evaluated numerically by introducing an increment in temperature and approximating the derivative as

$$\frac{\partial x_v}{\partial x_l} = \frac{x_{v,T+\Delta T} - x_{v,T}}{x_{l,T+\Delta T} - x_{l,T}} \qquad 3.69$$

The appropriate size for the increment in temperature depends on the accuracy required and was chosen as 0.005°C for this calculation. The equations can then be integrated numerically.

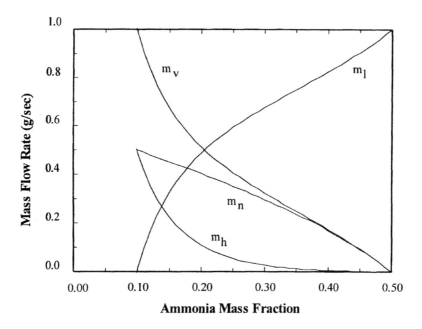

Figure 3.8 Mass transfer in total evaporation process

These equations represent a coupled set of first order ordinary differential equations and the boundary conditions define it as an initial value problem. Thus, an integration scheme such as Runge-Kutta is appropriate. The size of the integration step will influence the accuracy of the final result. The result obtained for an integration step of $\Delta x_l = 0.01$ is shown in Figure 3.8.

The total heat transfer needed can be obtained simply from an overall energy balance as

$$Q = m_{in}(h_v - h_l) \qquad\qquad 3.70$$

The heat flux required to evaporate all of the liquid varies along the length of the tube. An energy balance on a differential section yields

$$dQ = (h_v - h_l)\,dm_v + m_l\,dh_l + m_v\,dh_v \qquad\qquad 3.71$$

By integrating Equation 3.71 over the entire process, the heat transfer profile given in Figure 3.9 is obtained.

3.1.8   Summary

The overview of mixture thermodynamics given here has concentrated on those issues of direct relevance to absorption cycle analysis. Not all of the issues will be addressed in a particular analysis. If the object is to simply calculate the performance of a cycle from a complete set of thermodynamic data, then much of this material will be superfluous. However, if a new working fluid is being used, if the cycle conditions extend beyond the range of available data or if a detailed component analysis is needed, then many of the topics treated will be found useful. An alternative analysis procedure useful when limited thermodynamic data are available is given in Alefeld and Radermacher (1994).

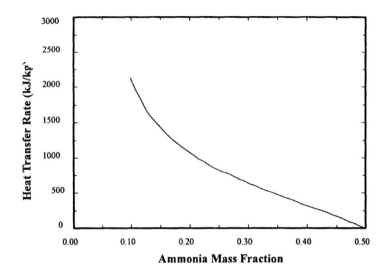

Figure 3.9 Heat transfer in total evaporation process

3.2   Graphical Perspective on Thermodynamic Properties of Absorption Working Fluids

The performance and efficiency of reversible cycles are independent of the properties of any working fluids. However, the performance and efficiency of a real machine are determined to a large degree by the properties of the working fluids. Both the first cost and the operating cost of an absorption system are strongly dependent on the working fluid properties. This chapter describes the properties of ammonia/water and water/lithium bromide mixtures which are the traditional absorption working fluids. Typical diagrams of fluid properties are discussed. Further, the determination of the properties for cycle calculations are explained in detail.

When designing an absorption system the most important thermodynamic variables to be considered are: pressure, temperature, mass fraction, enthalpy, specific volume and entropy. To display all variables, a multidimensional diagram is required. This is not practical. Several two-dimensional diagrams are in common use. These diagrams show any two variables on their axes and display other variables as sets of curves of constant properties such as isobars and isotherms. Usually, T-s, ln(P)-h or h-s diagrams are used for design calculations of cycles with pure fluids. However, in absorption processes, the additional variable, composition, must be considered. Historically, enthalpy-mass fraction diagrams (h-x diagrams) were preferred with temperature and pressure as parameters. Figures 3.10 and 3.11 are examples. They will be discussed later in full detail.

For the sake of simplicity, the explanation focuses on two-component (binary) mixtures only. From a thermodynamic point of view, a two-component mixture possesses one additional degree of freedom as compared to a pure fluid, the mass fraction. The mass fraction may be defined in many different ways including mole fraction, concentration, etc. In this text, the mass fraction is used throughout, defined as

$$x = \frac{mass \ of \ one \ component \ [kg]}{total \ mass \ of \ both \ components \ [kg]} \qquad\qquad 3.72$$

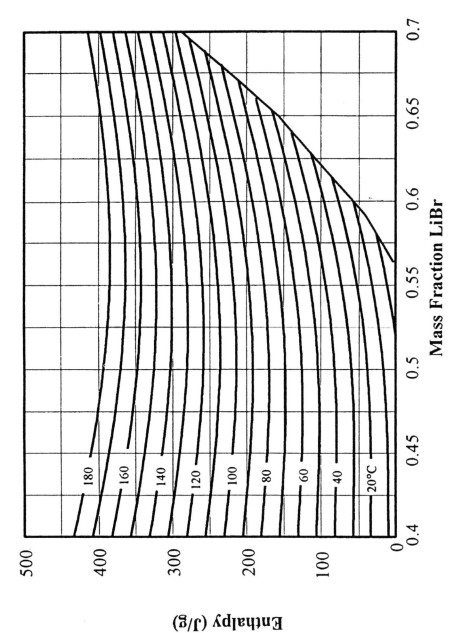

Figure 3.10 Enthalpy-mass fraction diagram for water/lithium bromide

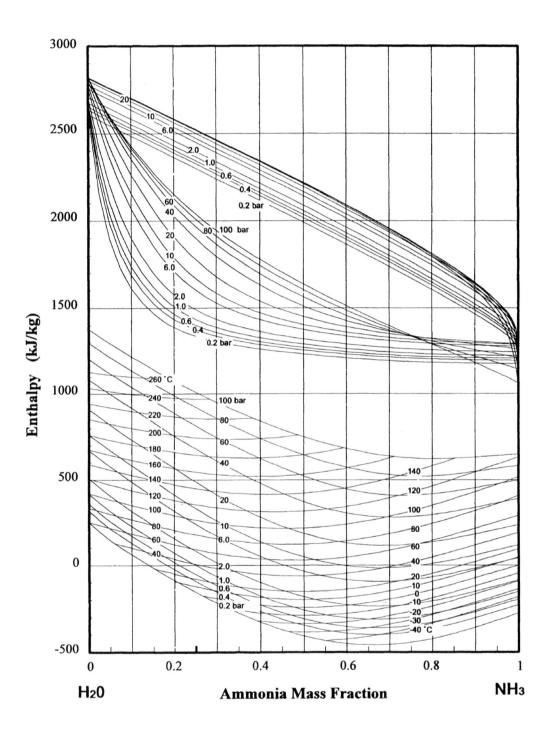

Figure 3.11 Enthalpy-concentration diagram for ammonia/water

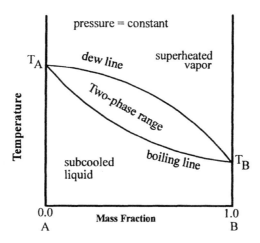

Figure 3.12 Schematic of temperature-concentration diagram

3.2.1 Temperature-Mass Fraction Diagram

When the liquid and vapor phases of a mixture coexist in equilibrium, the saturation temperature varies with the mass fraction even though the pressure is constant. This is in contrast to a pure fluid. Figure 3.12 shows a schematic of a temperature-mass fraction diagram (T-x diagram) for a mixture of two components, A and B, at constant pressure. The mass fraction axis ranges from 0 (only component A is present) to 1.0 (only component B is present). The area below the boiling line represents subcooled liquid. The area above the dew line represents superheated vapor. The area enclosed by the boiling and dew lines is the two-phase region. The boiling point for a mixture of mass fraction x is located on the boiling line at that mass fraction. The boiling line indicates the temperature at which the first vapor bubble is formed for the specified pressure and mass fraction. The boiling points of the pure components, T_A and T_B, are found on the respective ordinates. In Figure 3.12, the boiling point of component A is higher than that of component B. The dew line indicates the temperature at which the first liquid droplet is formed when a gas mixture of a given mass fraction is cooled.

To demonstrate the use of the diagram, a slow (equilibrium) evaporation process at constant pressure in a closed system is discussed as an example in Figure 3.13. The process begins with subcooled liquid at point 1'. Points with a single prime denote the liquid phase; points with a double prime, the vapor phase. As the mixture is heated, the temperature increases and the boiling line is reached. This is point 2'. Here the first vapor bubble forms. The mass fraction of the first vapor formed is in thermal equilibrium with the liquid found at point 2'. The vapor is enriched in component B as compared to the liquid. Its mass fraction is $x_{2''}$. This is a consequence of the fact that, at the same temperature, component B has a higher vapor pressure than component A.

As the heating process continues the evaporation process proceeds to point 3 where the mass fraction of the vapor in equilibrium with the remaining liquid is represented by point 3". The mass fraction of the liquid is now indicated by 3'. At this point, the amount of component B in the remaining liquid has been reduced as compared to point 2' while the vapor is enriched in component B. However, the vapor contains a lower fraction of component B than at point 2" and more of component A. As the evaporation process proceeds, the state points of the liquid and vapor phases continue to follow the boiling and dew point lines. When point 4" is reached, the evaporation process is completed. The vapor has the same mass fraction as the

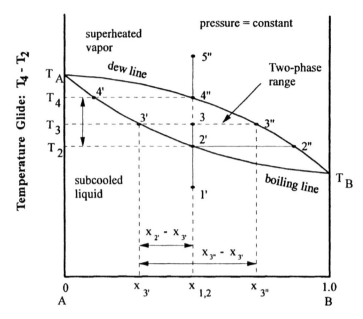

Figure 3.13 Evaporation process in temperature-concentration diagram

original subcooled liquid and the mass fraction of the last liquid droplet is indicated by point 4'. Further heating produces superheated vapor at point 5".

During the constant pressure evaporation process, the saturation temperature changed from T_2 to T_4. The temperature difference $(T_4 - T_2)$ is termed "temperature glide". Vapor quality x_q (defined as the ratio of mass of vapor over total mass) at point 3 can be calculated based on

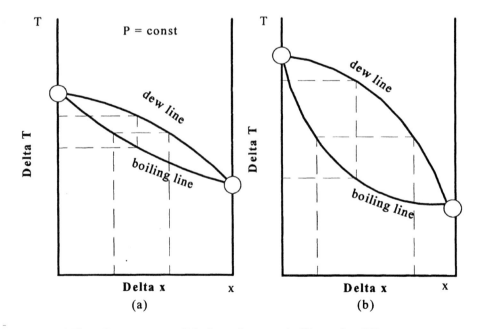

Figure 3.14 Size of temperature glide dependence on boiling point difference

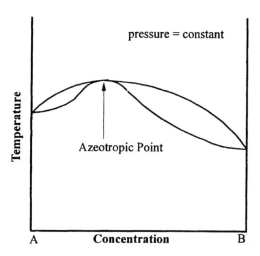

Figure 3.15 Schematic of the temperature-concentration diagram for an azeotropic mixture

a mass balance for the mixture and one pure component.

$$m_3{}'' + m_3{}' = m_2{}' \qquad\qquad 3.72$$

$$m_3{}'' x_3{}'' + m_3{}' x_3{}' = m_2{}' x_2{}' \qquad\qquad 3.73$$

After elimination of $m_3{}'$, the vapor quality is obtained as:

$$x_q = \frac{m_3{}''}{m_2{}'} = \frac{(x_2{}'-x_3{}')}{(x_3{}''-x_3{}')} \qquad\qquad 3.74$$

The vapor quality at state 3 is represented in Figure 3.13 by the ratio $(x_3 - x_3')/(x_3'' - x_3')$ and can be expressed as a function of mass fractions only.

It is noted that the temperature glide increases with increasing difference between the boiling points of the two pure components. Figure 3.14a shows a temperature-mass fraction diagram for a mixture where the difference in boiling points is relatively small, thus the temperature glide delta. This is relatively small as well. Figure 3.14b shows the temperature-mass fraction diagram of a mixture where the difference of the boiling points is large. Accordingly, the temperature glide marked as Delta T is large as well. The size of the temperature glide is also a function of the mass fraction. For small and large x the glide is generally smaller than for intermediate values of x.

The mixtures in Figures 3.12 to 3.14 are traditionally termed a "non-azeotropic mixture" or, in more recent literature a "zeotropic mixture". The name implies that in phase equilibrium the mass fractions of the vapor and liquid phases are always different. Some fluids form azeotropic mixtures, for example, a mixture of R12 and R152a or a mixture of water and ethanol. For an azeotropic mixture the mass fractions of the liquid and vapor phase is identical at a certain pressure and temperature as shown in Figure 3.15. This state is called the azeotropic point. The temperature glide is zero at this point. At all other mass fractions the mixture exhibits zeotropic behavior. The difference in mass fraction between the liquid and

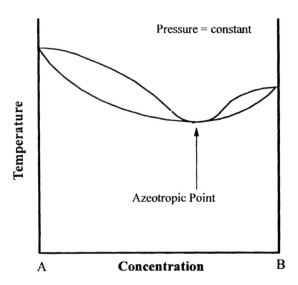

Figure 3.16 Schematic of the temperature-concentration diagram for an azeotropic mixture

vapor phases changes its sign when the overall mass fraction varies from a value less than the azeotropic mass fraction to a value larger than the azeotropic mass fraction. There are two types of azeotropes. These are distinguished from one another by the location of the boiling point at the azeotropic mass fraction relative to the boiling points of the pure fluids. The boiling point can either be higher than the boiling point of either of the two constituents of the mixture or lower than the boiling point of either of the two constituents as illustrated in Figure 3.16.

3.2.2 Pressure-Temperature Diagram

For preliminary investigations and comparisons of working fluids, the pressure-temperature diagram (or vapor pressure diagram as it is often referred to) turns out to be very helpful. Usually, the logarithm of the pressure is plotted versus the negative reciprocal of the temperature. This diagram is referred to at times as the ln(P), (-1)/T diagram. The advantage of this representation is the fact that the plots of saturation temperature vs. saturation pressure are almost straight lines for most fluids and fluid mixtures when the mass fraction is constant. These plots are referred to as vapor pressure curves. Figure 3.17 shows vapor pressure curves for a number of refrigerants. The following statements in this paragraph are valid for pure fluids only. The area to the left of the vapor pressure curve of one particular fluid represents higher pressures and/or lower temperatures than saturation values and represents subcooled liquid. The area to the right represents in contrast higher temperatures and/or lower pressures and represents superheated vapor. The two-phase range is represented by the vapor pressure line itself. In some representations, the vapor pressure curves are drawn exactly as straight lines. In those cases the scale of the pressure axes is adjusted accordingly. These diagrams were originally developed by Dühring and are referred to as Dühring plots.

For mixtures of fluids, curves similar to vapor pressure curves can be found by plotting the vapor pressure vs. -1/T for isosteres (lines of constant mass fraction) of the saturated liquid phase. Figure 3.18 shows, as an example, water/lithium bromide and Figure 3.19, ammonia/water. The space between the vapor pressure curves of the pure constituents of the mixture is termed the "solution field". The pressure-temperature diagrams of solutions are used extensively for the representation of cycle configurations.   Azeotropic mixtures yield vapor pressure lines as similar to those for pure components shown in Figure 3.17 (for example, R502, a mixture of R22 and R115). However, while for zeotropic mixtures all vapor pressure curves for mass fractions x with $0 < x < 1.0$ are located in between the two pure components, the vapor pressure curve for the azeotropic mass fraction of an azeotropic mixture is located outside the range limited by the vapor pressure curves of the pure components.

3.2.3 Pressure-Enthalpy Diagrams

Figure 3.21 displays the pressure-enthalpy diagram (ln(P)-h diagram) for ammonia. While this diagram is traditionally used in the refrigeration field for the evaluation of vapor compression cycles, it is of importance here whenever data for pure, superheated ammonia are required for cycle calculations. None of the other diagrams traditionally used in absorption technology analysis show data for superheated vapor since the additional variable complicates the presentation, rendering it incomprehensible on a two-dimensional plot.

The two-phase range is found under the vapor dome, the area where the isotherms are horizontal. To the left of the two-phase range is the subcooled liquid and to the right, the superheated vapor. The lines for the state points of saturated liquid and saturated vapor converge at the critical point.

3.2.4 Temperature-Entropy Diagrams

Figure 3.20 displays a temperature-entropy diagram for water. Traditionally, this diagram is used for representing the thermodynamic properties of water for the design of power plants. It is introduced here, since it provides information about the properties of superheated steam which are necessary to analyze absorption heat pumps that use the water/lithium bromide mixture. Both diagrams, the T-s diagram and the ln(P)-h diagram are representations of the properties of a pure fluid. The only difference is the choice of independent variables. The location of the ranges for the various phases is very similar as for the pressure-enthalpy diagram.

3.2.5 The Enthalpy-Mass Fraction Diagram

The enthalpy-mass fraction diagram (h-x diagram) was originally introduced by Merkel and Bosnjakovic [Niebergall, 1959]. It found wide applications for the design of absorption heat pumps and distillation equipment. It provides information about enthalpies, composition of the liquid and vapor phases, temperature and pressure. All lines on this diagram represent saturation properties.

Enthalpy-composition diagrams for water/lithium bromide and ammonia/water are given in Figures 3.10 and 3.11. Figure 3.22 shows a schematic of a constant pressure h-x diagram indicating three regions. The lower region displays properties of the liquid phase. Figure 3.11 shows sets of isotherms and isobars in this region. In Figure 3.10 only isotherms are shown for the sake of simplicity. Although the isobars and isotherms represent saturated liquid, they are also approximately valid for subcooled liquid, assuming that the enthalpy of the liquid phase is independent of the pressure. This assumption is usually fulfilled quite well when the temperature is below 0.9 of the critical temperature.

The middle region of the diagram, Figure 3.22, represents the two-phase region. This area is bordered by the boiling line and dew line. These two lines are isobars. The enthalpy difference at the endpoints ($x = 0$ and $x = 1.0$), between the boiling and dew line, represents the latent heat of evaporation of the pure fluids. Within the two-phase region, the equilibrium

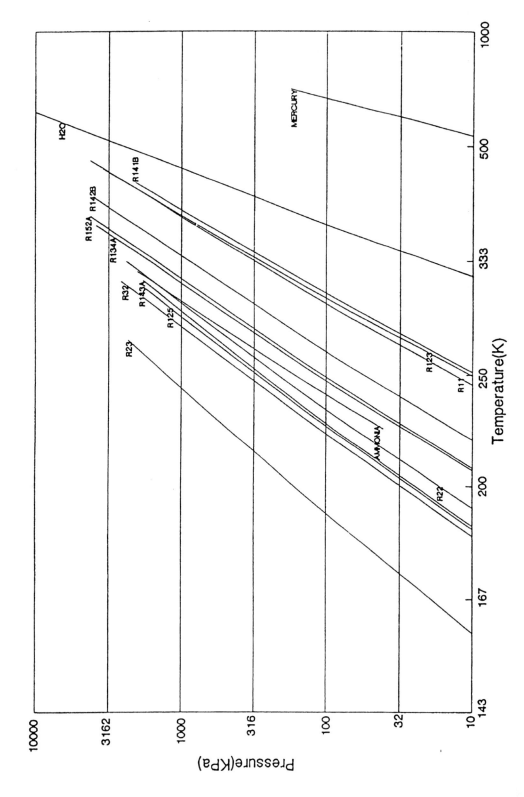

Figure 3.17 Pressure-temperature diagram for several pure fluids

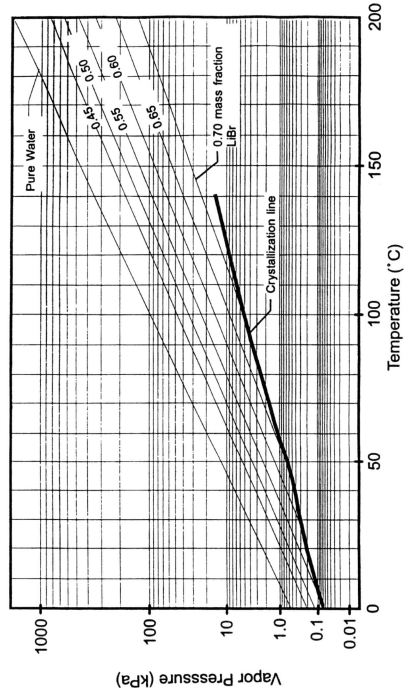

Figure 3.18 Pressure-temperature diagram for water/lithium bromide

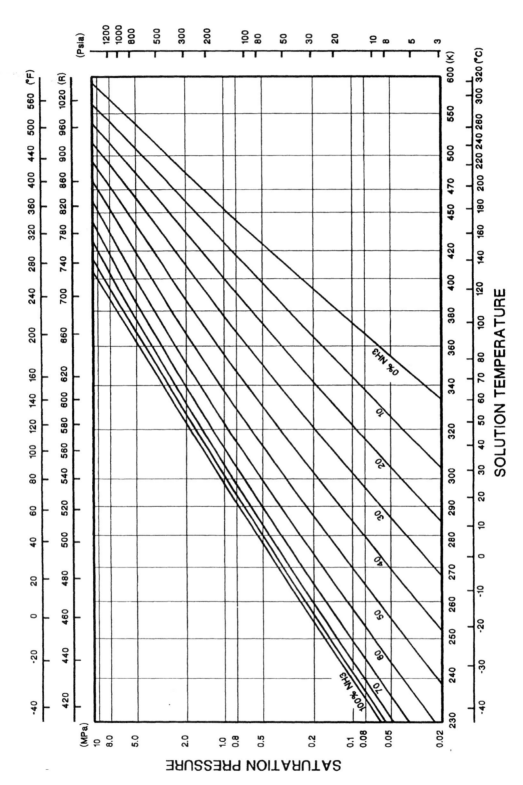

Figure 3.19 Pressure-temperature diagram for ammonia/water

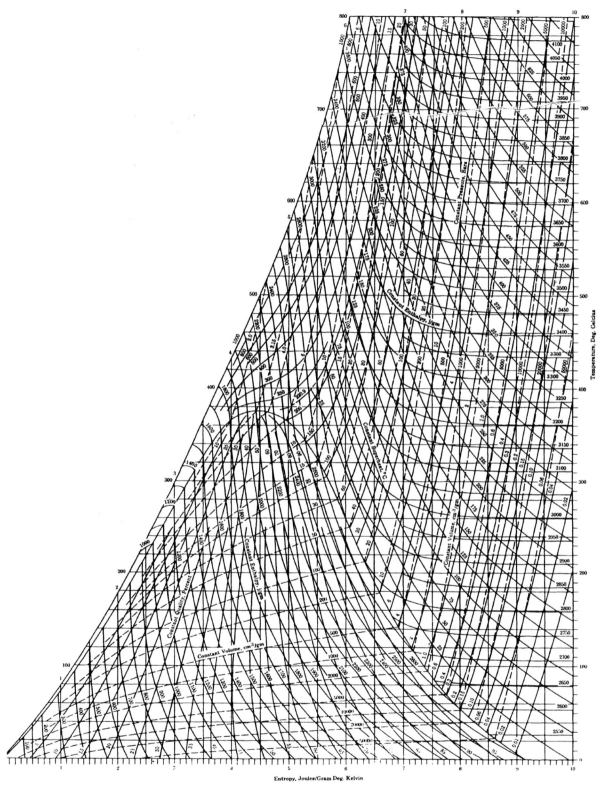

Figure 3.20 Temperature-entropy diagram for water. Source: Keenan, J.H., Keyes, F.G., Hill, P.G., Moore, J.G., 1969, <u>Steam Tables</u>, Wiley, New York, Reprinted by permission of John Wiley & Sons, Inc.

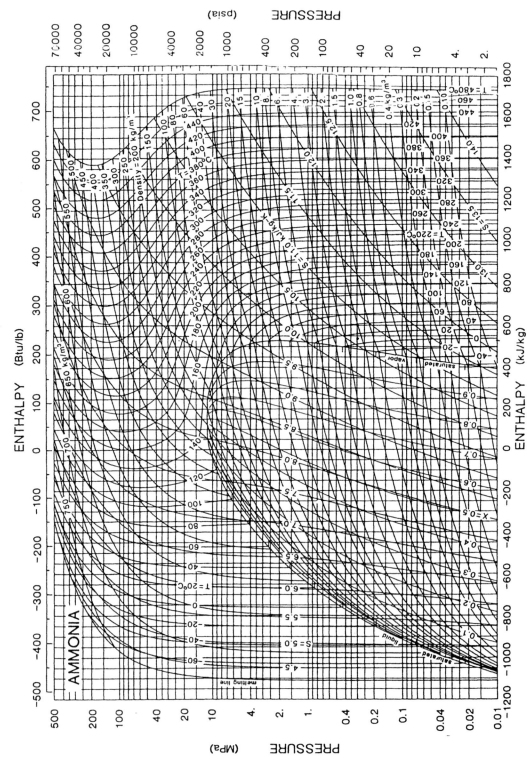

Figure 3.21 Pressure-enthalpy diagram for ammonia. Reprinted with permission from the American Society of Heating, Refrigerating and Air-Conditioning Engineers from the 1993 ASHRAE Handbook - Fundamentals.

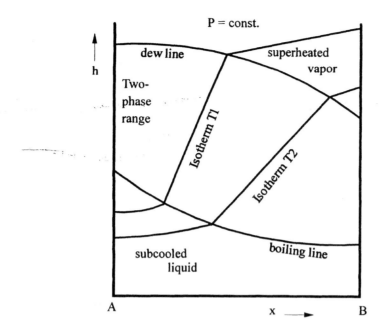

Figure 3.22 Schematic of an enthalpy-concentration diagram

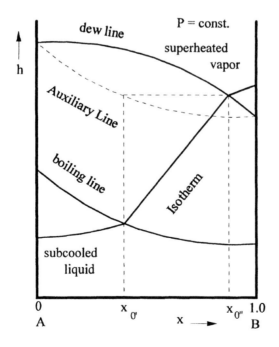

Figure 3.23 Schematic for finding the vapor concentration

vapor and liquid states can be connected by a tie line. Tie lines are isotherms and isobars. Two examples of typical isotherms are shown in Figure 3.22. These two examples are valid only for the pair of dew and boiling line (isobars) shown.   For other pressure levels, other pairs of dew and boiling lines exist. Each of these pairs has its own set of tie lines. The tie lines in the two-phase region are not shown for the sake of clarity.

The upper region of the diagram is the superheated vapor area. In this area the isotherms are usually almost straight lines since the heat of mixing of gases is negligible at low pressures. Again, each dew line has its own set of isotherms. They are usually not displayed to avoid crowding.

In Figure 3.11, the vapor mass fraction in equilibrium with a saturated liquid of a given mass fraction can be determined by finding the intersection of the liquid isostere $x_{0'}$ with the auxiliary line in the two-phase region. This construction determines the enthalpy of the saturated vapor in equilibrium with the liquid phase. From there, one follows a line of constant enthalpy (which is the enthalpy of the vapor) to the dew line. The intersection with the dew line indicates the vapor mass fraction $x_{0''}$. This procedure is shown in Figure 3.23.   In Figure 3.11 the isotherms (tie lines) in the two-phase region are not shown. Their endpoint on the dew line can be constructed with the help of the auxiliary lines.

HOMEWORK PROBLEMS

3.1   Derive Equations 3.20 and 3.21 from Equation 3.19.

3.2   Consider a mixture of ammonia and water. Evaluate the chemical potential of NH_3 in both the liquid and vapor phases at equilibrium for the state at $T = 100\,°C$ and $P = 10$ bar.

3.3   Repeat #2 for H_2O.

3.4   Perform the analysis in Section 3.1.6 in terms of mole fraction.

3.5   The mixture molecular weight, M_{eff}, is defined by

$$n\, M_{eff} \;=\; m$$

     a)   Find an expression for the mixture molecular weight in terms of the component molecular weights and the mole fraction.
     b)   Repeat for the mass fraction.

3.6   Consider NH_3/H_2O liquid at $P = 10$ bar, $T = 50°C$ and $x = 0.5$ mass fraction.
     a)   Determine the mixture specific enthalpy.
     b)   Convert the mixture specific enthalpy to molar units (kJ/kmole).
     c)   Determine the partial molal enthalpy of each of the components in mass units.
     d)   Determine the enthalpy of mixing.
     e)   Convert the partial molal enthalpy values to molar units.
     f)   Convert the enthalpy of mixing to molar units.

3.7   Consider the properties of aqueous lithium bromide along a 50°C saturated isotherm. Since the states are all saturated, the state of the equilibrium vapor is fully known. Since it is essentially pure water vapor, the chemical potential is known. Thus, the chemical potential of the water in the liquid is also known. From this information, determine the chemical potential variation of the lithium bromide in the liquid. Hint: Use the Gibbs-Duhem equation.

3.8   Write a function to compute the specific heat at constant pressure for ammonia/water and water/lithium bromide.

3.9   Evaluate the chemical potential of LiBr in both the liquid and vapor phases at equilibrium for the state at $T = 50°C$ and $P = 4$ kPa. Note that this problem is a natural follow-on problem to #3.7.

3.10   Repeat #9 for H_2O.

3.11   Check the thermodynamic consistency of the data for both ammonia/water and water/lithium bromide by integrating Equation 3.15 and comparing the result with the enthalpy differences obtained directly. It is of interest to perform this check at several points throughout the domain. To make the problem more specific consider the following specifications:

a. Ammonia/water from $(T,p,x) = 100°C$, 10 bar and 0.5 mass fraction ammonia to 120°C, 11 bar and 0.5 mass fraction ammonia.

b. Water/lithium bromide from $(T,p,x) = 100°C$, 1 kPa and 0.6 mass fraction lithium bromide to 120°C, 2 kPa and 0.6 mass fraction lithium bromide.

3.12   Generate an enthalpy-mass fraction diagram for water/lithium bromide.

3.13   Generate a pressure-temperature-mass fraction diagram for water/lithium bromide.

3.14   Generate an enthalpy-mass fraction diagram for ammonia/water.

3.15   Generate a pressure-temperature-mass fraction diagram for ammonia/water.

Chapter 4

THERMODYNAMIC PROCESSES WITH MIXTURES

In this chapter thermodynamic processes involving mixtures are discussed with emphasis on how mixtures change the operation of equipment as compared to pure fluids. As shown in Chapter 3.1 one of the important changes is the introduction of a temperature glide, i.e., non-isothermal evaporation, although the pressure is constant.

4.1   Mixing of Fluids and the Heat of Mixing

In order to explain the phenomenon of the heat of mixing, the following experiment is discussed. As shown in Figure 4.1 two fluid streams 1 and 2 are entering a mixing chamber. The mass flow rate of each stream, its temperature, and enthalpy are assumed to be known. It is further assumed that the pressure of both fluid streams is the same. The product of this mixing process is leaving the mixing chamber as stream 3 at the same pressure as the two incoming streams. For the sake of simplicity, it is assumed that the two entering streams are pure fluids, for example, ammonia and water. The mixing chamber is adiabatic. The following mass balance must apply:

$$m_1 + m_2 = m_3 \qquad\qquad 4.1$$

When Equation 4.1 is divided by m_3 and it is observed that the ratios m_i/m_3 ($i = 1,2$) are mass fractions, then the following equation is obtained:

$$x_1 + x_2 = 1 \qquad\qquad 4.2$$

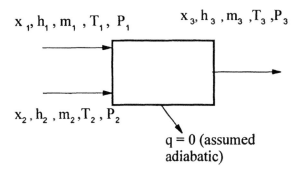

Figure 4.1 Mixing of two fluid streams

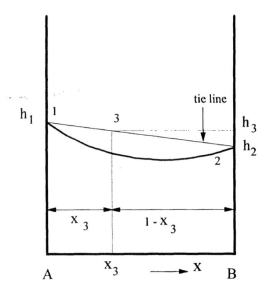

Figure 4.2 Mixing process in enthalpy-mass fraction diagram

Here the definition of the mass fraction was used, Equation 3.72. For the energy balance the following equation is obtained (adiabatic):

$$h_1 m_1 + h_2 m_2 = m_3 h_3 \qquad\qquad 4.3$$

or with Equation 4.2

$$h_1 x_1 + h_2 x_2 = h_3 \qquad\qquad 4.4$$

When all the properties of streams 1 and 2 are known, then the enthalpy h_3 can be found quite readily from the enthalpy-mass fraction diagram, Figure 4.2. According to Equation 4.4, h_3 is located on a straight line (tie line) that connects the state points of the two incoming streams, 1 and 2. h_3 is located where the line of constant concentration for x_3 intersects the tie line from 1 to 2. This point divides the tie line in the ratio of the mass fraction of the exiting stream.

If there were no heat of mixing (i.e., if the mixture of A in B were an ideal solution) then the tie line in Figure 4.2 would be an isobar. However, for most fluids, there exists a heat of mixing. For typical absorption fluid pairs, the heat of mixing is negative and there is an energy release upon mixing. The saturated isobar for the fluid is no longer a straight line but curved downward as in Figure 4.2. Nevertheless, for an adiabatic process, the construction using the tie line to find h_3 is still valid. The energy balance of Equation 4.4 is independent of the heat of mixing (i.e. the tie line construction is independent of the fluid properties). Once point 3 is located on the enthalpy-mass fraction diagram, the temperature of point 3 can be determined from the isotherm passing through that point.

The construction using the tie line is valid under all circumstances of an adiabatic mixing process. The incoming streams may already be mixtures of A and B themselves, and/or they may be in different phases or in the two-phase region.

To completely describe the mixing process for the case in which the incoming streams are binary mixtures an additional mass balance is required to account for all species. Equation 4.1 above is an overall mass balance. The second equation is a mass balance for one component, for example, B.

$$m_1 x_1 + m_2 x_2 = m_3 x_3 \qquad 4.5$$

After dividing by m_3 a mass flow rate ratio f can be introduced

$$f = \frac{m_1}{m_3} \qquad 4.6$$

and we obtain

$$f x_1 + (1 - f) x_2 = x_3 \qquad 4.7$$

with

$$f = \frac{x_3 - x_2}{x_1 - x_2} \qquad 4.8$$

and the energy balance yields

$$f h_1 + (1 - f) h_2 = h_3 \qquad 4.9$$

Equations 4.7 and 4.9 represent the graphical construction shown in Figure 4.2 generalized to the case where the inlet streams have arbitrary mass fraction.

Example 4.1 Adiabatic Mixing with Water/Lithium Bromide

Consider an adiabatic mixing process where two solution streams are mixed at the same temperature. Assume that the mixing process occurs isobarically at a pressure of 1 bar. The data for the two streams are given below. The object is to determine the outlet state. File: ex4_1.ees.

	T (°C)	X (% LiBr)	Mass Flow Rate (kg/sec)
Stream 1	50	50	1.5
Stream 2	50	60	7.5

Schematic:

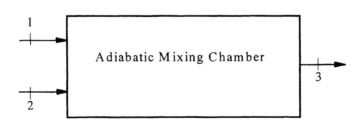

Analysis: An overall mass balance yields

$$m_3 = m_1 + m_2$$

$$= 9.0 \ kg/sec$$

A mass balance on the LiBr yields

$$x_3 = \frac{m_1 x_1 + m_2 x_2}{m_3}$$

$$= 0.58333 \ mass \ fraction \ LiBr$$

Based on the known data for the input streams, the enthalpy values can be obtained from an enthalpy chart or an appropriate equation. The results are $h_1 = 103.92$ J/g, $h_2 = 134.99$ J/g. An energy balance on the chamber then yields the enthalpy of the exiting stream as

$$h_3 = \frac{m_1 h_1 + m_2 h_2}{m_3}$$

$$= 129.81 \ J/g$$

Technically, the state is now totally determined since three properties are known. However, it is of interest to determine the temperature. This is done by a reverse process to the one used to find the inlet enthalpy values. The result is $T_3 = 51.4°C$.

Observations: The mass fraction of the outlet state lies between the mass fractions of the two inlet streams. This is also true of the enthalpy. The temperature of the outlet stream is higher than the temperature of either of the inlet streams. This comes about because of the exothermic nature of the mixing process whereby some of the internal energy associated with the LiBr-H_2O interactions in the inlet streams is liberated and shows up as an increase in the thermal energy (i.e., as a temperature increase).

4.2   Specific Heat of Mixtures

The specific heat of a mixture can be derived as follows. Assume the experiment of Figure 4.1 is now conducted as an isothermal process. Then q is the amount of energy (the heat of mixing) that has to be removed to maintain a constant temperature. The energy balance yields

$$f h_1 + (1 - f) h_2 - h_3 = q \qquad \qquad \textbf{4.10}$$

By taking the derivative of Equation 4.10 with respect to temperature, all enthalpies are expressed as specific heats at constant pressure

$$f c_{p1} + (1 - f) c_{p2} - c_{p3} = \left. \frac{dq}{dT} \right|_{P,x} \qquad \qquad \textbf{4.11}$$

Solving for the specific heat of the mixture, c_{p3}, the following expression is obtained.

$$c_{p3} = f c_{p1} + (1 - f) c_{p2} - \left. \frac{dq}{dT} \right|_{P,x} \qquad \qquad \textbf{4.12}$$

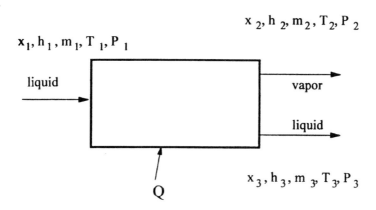

Figure 4.3 Desorption process

Thus, the specific heat of the mixture is equal to the mass fraction-weighted average of the specific heats of the constituents plus a so-called mixing term that accounts for the change of the heat of mixing with temperature. Equation 4.12 is very important for the construction of the single-phase regions of enthalpy-mass fraction diagrams.

4.3   Desorption

The term "desorption" describes the generation of vapor from the condensed phase (liquid or solid) of a mixture of two or more components. The term implies that the vapor contains predominantly one component in contrast to evaporation where all components are assumed to vaporize. A typical example is the desorption of water out of a water/lithium bromide mixture or of ammonia out of an ammonia/water mixture.

Figure 4.3 shows a desorption process in a steady-state, steady flow configuration. By adding thermal energy to the entering stream 1, a vapor stream 2, is generated while any remaining liquid leaves as stream 3. It is assumed that the properties of all three streams are known. The mass balance for the overall mass flow is written as follows

$$m_2 + m_3 = m_1 \qquad\qquad 4.13$$

and for one component as

$$m_2 x_2 + m_3 x_3 = m_1 x_1 \qquad\qquad 4.14$$

After dividing by the vapor mass flow rate (m_2) the solution circulation ratio, f, can be introduced

$$f = \frac{m_1}{m_2} \qquad\qquad 4.15$$

By introducing f, the following equations are all based on one unit of mass of refrigerant vapor

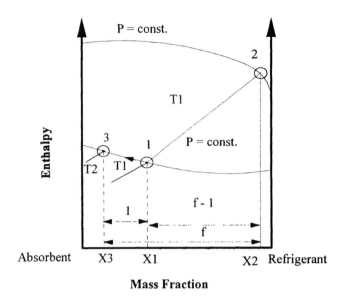

Figure 4.4 Desorption process in h-x diagram

produced in the desorption process. As will be shown later, this greatly simplifies the understanding of the interdependence of the variables involved. Rewriting the mass balance, Equation 4.14, using f yields

$$fx_1 - (f - 1)x_3 = x_2 \qquad\qquad\textbf{4.16}$$

which can be arranged as

$$f = \frac{x_2 - x_3}{x_1 - x_3} \qquad\qquad\textbf{4.17}$$

It should be noted that f depends only on the mass fractions of the entering and leaving streams.
    The energy balance on the system in Figure 4.3 yields

$$q + fh_1 = h_2 + (f - 1)h_3 \qquad\qquad\textbf{4.18}$$

The energy balance can be rearranged by solving for q to obtain

$$q = h_2 - h_3 + f(h_3 - h_1) \qquad\qquad\textbf{4.19}$$

This equation can be understood in the following way. The heat supplied for the desorption process, q, is used in two contributions $(h_2 - h_3)$ and $f(h_3 - h_1)$. The first contribution is an enthalpy difference between a liquid and a vapor phase. This term can be seen as accounting for the actual phase change process. The second contribution consists of the enthalpy difference between the liquid streams entering and leaving the desorber multiplied by the flow ratio of the liquid streams. This term describes the heating of the remaining liquid. Referring to Figure 3.13, it can be seen that the temperature of the remaining liquid is increasing as such an evaporation process progresses.
    Figure 4.4 shows the desorption process in an h-x diagram. It is assumed here that all

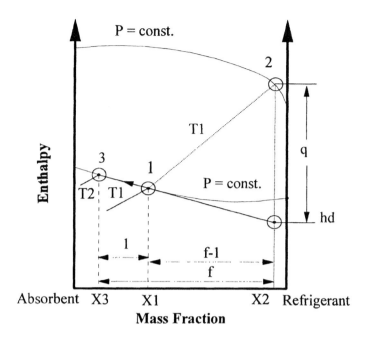

Figure 4.5 Graphic representation of the desorber heat requirement

entering and leaving fluid streams are saturated and that the desorption process occurs at constant pressure. Since it is assumed that at least two properties of each stream are known, the state points can be plotted. For Figure 4.4 it is also assumed that the vapor leaving the desorber is in equilibrium with the incoming liquid stream. However, the correctness of this assumption depends strongly on the design of the actual heat exchanger. In Figure 4.4, the isotherms T_1 and T_2 are shown. The isotherm T_1 continues through the two-phase area to the vapor state 2 which is assumed to be in equilibrium with the incoming liquid stream. At the bottom of the h-x diagram in Figure 4.4, the relationship between the solution circulation ratio f and the mass fractions is shown by applying the lever rule as follows. When, according to Equation 4.17, f is proportional to the difference $x_2 - x_3$ then f - 1 is proportional to the difference $x_2 - x_1$. This can be shown by subtracting 1.0 from both sides of Equation 4.17 and by using that 1.0 can be written as $(x_1 - x_3)/(x_1 - x_3)$ for the right-hand side. This graphical representation is an application of the lever rule for Equation 4.17.

The amount of energy required for the desorption process can be determined by a graphical procedure. For this purpose, Equation 4.19 is rewritten as

$$q = h_2 - h_d \qquad\qquad \textbf{4.20}$$

with

$$h_d = h_3 - f(h_3 - h_1) \qquad\qquad \textbf{4.21}$$

Equation 4.21 can be rearranged to yield

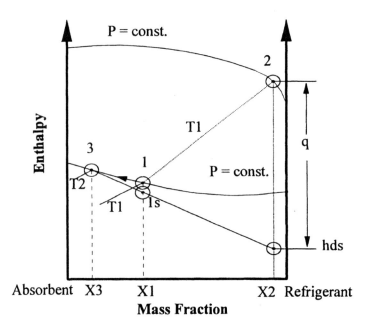

Figure 4.6 Desorption process in h-x diagram with subcooled inlet stream

$$\frac{(h_d - h_3)}{(h_1 - h_3)} = f \qquad\qquad 4.22$$

Using Equations 4.17, 4.20 to 4.22 and applying the lever rule, the heat of desorption can be plotted on the h-x diagram as shown in Figure 4.5. The desorber heat is the difference between the enthalpy of the leaving vapor stream and h_d. h_d is found by using the relationship between the differences in concentration and enthalpy as shown in Equations 4.21 and 4.17. h_d is found at the intersection of the extension of the tie line between points 1 and 3 with the isostere of the vapor. The vertical distance between the isobars of the liquid and vapor phase would yield an incorrect value. Because the enthalpy values in the diagram are valid per unit of mass of the fluid at a particular state point, the fact that for one unit of mass of desorbed vapor a liquid stream of the flow rate f (f > 1.0) is introduced into the desorber and heated must be accounted for appropriately.

The same method yields correct results independent of the state of the streams. The entering solution may be subcooled or the vapor may not be in equilibrium with any of the fluid streams. As long as the respective state points are known, q can be found either by Equation 4.19 or the graphical method. Figure 4.6 shows an example where the incoming liquid is subcooled, state point 1s. The value of the corresponding h_d, h_{ds}, decreases considerably, increasing the desorber heat requirement. This is particularly important for absorption systems because this is a common occurrence and the desorber heat increases considerably as a comparison of Figures 4.5 and 4.6 reveals.

Example 4.2 Desorption

The heat of desorption is calculated for the process in Figure 4.3. The inlet conditions of the rich liquid stream 1 are given as follows: T(1) = 370 K, x(1) = 0.30 and the fluid is saturated. The outlet conditions for the poor solution are given as x(3) = 0.20 and the solution is saturated. It is assumed that the vapor is in equilibrium with the incoming liquid. File: ex4_2.ees.

Solution: Using the enthalpy-mass fraction diagram or suitable software, the conditions of all state points can be determined as shown in the following Table 4.1. With these properties and

Table 4.1 Properties of ammonia/water at the state points according to Figure 4.3

	h (kJ/kg)	P (bar)	Quality	T (K)	x
1	231.8	8.507	0.000	370.000	0.300
2	1566.6	8.507	1.000	370.000	0.925
3	382.1	8.507	0.000	393.439	0.200

Equations 4.17 and 4.19 the mass flow rate of the rich solution f per kilogram of vapor generated as well as the amount of heat required can be calculated. f amounts to 7.25. This means further that the flow rate of the weak solution is 6.25 (Equation 4.13). The heat requirement is 2274 kJ/(kg of vapor).

Example 4.3 Desorption of Water from Aqueous Lithium Bromide

Consider a desorber operating at steady state with the following operating data. Determine the heat input needed. Assume pressure losses are negligible and that the pressure is 10 kPa throughout. File: ex4_3.ees.

	x (mass fraction)	T (°C)	Mass flow rate (kg/sec)
Point 3	0.55	50	1.0
Point 4	0.60	Saturated	0.85

Schematic:

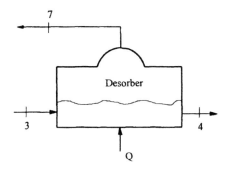

Assumptions: The exiting solution stream at 4 is saturated.

The vapor exits the desorber at the saturation temperature of the entering solution stream. Isobaric process

Analysis: The vapor mass flow rate is obtained from an overall mass balance.

$$m_7 = m_3 - m_4$$

$$= 0.15 \ kg/sec$$

The enthalpy of the solution streams are obtained from property charts or equations. Since point 4 is a saturated state, it can be read exactly from the chart. Point 3 is a subcooled state. Using the subcooled approximation, the enthalpy is read from the chart at the same temperature and mass fraction. The water vapor is superheated and the properties are taken from a pure water data source. The results are $h_3 = 114.33$ J/g, $h_4 = 213.72$ J/g and $h_7 = 2650.2$ J/g. Based on these enthalpy values, the energy balance on the desorber yields

$$Q = m_7 h_7 + m_4 h_4 - m_3 h_3$$

$$= 464.86 \ kW$$

The energy required to raise the inlet stream temperature up to the saturation temperature can also be calculated. Designate the saturation state corresponding to the mass fraction of stream 3 as 3s. The saturation temperature can be obtained from a Dühring plot or an appropriate equation as $T_{3s} = 80.01°C$. The enthalpy is then $h_{3s} = 176.49$ J/g. The sensible heat required is obtained from an energy balance on the inlet stream as

$$Q_{sen} = m_3 (h_{3s} - h_3)$$

$$= 62.16 \ kW$$

Then the remainder is the heat of vaporization of the working fluid

$$Q_{vap} = Q - Q_{sen}$$

$$= 402.70 \ kW$$

For comparison, consider the heat which would be required to vaporize stream 7 at the same pressure ($h_{fg} = 2393.2$ J/g)

$$Q_{fg} = m_7 h_{fg}$$

$$= 358.98 \ kW$$

Observations:

The heat requirement can be broken down into two categories: sensible heat and heat of vaporization of the working fluid. The sensible heat requirement for this case is approximately 13% of the total heat requirement. This is fairly typical in practice. The heat of vaporization can be viewed as consisting of the heat of vaporization of pure water and the heat of mixing of the liquid solution. A rough comparison can be made with the heat of vaporization of water at the same pressure. This indicates that the heat of mixing is approximately 11% of the heat of vaporization. This is fairly typical for water/lithium bromide.

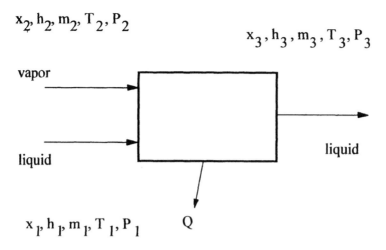

x_2, h_2, m_2, T_2, P_2

x_3, h_3, m_3, T_3, P_3

vapor

liquid

liquid

x_1, h_1, m_1, T_1, P_1 Q

Figure 4.7 Absorption process

4.4   Absorption

The term "absorption" describes the transfer of a binary vapor into the absorbent of a liquid or solid state. It is similar to "condensation" in the sense that a phase change occurs from a vapor state to a liquid state. However, "absorption" implies that there is already a condensed phase present at the absorber inlet. Figure 4.7 shows a schematic of an absorber. The vapor stream 2 is entering the absorber together with the liquid stream 1. After the vapor is absorbed into the condensed phase within the chamber and the heat of absorption released, the product stream 3 is leaving. Again, mass and energy balances can be applied to determine the amount of absorber heat. The analysis procedure is analogous to the one for the desorber. As a final result we obtain for the absorber

$$q = h_2 - h_1 + f(h_1 - h_3) \qquad\qquad 4.23$$

with q representing the amount of heat released per unit of mass of vapor absorbed and f the solution circulation ratio as defined as $f = m_3/m_2$. Again there are two terms, one for the phase change of the vapor and a second one that represents the cooling of the solution. The absorption process is shown in an h-x diagram in Figure 4.8. It is assumed that the liquid streams 1 and 3 are saturated. The vapor state point 2 represents vapor of the same pressure as the liquid streams; however, this vapor is not necessarily in thermodynamic equilibrium since it is supplied from some unknown source. In fact, the vapor could be superheated (2 would be located above the isobar), contain liquid droplets (2 would be located in the two-phase region) and can have any mass fraction, different from the one indicated in Figure 4.8. The amount of energy released in the absorber can be shown graphically as displayed in Figure 4.9. For this purpose Equation 4.23 is rearranged introducing h_a.

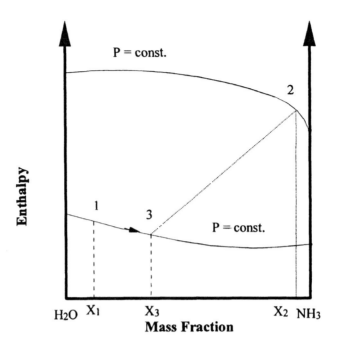

Figure 4.8 Absorption in h-x diagram

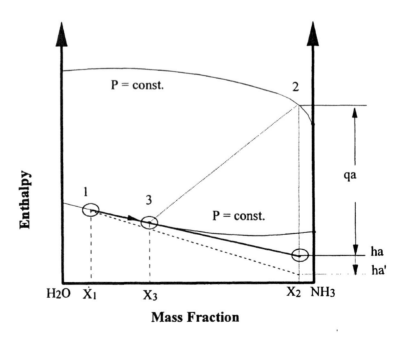

Figure 4.9 Absorption process in h-x diagram with subcooled outlet stream

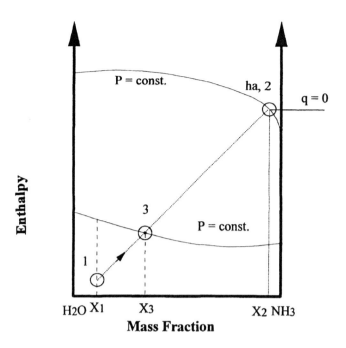

Figure 4.10 Adiabatic absorption

$$q = h_2 - h_a \qquad\qquad 4.24$$

with

$$h_a = h_1 - f(h_1 - h_3) \qquad\qquad 4.25$$

Solving Equation 4.25 for f yields again an expression that contains only differences in enthalpy. By expressing f also in terms of mass fraction, the lever rule can be applied as shown in Figure 4.9. The way of constructing the distance that represents the absorber heat q is quite similar as for the desorber.

The dashed line (long dashes) in Figure 4.9 represents a case in which the liquid stream 3 is subcooled. The change in the absorber heat with h_a is indicated on Figure 4.9. In a similar way, the reader can study how the absorber heat changes when the incoming vapor is superheated, for example.

In some absorption heat pumps a so-called adiabatic absorber is employed. The incoming liquid stream is subcooled significantly. During the absorption process the stream is heated until saturation is reached. Then the absorption process ceases to operate. There is no heat released to the outside, $q = 0$. In an enthalpy-mass fraction diagram, Figure 4.10, it can be seen that h_a becomes equal to h_2. Accordingly, the tie line has to be the same as the line connecting the vapor state point 2 with the liquid state point 1. The intersection of this tie line with the isobar determines the endpoint of the absorption process (point 3).

Example 4.4 Absorber

The heat of absorption is calculated for the process in Figure 4.7. The inlet conditions of the liquid stream 1 are given as follows: T(1) = 310 K, x(1) = 0.20 and the fluid is saturated. The outlet conditions for the rich solution are given as x(3) = 0.30 and the solution is saturated. It is assumed that the vapor originates from an evaporator of the same pressure and has a mass fraction of 0.999. File: ex4_4.ees.

Solution : Using the enthalpy-mass fraction diagram or suitable software, the conditions of all state points can be determined as shown in the following Table 4.2.

Table 4.2 Properties of ammonia/water at the state points according to Figure 4.7

	h (kJ/kg)	P (bar)	Quality	T (K)	x
1	22.1	0.631	0.000	310.0	0.200
2	1227.3	0.631	1.000	257.5	0.999
3	-111.9	0.631	0.000	291.2	0.300

With these properties and Equations 4.17 and 4.23 the mass flow rate of the rich solution f per kg of vapor generated as well as the amount of heat required can be calculated. f is found to be 7.99. This means further that the flow rate of the weak solution is 6.99 (Equation 4.13). The heat requirement is 2275 kJ/(kg of vapor).

Example 4.5 Absorption of Water Vapor into Aqueous Lithium Bromide

Consider a water/lithium bromide absorber operating at steady state with the following operating data. Determine the heat input needed. Assume pressure losses are negligible and that the pressure is 1 kPa throughout. File: ex4_5.ees.

	x (mass fraction)	T (°C)	Mass flow rate (kg/sec)
Point 6	0.60*	Saturated	0.85
Point 1	0.55	Saturated	1.0

* - The overall composition of the two-phase state is 0.6.

Assumptions: The exiting solution stream at 1 is saturated.
                    Isobaric process

Analysis: The vapor mass flow rate is obtained from an overall mass balance

$$m_{10} = m_1 - m_6$$

$$= 0.15 \ kg/sec$$

The enthalpy of the solution stream at 1 is obtained from property charts or equations. Since point 1 is a saturated state, it can be read exactly from the chart, resulting in $h_1 = 85.25$ J/g. Point 6 is a two-phase state arising from the solution partially flashing as it passes through the expansion valve. It was assumed that the temperature of the solution entering the valve is 50°C.

Schematic:

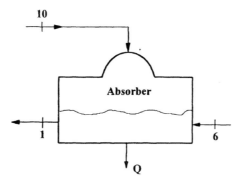

This results in an enthalpy of 134.99 J/g. The process in the valve is assumed to be adiabatic, yielding $h_6 = 134.99$ J/g. Based on the assumed temperature of the vapor stream the enthalpy is read from the steam tables as $h_{10} = 2520.7$ J/g. Based on these enthalpy values, the energy balance on the absorber yields

$$Q = m_6 h_6 + m_{10} h_{10} - m_1 h_1$$

$$= 407.60 \ kW$$

Based on the given conditions at 6, the vapor quality is 0.28% and the temperature is 45.97°C. Because a portion of the water in the solution flashes, the liquid mass fraction entering the absorber is 0.6017.

Observations: The heat requirement in the absorber is approximately equal to the heat requirement in the desorber (of course, the direction is opposite). The heat requirement is larger than one would estimate from pure water consideration due to the heat of mixing in the solution. The heat of mixing increases the heat requirement in both the desorber and the absorber.

4.5   Condensation and Evaporation

The terms "condensation" and "evaporation" refer to the phase change of a pure fluid or a mixture where the process is complete, i.e., there is only vapor entering the condenser (no liquid stream as in an absorber) and there is only vapor leaving the evaporator. In an enthalpy mass fraction diagram, the state points describing the beginning and end of the process are located on the same line of constant mass fraction. The amount of energy released or absorbed in this phase change process is calculated as follows

$$q = h_{out} - h_{in} \qquad\qquad 4.26$$

h_{out} and h_{in} represent the enthalpies of the leaving and entering fluid streams. In the case of condensation the entering vapor may be superheated, in which case h_{in} is the enthalpy of superheated vapor. Further, the leaving condensate may be subcooled liquid. In the case of the evaporator the vapor at the exit may be saturated, be superheated or contain liquid droplets depending on the design of the system.

Example 4.6 Condensation and Evaporation of Water

Consider a condenser and evaporator connected by an expansion valve. Based on given pressures and flow rate, determine the heat transfer rates in the components. The high pressure is 10 kPa and the low pressure is 1 kPa. The mass flow rate of refrigerant into the condenser is 0.15 kg/sec and the inlet temperature is 50°C. File: ex4_6.ees.

Assumptions: Outlet from condenser is saturated liquid. Outlet from evaporator is saturated vapor. Throttling valve is adiabatic.

Schematic:

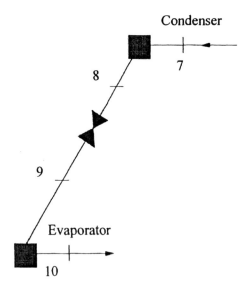

Analysis: Incoming vapor is superheated and the enthalpy can be obtained from the steam tables as $h_7 = 2593.0$ J/g. The condenser exit is saturated liquid yielding $h_8 = 191.81$ J/g. An energy balance on the condenser yields

$$Q_c = m_7(h_7 - h_8)$$

$$= 360.18 \ kW$$

Based on an adiabatic throttling model we get $h_9 = h_8$. The enthalpy of the saturated vapor at point 10 is $h_{10} = 2514.7$ J/g. An energy balance on the evaporator yields

$$Q_e \;=\; m_7(h_{10} - h_9)$$

$$=\; 348.43 \;\; kW$$

Observations: The load in the condenser is slightly higher than that in the evaporator due primarily to superheating of the inlet vapor to the condenser. The condenser and evaporator loads are approximately 10% less than the corresponding generator and absorber loads. This difference is largely due to heat of mixing effects in the solution which are not present in the pure fluid.

4.6   Compression

The compression of a vapor is often assumed to be isentropic to simplify the analysis and in most instances the process occurs entirely in the vapor phase. Only occasionally so-called wet compression is encountered. For the appropriate representation of the compression process, lines of constant entropy are required in the vapor phase. These are not available in an enthalpy-mass fraction diagram. Usually the temperature-entropy diagram or the pressure-enthalpy diagrams are used. Figure 3.21 shows the pressure-enthalpy diagram for ammonia and an isentropic compression process is shown as a solid line. The compression work is calculated using an energy balance as shown next.

$$w \;=\; h_{out} - h_{in} \tag{4.27}$$

h_{out} and h_{in} represent the enthalpies of the entering and leaving fluid streams and Equation 4.27 is valid for the mass flow rate of 1 kg/sec.

4.7   Pumping

In absorption systems, liquid pumps are used for two purposes. The first is for circulating liquid streams through heat exchangers for increased heat and mass transfer and the second is to convey liquid from the low pressure side to the high pressure side. In general terms, for pure fluid, the enthalpy difference can be expressed as

$$dh \;=\; c_P dT + [v - T\left(\frac{\partial v}{\partial T}\right)_P] dP \tag{4.28}$$

For incompressible flow, this equation converts to

$$dh \;=\; c_P dT + v dP \tag{4.29}$$

Also the following equation holds for the entropy change of a pure fluid.

$$ds \;=\; \frac{c_P}{T} dT - \left(\frac{\partial v}{\partial T}\right)_P dP \tag{4.30}$$

Assuming that the compression process is isentropic, ds = 0. Then, since the fluid is incompressible (dv = 0), it is concluded that it is also isothermal, dT = 0, based on Equation 4.30. For an isothermal and incompressible process, Equation 4.29 shows that the enthalpy change is just the volume multiplied by the pressure change. When Equation 4.29 is integrated

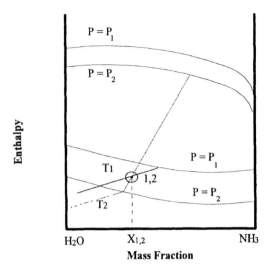

Figure 4.11 Throttling process

over a finite process and inserted into an energy balance on the pump, the pump work is calculated as

$$w = (p_{high} - p_{low})\frac{v\,m}{\eta_p}$$ 4.31

$p_{high} - p_{low}$ represents the pressure difference across the pump, v is the specific volume of the liquid, m the mass flow rate and η_p the pump efficiency.

4.8   Throttling

In absorption systems throttling devices such as expansion valves, orifice plates, and capillary tubes are used to reduce the pressure of a liquid stream passing from the high pressure side to the low pressure side. The throttling process is assumed to be adiabatic, resulting in a constant enthalpy process. Since the enthalpy and the overall mass fraction of the stream do not change while passing a throttle, the state point on the enthalpy mass fraction diagram does not change either, Figure 4.11. When 1 represents the entering stream and 2 the leaving stream, then they occupy the same position. However, what do change are the other variables like temperature and pressure. Before the throttle process, point 1 is subcooled liquid at pressure P_1 and temperature T_1. After the expansion process, the pressure has dropped to P_2 a value for which the fluid is now in the two-phase range. The new temperature, T_2 is indicated by the respective isotherm, Figure 4.11. It can be said that the state point location does not change during a throttling process, but that rather the diagram changes around it, since a different set of isobars represents the fluid state after the process.

When a liquid is throttled the outlet state can be subcooled or a portion of the liquid can flash (vaporize) as it passes through the throttle. The end result depends on the particular conditions. In general, when the inlet stream is more highly subcooled, the outlet stream will have less flash vapor. Assuming that the process is adiabatic, the amount of flash gas and the vapor quality of the outlet stream can be readily calculated as shown in the following three examples.

Example 4.7   Throttling Process with Pure Fluid

Determine the outlet state from a throttling valve where water is throttled from a condenser to an evaporator. Assume the pressures are fixed at 10 and 1 kPa and that the inlet temperature ranges from the saturation temperature down to sufficient subcooling to ensure that the outlet state is still subcooled. File: ex4_7.ees.

Solution: The inlet state is completely specified once the pressure and temperature are specified. For a pressure of 10 kPa, the saturation temperature is 45.82°C. At the saturated state, the enthalpy of the water is 191.8 J/g. For an adiabatic throttle, the energy balance is just $\Delta h = 0$. This implies that the enthalpy at the outlet equals the enthalpy at the inlet. Thus, for the outlet state, the enthalpy and the pressure are known. Since two properties completely define state for a pure fluid, the other properties of interest including the temperature and vapor quality can be determined. This determination typically requires an iterative process using the property routines. This calculation was done for a series of temperatures and the result is given in the following table.

T_{in} (°C)	T_{out} (°C)	Q (fraction)	h (J/g)
45.817	6.971	0.065	191.8
45.0	6.971	0.064	188.4
35.0	6.971	0.047	146.6
25.0	6.971	0.030	104.8
15.0	6.971	0.014	62.9
5.0	5.002		21.0

Observations: The last run, at an inlet temperature of 5°C, results in a subcooled state at the outlet because the inlet enthalpy is sufficiently low that there is not enough energy in the stream to cause vaporization. For that case, the temperature actually rises slightly as the liquid passes through the throttle. The temperature rise is due to viscous friction effects. In all other cases, a portion of the mass flashes as it passes through the throttle with a maximum of 6.5% when the liquid enters in a saturated state. For all the cases where the outlet is two phase, the temperature is the same as determined by the given outlet pressure.

Example 4.8   Throttling Aqueous Lithium Bromide

Determine the outlet state from a throttling valve where aqueous lithium bromide is throttled. Assume the pressures are fixed at 10 and 1 kPa, the solution mass fraction is 0.6 and the inlet temperature ranges from the saturation temperature down to sufficient subcooling to ensure that the outlet state is still subcooled. File: ex4_8.ees.

Solution: This example is similar to Example 4.7 with the additional complexity that the fluid is a binary mixture. It is a special case of a binary mixture for which the vapor composition is known. The basic energy balance on the throttle remains the same. The iterative solution process is also quite similar. It basically involves assuming an outlet condition and calculating all properties to see if the energy balance is satisfied. If it is not, the properties must be adjusted in the direction required to bring it more closely in balance. The iteration is terminated when sufficient accuracy is obtained. At the conditions specified for the inlet, the saturation temperature of·aqueous lithium bromide is 90.8°C and the enthalpy is 213.7 J/g. The outlet conditions for a range of inlet temperature are indicated in the following table.

T_{in} (°C)	T_{out} (°C)	Q (fraction)	h (J/g)	$x_{l,out}$ (% LiBr)
90.793	49.255	0.0288	213.7	61.781
90.0	49.190	0.0283	212.2	61.749
80.0	48.369	0.0220	192.9	61.348
70.0	47.559	0.0156	173.6	60.950
60.0	46.761	0.0092	154.3	60.558
50.0	45.974	0.0028	135.0	60.169
45.0	45.0		125.4	60.0

Observations: The last run, with an inlet temperature of 45°C, resulted in no flashing across the throttle. In this case, there is no temperature increase predicted for the throttling of the essentially incompressible liquid because the enthalpy function being used is independent of pressure. This is an internal inconsistency in the thermodynamic data for aqueous lithium bromide but it contributes only a small error in most calculations.

For a binary mixture, the vapor which flashes causes the mass fraction of the remaining liquid to change as it passes through the throttle. This can be seen in the last column where the liquid mass fraction at the outlet is tabulated. It can also be seen in the fact that the temperature at the outlet changes according to how much vapor flashes.

A useful relationship that can be derived from mass balance considerations only is

$$Q = \frac{x_{l,out} - x_{in}}{x_{l,out}}$$

where Q is the vapor fraction (i.e., the vapor quality), x_{in} is the inlet mass fraction and $x_{l,out}$ is the outlet liquid mass fraction.

Example 4.9   Throttling of Liquid Ammonia/Water

Determine the outlet state from a throttling valve where liquid ammonia/water is throttled. Assume the pressures are fixed at 1000 and 500 kPa, the solution mass fraction is 0.3 and the inlet temperature ranges from the saturation temperature down to sufficient subcooling to ensure that the outlet state is still subcooled. File: ex4_9.ees.

Solution: The solution here is identical in concept to the preceding two examples. One difference here is that the vapor mass fraction is not known and must be included in the iterative calculation. It is assumed that the vapor and liquid at the outlet are in equilibrium. The results are indicated in the table below.

T_{in} (°C)	T_{out} (°C)	Q (fraction)	h (J/g)	$x_{l,out}$	$x_{v,out}$
103.529	84.22	0.053	262.103	0.266	0.918
100.0	83.24	0.046	245.956	0.270	0.922
90.0	80.43	0.026	201.123	0.283	0.932
80.0	77.74	0.006	156.844	0.296	0.941
70.0	70.08		113.025	0.300	

Observations: The overall observations are similar to those in the preceding two examples. The ammonia/water property routines used here do include a pressure term in the enthalpy determination and thus, the throttling of a liquid shows a temperature increase as shown in the last case tabulated. For ammonia/water, the ammonia mass fraction decreases as vapor flashes because the mass fraction is defined in terms of the volatile component. The opposite trend is seen for aqueous lithium bromide since the mass fraction is defined in terms of the less volatile component (i.e., the LiBr).

Once again, a useful relationship can be derived from mass balance considerations only to yield

$$Q = \frac{x_{in} - x_{l,out}}{x_{v,out} - x_{l,out}}$$

where Q is the vapor fraction (i.e., the vapor quality), x_{in} is the inlet mass fraction, $x_{l,out}$ is the outlet liquid mass fraction and $x_{v,out}$ is the outlet vapor mass fraction.

4.9   Ammonia Purification

There are two ways to remove water vapor from an ammonia/water vapor mixture. The first is partial condensation and the second involves a counter-flow purification column. In partial condensation, the vapor is cooled so that a small fraction condenses. The condensate, also called the reflux, contains a significant amount of water leaving the remaining vapor with a higher ammonia concentration. A counter-flow purification column requires liquid reflux at the top of a column. In this way, the two mechanisms work together to achieve the purification.

4.9.1   Reflux Cooling
In order to purify the vapor, it is brought into contact with a cooled surface in the reflux cooler. The surface temperature is kept below the dew point of the vapor stream and a portion of the vapor condenses. This portion is enriched in water and returns as reflux to the desorber. As a result, the remaining vapor stream is enriched in ammonia. Figure 4.12 shows a reflux cooler mounted on top of a desorber. The control volume is shown as a dotted line. The fluid streams leaving and returning to the desorber from the reflux cooler have the subscripts va and la for the vapor phase and liquid phase, respectively. A mass balance yields

$$m_{va} - m_{la} = m_7 \tag{4.32}$$

A mass balance for the ammonia yields

$$m_{va} x_{va} - m_{la} x_{la} = x_7 m_7 \tag{4.33}$$

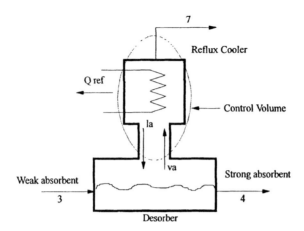

Figure 4.12 Reflux cooler

The mass of the reflux m_{la} can be calculated as

$$m_{la} = m_7 \frac{(x_7 - x_{va})}{(x_{va} - x_{la})}$$ 4.34

Thus when the amount of vapor leaving the reflux cooler at the top as product is set to $m_7 = 1$ then the amount of reflux depends only on the mass fraction of the streams entering and leaving the reflux cooler. An energy balance yields for the control volume

$$(m_7 + m_{la})h_{va} = m_7 h_7 + m_{la} h_{la} + Q_{ref}$$ 4.35

and the heat to be removed from the reflux cooler can be calculated as

$$Q_{ref} = m_7(h_{va} - h_7) + m_{la}(h_{va} - h_{la})$$ 4.36

Then on a per kilogram of rectified vapor (at point 7) basis Equation 4.36 becomes

$$q_{ref} = h_{va} - h_7 + \frac{x_7 - x_{va}}{x_{va} - x_{la}}(h_{va} - h_{la})$$ 4.37

where Equation 4.34 has been introduced. This equation reveals that the amount of heat removed is composed of two terms. The first accounts for the vapor stream being cooled and the second for a portion of the vapor to be condensed.

Now it is necessary to find all state points. Assume that the vapor leaving the desorber is in equilibrium with the solution of a mass fraction that is the algebraic average of the mass fraction of the weak and strong absorbent ($x_3 - x_4$)/2. Further assume that the entire process occurs at constant pressure and that the temperature of the vapor leaving at 7 is known. Figure 4.13 shows an enthalpy-mass fraction diagram that reveals the state points. It can further be observed that Equation 4.37 can be rewritten introducing the pole for reflux cooling. The enthalpy of this pole is defined as

$$q_{ref} = h_{pole} - h_7$$ 4.38

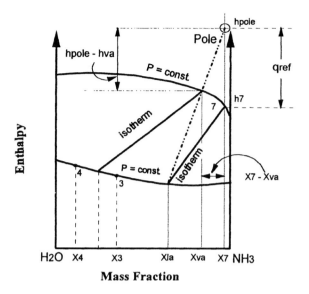

Figure 4.13 Reflux cooling process in an enthalpy-mass fraction diagram

A more useful form is

$$h_{pole} = h_{va} + \frac{x_7 - x_{va}}{x_{va} - x_{la}}(h_{va} - h_{la})$$ 4.39

Equation 4.39 can be rewritten as

$$\frac{(h_{pole} - h_{va})}{(h_{va} - h_{la})} = \frac{(x_7 - x_{va})}{(x_{va} - x_{la})}$$ 4.40

This expression can be interpreted graphically as shown in Figure 4.13 by using the lever rule. The differences in h and x are shown in Figure 4.13 and the pole is found at the intersection of the extension of the isotherm of the vapor leaving the desorber with the isostere of the vapor leaving the reflux cooler. The distance (h_{pole} - h_7) represents the heat removed by the reflux cooler per unit mass of vapor leaving at point 7.

Example 4.10 Reflux Cooler

Consider a reflux cooler, as given in Figure 4.12, designed to purify an ammonia/water vapor stream. It is assumed that the mass fraction of the purified vapor is 0.999, that the vapor leaving the desorber is in equilibrium with the incoming saturated solution of $x_3 = 0.50$ and $T_3 = 100°C$. The process is isobaric. Determine the cooling required in the reflux cooler. File: ex4_10.ees.

Solution: Using appropriate software or the enthalpy-mass fraction diagrams in Chapter 3, the state points of the fluid streams at the solution inlet into the desorber (3), the purified vapor leaving the reflux cooler (7), the vapor leaving the desorber (va in Equation 4.32; 11 in EES file) and, the liquid returning from the reflux cooler to the desorber (la in Equation 4.32; 12 in

EES file), can be determined as shown next. For (3) the temperature and concentration are known as well as the fact that the solution is saturated. Thus the pressure can be read off a chart, the enthalpy-mass fraction diagram, for example. Further the composition of the vapor in equilibrium with this solution can be determined using the same chart. Thus point (va) is found. Point (7) is assumed to be saturated vapor at the same pressure (no pressure loss in the reflux cooler) as 3 at a concentration of 0.999. The saturated liquid at (la) must be in equilibrium with the vapor at (7) and at the same pressure. The following Table 4.3 of property data results.

Table 4.3 Property data for Example 4.10

Point	h (kJ/kg)	P (bar)	Qu	T (^{o}C)	x
3	218	22.9	0.0	373.0	0.5
7	1319	22.9	1.0	335.2	0.999
va,11	1462	22.9	1.0	373.0	0.979
la,12	176	22.9	0.0	335.2	0.83

Based on the data in Table 4.3 the amount of reflux, m_{la}, can be calculated per 1 kg of vapor and the amount of heat to be removed from the reflux cooler, q_{ref} using the above equations. m_{la} amounts to 0.135 kg/(kg of vapor), and the q_{ref} to 317 kJ/kg.

4.9.2   Rectification

Whenever reflux cooling does not provide vapor of the desired purity or the amount of energy released becomes large compared to the desorber heat, it is better to employ rectification using a rectification column. Figure 4.14 shows a rectification column (top portion) combined with a desorber (bottom portion). State point 1 represents the liquid feed of ammonia/water mixture entering the desorber. The feed enters the analyzer section first in counterflow to vapor rising from the reboiler and then falls into the reboiler where heat is added to produce a vapor stream. The remaining solution leaves the desorber at 2. The vapor rises from the reboiler in counter-flow to the falling liquid solution. In this way heat and mass exchange between vapor and liquid streams are enabled. The vapor rises further through the column in counter-flow to a liquid stream that is the reflux which is condensed at the top of the column in the reflux cooler. This condensate has a relatively high concentration of water as compared to the vapor. Nevertheless, its ammonia content is much higher than that of the solution at 1. The rising vapor is in heat and mass exchange with the reflux. This process is enhanced by using a packing material or set of plates that prolong the contact time between vapor and liquid and enhance the mixing. The horizontal baffles in the column in Figure 4.14 symbolize the plates. Thus, as the vapor rises, its temperature is reduced which for a constant pressure equilibrium process requires that the amount of water is reduced as well. A portion of the water content of the vapor is transferred to the reflux releasing heat of condensation. This heat is used to evaporate a certain amount of reflux which is almost pure ammonia. The temperature of the liquid stream and its flow rate gradually increase as it flows toward the bottom of the column.

Figure 4.14 shows a control volume within the column that includes just one plate.

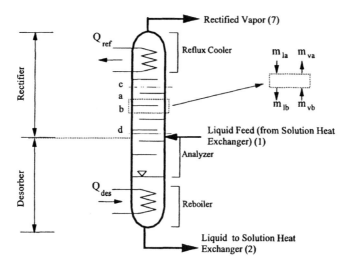

Figure 4.14 Distillation column

Figure 4.14 shows a control volume within the column that includes just one plate. Applying a mass balance yields

$$m_{la} + m_{vb} = m_{lb} + m_{va} \qquad 4.41$$

The subscripts a and b stand for the upper or lower surface of the control volume and the subscripts l and v for the liquid and vapor flow rates, respectively. Equation 4.41 can be rearranged to yield

$$m_{va} - m_{la} = m_{vb} - m_{lb} \qquad 4.42$$

Such a mass balance must hold for all control volumes within the column that do not straddle the feed point. Applying the mass balance to the top of the column around the reflux cooling coil, reveals that the difference between the vapor and liquid streams is the amount of vapor leaving the column.

$$m_v - m_l = constant = m_7 \qquad 4.43$$

This is the case for any cross section of the column above the feed. Applying the mass balance for one component of the mixture, e.g., ammonia, the following is obtained.

$$m_{la}x_{la} + m_{vb}x_{vb} = m_{lb}x_{lb} + m_{va}x_{va} \qquad 4.44$$

Applying the mass balance for any control volume above the feed point, it again turns out that the difference between the liquid and vapor streams is the amount of ammonia leaving the column at the top.

$$m_v x_v - m_l x_l = m_7 x_7 \qquad 4.45$$

Applying an energy balance to the control volume of Figure 4.14 yields the following equation, assuming that the column operates adiabatically:

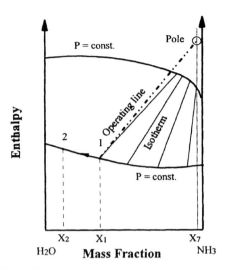

Figure 4.15 The pole of rectification

$$m_{la}h_{la} + m_{vb}h_{vb} = m_{lb}h_{lb} + m_{va}h_{va} \qquad \textbf{4.46}$$

This equation holds for all control volumes above the feed point and can be generalized to

$$m_{v}h_{v} - m_{l}h_{l} = constant \qquad \textbf{4.47}$$

Applying the energy balance to the top of the column around the reflux cooler, the following expression is obtained

$$m_{vc}h_{vc} = m_{lc}h_{lc} + m_{7}h_{7} + Q_{rec} \qquad \textbf{4.48}$$

The subscript c denotes an arbitrary cross section in the column above the feed point. The term Q_{rec} denotes the amount of energy that has to be removed by the reflux cooler. It is termed the heat of rectification.

In order to determine the amount of reflux at an arbitrary position in the column above the feed point, a mass balance for the total mass flow rate and for one component, the ammonia, is employed for the entire rectification column above the point of interest, d. The total mass balance is

$$m_{ld} + m_{7} = m_{vd} \qquad \textbf{4.49}$$

and the mass balance for ammonia

$$m_{ld}x_{ld} + m_{7}x_{7} = m_{vd}x_{vd} \qquad \textbf{4.50}$$

The subscript d designates an arbitrary control surface in the column above the feed point. Combining Equations 4.49 and 4.50 and eliminating m_{vd} yields

$$m_{ld} = \frac{(x_{7} - x_{vd})}{(x_{vd} - x_{ld})}m_{7} \qquad \textbf{4.51}$$

With Equation 4.51 the heat of rectification $Q_{rec}/m_7 = q_{rec}$ can be written as

$$q_{rec} = h_{vd} - h_7 + \frac{x_7 - x_{vd}}{x_{vd} - x_{ld}}(h_{vd} - h_{ld})$$ 4.52

The first term on the right-hand side of this equation represents the difference in the enthalpies of the vapor streams entering and leaving the column. The second term represents the heat of condensation that has to be removed to generate the reflux m_{ld}.

The challenge is now to determine the properties of the state point of the reflux at the control surface d. For this purpose an enthalpy-mass fraction diagram is considered, Figure 4.15. The operating line (it connects two state points that exist at one cross section during the operation of the column) connects the state points of the incoming solution at 1 with the state point of the vapor that is entering the column. This operating line can, at best, have the same slope as the isotherm, meaning that the vapor and liquid are in thermodynamic equilibrium. However, in a real column the vapor contains less ammonia than under equilibrium conditions. Thus the operating line for an actual process will be steeper than the isotherm (a violation of this rule would be a violation of the Second Law). This statement holds for each cross section of the column. The vapor leaving each column cross section cannot be enriched beyond the equilibrium state. Accordingly, the operating line in each cross section cannot be "flatter" than the respective isotherm (i.e., these two lines cannot intersect). In Figure 4.15 only one operating line is shown corresponding to the control surface d of Figure 4.14. However, there are additional isotherms shown in the two-phase region which represent equilibrium states at other plates within the column. The respective operating lines have to have a slope that is at least as steep or steeper than that of the respective isotherms.

To better understand the heat of rectification, define the pole of rectification such that

$$q_{rec} = h_{pole} - h_7$$ 4.53

which implies

$$h_{pole} = h_{vd} + \frac{m_{ld}}{m_7}(h_{vd} - h_{ld})$$

$$= h_{vd} + \frac{x_7 - x_{vd}}{x_{vd} - x_{ld}}(h_{vd} - h_{ld})$$ 4.54

Equation 4.54 can be rewritten as follows to clarify the graphical interpretation

$$\frac{h_{pole} - h_{vd}}{h_{vd} - h_{ld}} = \frac{x_7 - x_{vd}}{x_{vd} - x_{ld}}$$ 4.55

The distances are indicated in Figure 4.16. Thus, according to Equation 4.53 the heat of rectification is represented by the enthalpy difference between the pole and the state point of the rectified vapor leaving the column. Thus the pole of rectification determines the amount of energy that has to be removed in order to achieve the desired vapor purity. The location of the pole is determined by the intersection of the operating lines with the isostere of the purified vapor. The slope of the operating lines cannot be smaller than that of the respective isotherm in a given column cross section in the two-phase range between x_{ld} and x_7. Thus, the heat of rectification is determined by the slope of the isotherm that provides the highest point of intersection. It is not known a priori where this isotherm occurs. It could be at either end of

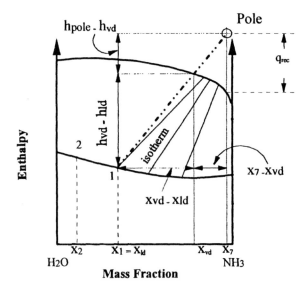

Figure 4.16 Determination of the pole of rectification

the column or at an intermediate cross section. Procedures for locating the pole are described below.

To determine how many theoretical plates the rectification column requires to achieve a given vapor purity, several methods are available. For fluid mixtures with a small difference in boiling points the McCabe and Thiele diagram is employed, the use of which requires simplifying assumptions that are not very well fulfilled with ammonia/water. Another procedure involves the enthalpy-mass fraction diagram and a graphical method by plotting out

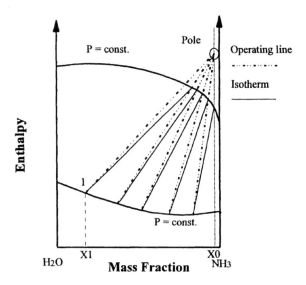

Figure 4.17 Determination of the number of theoretical plates

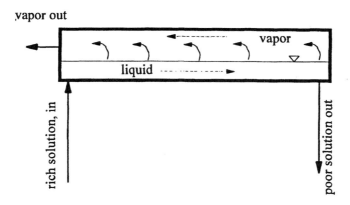

Figure 4.18 Counter-flow desorber to minimize need for rectification

all operating lines and respective isotherms. Computer simulation is the method of choice today. The procedure is the following. Figure 4.17 shows the isotherms and respective operating lines. All the operating lines have to intersect in the pole in order to satisfy the energy and mass balances. It is assumed here that the intersection of each operating line with the saturated vapor isobar determines the liquid concentration of the next higher plate in the column which in turn determines the equilibrium vapor concentration that can at best be achieved. Each operating line represents one plate. Proceeding in this way from operating line to isotherm to operating line across the entire two-phase range between x_1 and x_7 the total number of plates can be determined. Of course, there will usually be a "fraction" of a plate left at the end, which always makes sense to round to the next higher integer.

The location of the pole and therefore the rectification heat can be chosen to a certain degree. If the pole is moved higher, i.e., rectification heat increases, the operating lines become steeper. The vapor- and liquid-phase mass fractions at each plate depart further from equilibrium. At the same time fewer operating lines (and therefore fewer plates) are required to achieve the same purification. When more heat is removed, the amount of reflux is increased which in turn provides more liquid that can participate in the heat and mass exchange process of the rectification column.

Last, the amount of heat that is required to operate the desorber q_{des} together with the rectifier is calculated. Based on mass balances for the total flow rates and that of the ammonia only and the energy balance for the entire column, the following equation is obtained.

$$q_{des} \ = \ h_7 \ - \ h_2 \ + f(h_2 \ - \ h_1) \ + \ q_{rec} \qquad\qquad 4.56$$

The terms are defined as $f = m_1/m_7$ and $q_{des} = Q_{des}/m_7$. q_{rec} is calculated according to Equation 4.52.

Equation 4.56 shows very clearly that the heat of rectification and therefore the need to rectify is a penalty because it increases the heat requirement of the desorber. Thus careful desorber design that limits the water content of the vapor by bringing the leaving vapor into contact with the entering rich solution is quite important (this is related to the analyzer section shown in Figure 4.20. Figure 4.18 symbolizes such a design. The boiling solution is flowing from left to right, while the vapor, in contact with the solution, flows in counter-flow from right to left. Thus an opportunity is provided for heat and mass exchange between vapor and solution which is similar to the process in the rectification column. Further, it is very important that the rich solution entering the desorber be as rich as possible. If the liquid supplied to the generator has a high water content, then the respective two-phase isotherm is quite steep, forcing the pole to very high enthalpies increasing rectification heat requirement dramatically.

The number of plates discussed above is a theoretical value. In an actual system, the heat and mass transfer coefficients and contact time play an important role in the effectiveness of the column. Therefore, there is frequently an efficiency included in the design procedure that either increases the number of plates without lowering the pole or increases the heat. Thus the amount of rectification heat calculated in Equation 4.52 is a minimum value.

Example 4.11 Rectifier

Analyze a rectifier for the following operating conditions. Assume the device operates at constant pressure. The vapor entering the column (control surface d), Figure 4.14, is in equilibrium with a saturated solution of a mass fraction of 0.20 at a temperature of 129.85°C, 1. The rectified vapor is assumed saturated at a mass fraction 0.994. Calculate the amount of reflux and heat released assuming that none (reflux cooler only), one, two, and three plates are used. File: ex4_11.ees.

Solution: Assumptions: no pressure drop, the vapor and liquid leaving each plate and the reflux cooler are in thermodynamic equilibrium.

Based on the information given, the states of the inlet vapor and the outlet vapor as well as of the reflux leaving the reflux coil can be evaluated. Then the equations for the equilibrium at each plate have to be solved. The latter involves an iterative process best conducted with appropriate software or the construction using the enthalpy-mass fraction diagram. The results for no plate, one, two, and three plates are summarized in the Tables 4.4 - 4.6 below.

Table 4.4 Result for rectifier with no plate (reflux cooler)

	h_l (kJ/kg)	h_v (kJ/kg)	Liquid Fraction	T_l (K)	T_v (K)	Vapor Fraction	x_l	x_v
1	425	1797		403.0	403.0	1.877	0.200	0.781
2	34.5	1386	0.877	333.3	333.3		0.538	0.994
Reflux = 0.877, heat of rectification = 1956 kJ/kg of rectified vapor, P = 10.6 bar							

State point 2 represents the liquid and vapor stream leaving the reflux cooler (i.e. it represents the equilibrium tie line for the reflux cooler). The liquid and vapor are in thermodynamic equilibrium and have the same pressure and temperature. The term "reflux" in the above table represents the reflux mass flow rate (leaving the reflux cooler) as a fraction of the rectified vapor flow rate leaving the column. The liquid and vapor fractions listed represent the mass flow rate of liquid or vapor assuming a unit mass flow rate of rectified vapor.

Table 4.5 Result for rectifier with one plate

	h_l (kJ/kg)	h_v (kJ/kg)	Liquid Fraction	T_l (K)	T_v (K)	Vapor Fraction	x_l	x_v
1	425	1797		403.0	403.0	1.400	0.200	0.781
2	349	1683	0.400	391.2	391.2	1.421	0.248	0.859
3	34.5	1386	0.421	333.3	333.3		0.538	0.994
Reflux = 0.421, heat of rectification = 990 kJ/kg of rectified vapor, P = 10.6 bar							

In this case state point 3 presents the liquid and vapor streams leaving the reflux cooler. State point 2 represents the liquid and vapor streams leaving the plate. The temperatures Tl_2 and Tv_2 are the same which is the temperature of the plate. Compared to the case with no plate, the amount of reflux and the amount of heat to be rejected are reduced significantly.

Table 4.6 Result for rectifier with two plates

	h_l (kJ/kg)	h_v (kJ/kg)	Liquid Fraction	T_l (K)	T_v (K)	Vapor Fraction	x_l	x_v
1	425	1797		403.0	403.0	1.371	0.200	0.781
2	414	1779	0.371	401.3	401.3	1.370	0.207	0.794
3	340	1672	0.370	389.8	389.8	1.390	0.254	0.866
4	34.5	1386	0.390	333.3	333.3		0.538	0.994
Reflux = 0.390, heat of rectification= 924.5 kJ/kg of rectified vapor, P = 10.6 bar							

State points 2 and 3 are the liquid and vapor streams leaving the plates. Again heat and reflux are reduced, but the reduction is smaller than before. It can be observed that the amount of water in the reflux increases with each plate as we proceed from the top of the column (reflux cooler) to the bottom. This change represents the amount of water (and some ammonia) that is being extracted from the vapor.

Table 4.7 Result for rectifier with three plates

	h_l (kJ/kg)	h_v (kJ/kg)	Liquid Fraction	T_l (K)	T_v (K)	Vapor Fraction	x_l	x_v
1	425	1797		403.0	403.0	1.367	0.200	0.781
2	423	1795	0.367	402.7	402.7	1.367	0.201	0.783
3	412	1776	0.367	401.0	401.0	1.366	0.208	0.796
4	339	1670	0.366	389.7	389.7	1.386	0.255	0.867
5	34.5	1386	0.386	333.3	333.3		0.538	0.994
Reflux = 0.386, heat of rectification = 916 kJ/kg of rectified vapor, P = 10.6 bar								

An additional plate further reduces reflux and heat but the effect is small. It becomes clear that the return for this additional complexity is diminishing rapidly. An inspection of Table 4.7 reveals that the effect of the lowest plates is very minimal (i.e. not much change in mass fraction is observed). Thus, it is found that under these conditions, an increase in the number of plates beyond the effective number is a wasted investment. This results from the characteristics of the mixture. The only way to reduce the rectifier heat further is to change the feed conditions or the mass fraction of the incoming vapor. Such topics are of interest for advanced cycle design. A more advanced treatment of distillation analysis can be found in Bogart (1981).

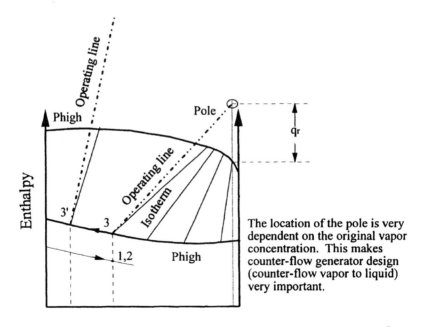

The location of the pole is very dependent on the original vapor concentration. This makes counter-flow generator design (counter-flow vapor to liquid) very important.

Figure 4.19 Sensitivity of the pole position to the feed concentration

As the example shows quite clearly, rectification does purify the vapor but has a limited effectiveness in reducing the amount of heat required. It should be noted that the rectification heat increases quite dramatically as the concentration of the solution entering the desorber/rectifier at point 1, Figure 4.15, decreases. Figure 4.19 shows that as the solution concentration decreases, the pole rises very high because of the increasing slope of the isotherms in the two-phase range. However, there is a second method that is quite effective in absorption systems. While a rectifier produces a purer vapor, a so-called analyzer produces a purer vapor leaving the generator. The analyzer has essentially the same appearance as the rectification column, except the feed stream, 3, enters close to the top. Figure 4.20 shows a schematic. The analysis of an analyzer is very similar to that of a rectifier. The reasoning and derivation of all the equations is essentially the same. There exists a pole which is now located on the line of constant concentration for the strong absorbent leaving the generator (Bogart, 1981).

4.10   Heat Exchangers

Heat exchange processes occur throughout absorption machines. The processes are complicated by coupled mass transfer and by the properties of the binary mixture working fluids. The purpose of the present section is to introduce the terminology that is used in this text to describe the heat exchange processes and to introduce simple heat exchanger models.

Coupled heat and mass transfer occurs in all of the major heat exchangers in an absorption machine except the solution heat exchanger. In the coupled heat/mass transfer processes, both heat and mass transfers occur simultaneously and the two processes are coupled by the internal energy flowing with the working fluid. In particular, the latent heat couples the processes since all the coupled processes involve phase change. It may be argued that a pure fluid evaporator or condenser, such as those that exist in water/lithium bromide systems, does not experience mass transfer since there are no concentration gradients in the system. However, a more general definition of mass transfer is a transfer of mass due to a chemical potential gradient. From this perspective, all the phase change devices experience both heat and mass transfer.

Although these components undergo both heat and mass transfer, it is often convenient to analyze them based on heat transfer alone. This is convenient because it allows the engineer to concentrate on a single set of variables which are temperature and heat transfer rate. Once these variables are specified for a given design, then the mass transfer aspects can be deduced from appropriate models. This heat transfer perspective works because both processes are coupled together so that if you know one then you know the other. However, a full appreciation of the design perspective must give full acknowledgment to the role of both heat and mass transfer since either transfer process can control the overall transfer process.

As an example of this, consider the absorption process. It has been found from numerous modeling and experimental studies that the absorption process, in both water/lithium bromide and ammonia/water, is controlled by the mass transfer resistance on the liquid side. This is because the refrigerant vapor which absorbs at the liquid interface does not effectively transfer into the bulk of the liquid. Instead, it tends to stay near the interface and to slow down the absorption of additional refrigerant. The energy released at the liquid interface causes the temperature to rise there and this energy must also transfer through the liquid film to the heat transfer surface. However, the overall process is controlled by the mass transfer characteristics in the sense that augmentation of the mass transfer mechanisms has a larger effect on the

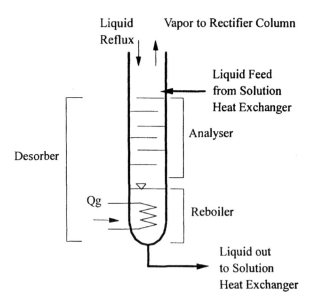

Figure 4.20 Analyzer

overall transfer process than augmentation of the heat transfer mechanisms.

In this section, a heat transfer perspective is taken to simplify the discussion. However, it must be kept in mind that this perspective is only appropriate for design purposes if the overall transfer characteristics of the device are already known in terms of the heat transfer characteristics. In the more general case, it is necessary to account for both of the transfer processes to determine the overall transfer characteristics.

Heat Exchanger Diagrams

A convenient format for visualizing the operating point of a heat exchanger is to plot the temperature versus cumulative heat transfer rate characteristics. Such diagrams are termed heat exchanger diagrams in this text. An example of such a diagram is provided as Figure 4.21. The temperature coordinate is simply the bulk temperature measured along the fluid stream as it passes through the heat exchanger. For a single pass device, this is a well-defined quantity. The cumulative heat transfer rate coordinate represents the rate of energy transferred between any two temperatures. Thus, the difference between the endpoints of the heat exchanger, as measured on the cumulative heat transfer rate coordinate, represents the total heat transfer rate in the heat exchanger. The zero point for the cumulative heat transfer rate scale is arbitrary but it is convenient to set it to zero at one end of the heat exchanger.

When a heat exchanger operating point is plotted on these coordinates, it is immediately clear whether the heat exchanger is well designed. A primary objective in heat exchanger design is to achieve a uniform temperature difference throughout the device. What this implies is that the vertical difference (i.e., the temperature difference) between the hot and cold sides of the heat exchanger should be uniform for the entire energy transfer. Such a uniform design yields the lowest entropy production (i.e., the lowest irreversibility) of any possible design for the same heat transfer resistance. This is why a uniform temperature difference design is referred to as a well-matched design. When temperature matching can be achieved in a heat exchanger design, it pays off in increased system performance.

Commonly it is difficult to achieve a well-matched design in absorption components. This is true primarily because of sensible heat effects and non-linear properties. One characteristic that can be identified for most real heat exchanger designs is a temperature pinch point. The temperature pinch is the point at which the hot and cold side temperatures have their closest approach. The pinch can usually be readily identified from a heat exchanger diagram. Frequently, but not always, the pinch point occurs at either the inlet or the outlet of the heat exchanger. The existence of a pinch point indicates the possibility of improvement of the heat exchanger design. If the two traces on the heat exchanger diagram are far from parallel, the pinch point acts as a bottleneck to the energy transfer.

Examples of heat exchanger diagrams for absorption components are included in Chapter 6 for each of the five major heat exchangers in a single-effect water/lithium bromide machine.

Heat Exchanger Models

Various heat exchanger models exist in the literature that are useful when modeling absorption machines. These include 1) pinch point specification, 2) UA models and 3) effectiveness models. The specification of the pinch point is a simple approach that does not require much discussion. However, it should be noted that such a simple model is an excellent place to start when modeling a new cycle since the complexity of the more realistic models can cause problems in the convergence of the overall cycle model.

When analyzing heat exchangers with phase change on one side, it is often convenient to employ simplifying assumptions to deal with superheated and subcooled streams. The first order assumption commonly made is to ignore the temperature effects associated with sensible heat. This implies that the solution side of the heat exchanger is represented by the saturation temperatures associated with the local streams. The result of such assumptions for each of the components in a typical cycle are discussed in Appendix C where examples of heat exchanger models are given for each of the components in Table C.3.

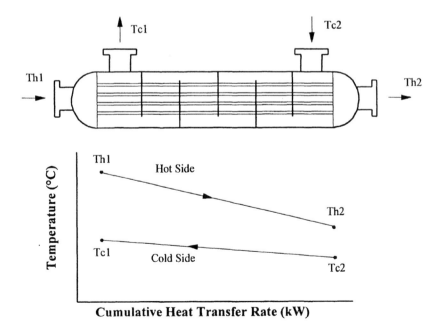

Figure 4.21 Typical counter-flow heat exchanger

UA Type Heat Exchanger Models

The product of the overall heat transfer coefficient, U, and the heat exchanger area, A, is a convenient way to specify the size and performance of a heat exchanger in a single parameter (i.e., the UA value). It is common to use the UA formulation along with the log mean temperature difference as follows.

$$Q = UA \, \Delta T_{lm} \qquad\qquad 4.57$$

with

$$\Delta T_{lm} = \frac{(T_{h,1} - T_{c,1}) - (T_{h,2} - T_{c,2})}{\ln \dfrac{T_{h,1} - T_{c,1}}{T_{h,2} - T_{c,2}}} \qquad\qquad 4.58$$

where h and c refer to the hot and cold sides, respectively. The subscripts 1 and 2 refer to either end of the device as shown in the heat exchanger diagram in Figure 4.21.

The difficulty in using a UA-type formulation comes from the logarithmic term which exhibits singular behavior if the temperatures are not in the correct ranges. This causes problems for iterative solution schemes that sometimes search over wide ranges of the variables to find a final solution. The UA formulation can cause such an iteration to fail before it reaches the answer. Another difficulty with this formulation is that the overall heat transfer coefficient may be a function of other variables such as flow rates, temperature and pressures. Thus, although it is convenient to view the UA product as a constant for modeling purposes, reality is more complex.

Effectiveness Type Heat Exchanger Models

The heat exchanger effectiveness is a useful way to define the performance of a heat exchanger. The effectiveness is defined as the ratio of the actual heat transfer to the maximum possible heat transfer for the given inlet conditions. This can be translated into symbols as

$$\varepsilon = \frac{Q_{act}}{Q_{max}} \qquad\qquad 4.59$$

Consider the heat exchanger shown schematically in Figure 4.21. The maximum possible heat transfer is taken to be the product of the minimum heat capacity rate multiplied by the temperature difference between the two inlet streams or

$$Q_{max} = (mc_p)_{min} \Delta T_{inlet}$$
$$= (mc_p)_{min}(T_{h1} - T_{c2}) \qquad\qquad 4.60$$

In general, the minimum heat capacity can occur on either the hot or cold sides of the heat exchanger.

Consider a case where the minimum heat capacity side is known to occur on the cold side of the heat exchanger. Then the effectiveness would be

$$\varepsilon = \frac{(m\,c_p)_{cold}(T_{c1} - T_{c2})}{(m\,c_p)_{cold}(T_{h1} - T_{c2})}$$

$$= \frac{T_{c1} - T_{c2}}{T_{h1} - T_{c2}}$$

4.61

The definition of the effectiveness in terms of only the temperatures makes it a very convenient heat exchanger performance parameter.

The definition of the effectiveness can be generalized to include heat exchangers that undergo a phase change on one side. For pure fluids, isobaric phase change processes imply no temperature changes on the phase change side and thus the effective specific heat is infinite and the minimum heat capacity must occur on the non-phase change side of the heat exchanger. For absorption components that involve a phase change and a temperature glide, it is not obvious how to proceed because the heat capacity is undefined in the two-phase region. In this text, effectiveness values are calculated for such components based on the assumption that the non-phase change side has the minimum heat capacity. When defined in this manner, the interpretation of the effectiveness loses rigor but it is still convenient and useful as a modeling scheme. As long as it is interpreted in the way it is defined, everything works smoothly.

HOMEWORK PROBLEMS

4.1   Determine the outlet state when two saturated vapor streams of ammonia/water are mixed in equal amounts adiabatically and isobarically at 10 bar. The two streams are at mass fractions of 0.7 and 0.9.

4.2   Consider a water/lithium bromide single-effect cycle operating between pressures of 0.8 and 8 kPa. Assume the solution circuit mass fractions are 0.55 and 0.6 and that the solution heat exchanger has an effectiveness of 50%. Assume that the streams leaving each of the components are saturated.
   a.)   Determine the heat transfer rate in each of the components
   b.)   Determine the solution circulation ratio
   c.)   Determine the cooling COP
   d.)   Determine the ideal pump work
   e.)   Determine the vapor quality at the outlet of the refrigerant expansion valve
   f.)   Determine the vapor quality at the outlet of the solution expansion valve

4.3   Utilize the properties of ammonia/water directly to determine the error introduced in the pump work prediction by the incompressible assumption. Use an isentropic process but evaluate the properties directly and compare the result against the approximation in Equation 4.31.

4.4   Evaluate the losses in the expansion valve by replacing it by a work producing machine such as a turbine. Model the turbine as isentropic and determine the work output. Compare this against the latent heat of the refrigerant to obtain a rough estimate of the magnitude of the losses in the expansion valve.
   a.)   Consider pure ammonia over the pressure range from 25 to 4 bar. Assume the expansion valve inlet is saturated liquid.
   b.)   Consider ammonia/water at as mass fraction of 0.99 under the same conditions as in a.).
   c.)   Consider pure water over the pressure range from 10 to 1 kPa. Assume the expansion valve inlet is saturated liquid.

4.5   Compute the reversible rectifier solution for the conditions of Example 4.11. The results should include the rectifier heat transfer and the flow rate of the reflux liquid out of the column.

4.6   Perform a parametric study, using the rectifier model of Example 4.11, to determine the influence of the key variables on the rectification heat and the reflux flow rate. For each of the following variables, consider a range and prepare a plot of rectification heat and reflux flow rate versus that variable.
   a.   Number of plates
   b.   Mass fraction of rectified vapor
   c.   Mass fraction of inlet vapor

4.7   Plot the h-x diagram for each of the 4 cases contained in Tables 4.4 to 4.7.

4.8   Compute h_{pole} from Equation 4.54 for each of the cases contained in Tables 4.4 to 4.7.

Chapter 5

OVERVIEW OF WATER/LITHIUM BROMIDE TECHNOLOGY

Aqueous lithium bromide is used as an absorption working fluid because it is one of the best choices found among hundreds of working fluids that have been considered. Although aqueous lithium bromide is the preferred choice for many applications, there are numerous limitations associated with the choice that need to be understood. The thermodynamic and transport properties of working fluids are discussed in some detail in Chapter 3 and that material will not be repeated here. However, there are additional properties of aqueous lithium bromide, beyond the thermodynamic and transport properties discussed previously, which impact the design and operation of a machine based on this working fluid. These properties and the limitations that they impose form the focus of this chapter.

5.1   Fundamentals of Operation

The fundamentals of operation of an absorption cycle using aqueous lithium bromide as the working fluid are discussed in this section. To keep the discussion simple, only the most basic cycle is considered. Advanced cycle design features are introduced in Chapters 6 to 8. The discussion is meant as an overview and as an introduction to the many design aspects that must be considered to successfully use the technology.

A block diagram of a single-effect machine is provided as Figure 5.1. The diagram is formatted as if it were superimposed on a Dühring plot of the working fluid. Thus, the positions of the components indicate the relative temperature, pressure and mass fraction. The machine consists of four components that exchange energy with the surroundings, one internal heat exchanger, two flow restrictors that are termed valves on the diagram for brevity, and a pump. The connecting piping between these devices is also important for design.

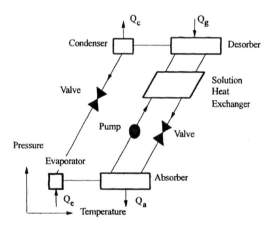

Figure 5.1 Single-effect absorption cycle schematic

5.1.1   Solution Circuit

(The solution circuit circulates between the desorber and absorber.) This liquid loop is pumped from the low pressure in the absorber to the high pressure in the desorber. As a first approximation, the entire machine can be considered to operate between two pressure levels. Of course, in actual operation, there will be pressure losses associated with: 1) flow through the components and 2) changes in elevation. These losses must be minimized by design. In its simplest form, an absorption machine is a two-pressure device with significant pressure changes occurring only in the flow restrictors and the pump.

The liquid solution is pumped into the desorber where heat is supplied by external means such as a combustion source or any other source with a sufficiently high temperature. The required temperature level is governed by the properties of the working fluid and the operation of the other components in the machine. For a typical single-effect aqueous lithium bromide machine the desorber heat must be supplied above a temperature of approximately 90°C (this value is a rule of thumb with actual requirements depending on the details of the application). When heat is applied to the solution, the volatile component (i.e., the refrigerant) is boiled off. In the case of aqueous lithium bromide, water is the refrigerant.

In dealing with mixtures, the relative volatility of the components is a property of major interest. In the case of aqueous lithium bromide the salt (lithium bromide) is essentially non-volatile and the relative volatility is effectively infinite. From a molecular viewpoint, we expect that some salt molecules may escape from the liquid surface and be present in the vapor. However, the escaping tendency is so small under the conditions encountered in an absorption machine that the vapor above the liquid solution is essentially pure water vapor (steam). This fact can be appreciated more fully by realizing that the normal boiling point of solid lithium bromide salt is 1282°C (Foote). Thus, the vapor pressure of the salt at typical absorption machine conditions is exceedingly low. From a thermodynamic standpoint, we will assume that there is no salt content in the vapor and that the properties of the vapor are those of pure water (i.e., steam). At high vapor velocities, liquid entrainment can also carry salt throughout the machine. Trace salt quantities are important from a corrosion perspective. The presence of trace amounts of salt contributes to accelerated corrosion throughout the vapor space.

When heat is applied to the solution in the desorber, vapor is "generated" and the vapor flows to the condenser. The remaining liquid solution exits the desorber and flows back to the absorber. The process in the desorber is a partial evaporation. Since the vapor leaving the desorber is essentially free of salt, the liquid solution becomes concentrated during the partial evaporation process. Thus, the solution flowing back to the absorber is a relatively concentrated salt solution (compared to that exiting the absorber). A number of terms are in common use to describe the concentrations in absorption systems. In general, the mass fraction is used as a concentration measure in this text. However, it may be of use to the reader to define the commonly used terms. The terms "rich" and "poor" are sometimes used but care must be taken to know to which component these terms refer. When using these terms, one must say, for example, that the solution is "rich in refrigerant". A similar set of terms is "strong" and "weak". Once again, one can refer to a solution as "strong in refrigerant" meaning a low mass fraction of lithium bromide. The reader is referred to (ASHRAE, 1993) for suggestions on how to standardize this type of terminology. For our purposes in this book, it was decided to avoid the ambiguity by simply using mass fraction of salt throughout.   water is the refrigerant

The concentrated salt solution leaving the desorber passes through a solution heat exchanger and exchanges energy with the solution leaving the absorber. This heat exchange process occurs between two liquid streams and involves only sensible heat (no phase change occurs in this device under normal conditions). The purpose of this internal heat exchange device is to reduce the external heat input requirement by utilizing energy available within the machine that would otherwise be wasted. By including the solution heat exchanger, the quantity of rejected heat is also reduced. Thus, the solution heat exchanger is a key component; the performance of this component has a major impact on the design of an absorption machine.

The solution stream leaving the desorber returns to the absorber. The stream gives up energy in the solution heat exchanger and typically arrives at the flow restrictor subcooled. As the liquid is throttled through the restrictor, some vapor usually evolves from the liquid. The two-phase stream then enters the absorber. In the absorber, the concentrated salt solution is brought into contact with vapor supplied by the evaporator. The absorption process occurs if the absorber is cooled by an external sink (for example, a flow from a cooling tower). As the vapor is absorbed, the liquid mass fraction is reduced to the level of the desorber input. Since vapor is absorbed into the solution, the mass flow rate of liquid leaving the absorber is greater than that of the liquid entering the absorber. The reverse is true for the desorber.

5.1.2 Refrigerant Loop

The refrigerant loop of an absorption machine is identical in function to the corresponding components in a vapor compression machine. The refrigerant loop takes the refrigerant vapor from the desorber and directs it to the condenser where it is liquified by rejecting heat to a sink. In a typical installation, the absorber and the condenser would reject heat to the same sink (i.e., approximately the same temperature level). The subcooled liquid leaving the condenser is throttled through the restrictor to the low pressure. This throttling process is typically accompanied by some vapor flashing. However, due to the high latent heat of water, the vapor quality leaving the restrictor is relatively low as compared to common refrigerants used in vapor compression systems. The two-phase refrigerant then enters the evaporator. Evaporation takes place, accompanied by heat transfer from the evaporator environment, due to the low pressure created by the absorber. Complete evaporation then implies that all of the refrigerant flow arrives at the absorber as vapor.

5.2   Crystallization and Absorber Cooling Requirements

The nature of salt solutions, such as aqueous LiBr, is that the salt component precipitates when the mass fraction of salt exceeds the solubility limit. The solubility limit is a strong function of mass fraction and temperature and a weak function of pressure. Furthermore, crystal nucleation is a process sensitive to the presence of nucleation sites. If no suitable nucleation sites are present, supersaturation can occur where the salt content of the liquid is greater than the solubility limit. Once crystals begin to form, the crystals themselves provide favorable nucleation sites and the crystals grow on themselves. The phenomenon of precipitation of salt from an aqueous solution can be readily observed by preparing a solution of 0.70 mass fraction LiBr. Precipitation of LiBr crystals can be observed by mixing the solution at 100°C and then cooling it to room temperature. At 100°C, the solution consists of a single-phase liquid. As the solution is slowly cooled, wispy white flakes begin to appear which appear to be only slightly more dense than the liquid.

In the case of LiBr, the precipitate observed is a solid hydrate phase. The phase diagram for aqueous LiBr is included as Figure 5.2. This phase diagram is a plot of temperature as a function of mass fraction and shows the various phase boundaries present in the system. It should be mentioned that the diagram was constructed based on data taken at atmospheric pressure. However, since pressure has only a weak effect in the pressure range of interest, these data are directly applicable. The two-phase regions adjacent to the liquid region consist of solid hydrate along with liquid solution. This is typical of the wet solid (slush) that can form in the absorption machine piping. The solid precipitate tends to cling to piping components and, if conditions are allowed to persist, the precipitate can completely clog the flowing system and stop the flow.

When the flow stops, additional cooling of the piping occurs due to heat loss to the surroundings and the wet solid slush in the pipes becomes even more solid. Visualization of the flowing solution prior to a crystallization shutdown typically shows floating solid. If filters are used in the system, the observation of floating solid is an indication of an imminent shutdown.

When such a flow stoppage occurs in an operating absorption machine, it tends to occur at the outlet of the solution heat exchanger where temperatures are relatively low and mass

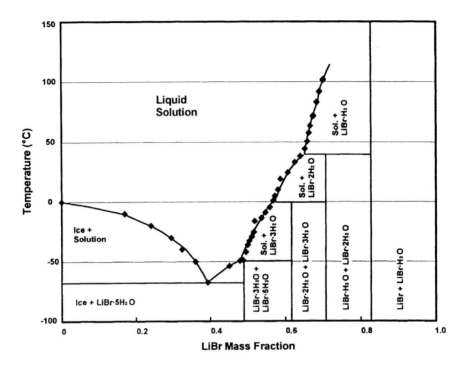

Figure 5.2 Aqueous lithium bromide phase diagram

fractions are high. The phase boundaries are usually included on enthalpy and Dühring plots to remind the designer of the proximity of this characteristic of the fluid (see Figure 3.10 and 3.18). Various methods can be used by the designer to minimize the possibility of crossing the phase boundary. The simplest method used is to ensure a sufficiently low temperature sink for cooling the absorber. Based on the properties of aqueous lithium bromide, low temperatures in the absorber require lower absorber solution concentration and thus tend to avoid the phase boundary. The crystallization characteristic of the fluid is the main obstacle to producing an air-cooled absorption machine based on LiBr. Air-cooled absorbers tend to run hotter than water-cooled units due to the relative heat transfer characteristics of the coolants. Particularly in hot climates, this presents a difficult design problem.

Water-cooled absorption machines generally can operate year round without crystallization problems. However, crystallization events require considerable time and effort to correct. Although the slushy solid that forms is wet, it is quite viscous. Thus, the preferred recovery procedure is to raise the temperature of the portion that is crystallized to a point where the viscosity is reduced sufficiently so that the pumps can circulate the solution. Once the solution is circulating, it can be easily diluted using water from the evaporator. It is highly desirable to avoid crystallization events. Thus, manufacturers generally include controls that sense the possibility of crystallization and take appropriate action to avoid the condition, by reducing heat input to the desorber or by diverting liquid water from the evaporator to the absorber and thus diluting the solution.

5.3   Corrosion and Materials Compatibility

In the presence of dissolved oxygen, aqueous LiBr is highly aggressive to many metals including carbon steel and copper. However, in the hermetic environment inside an absorption machine, very little oxygen is present and corrosion rates are much slower. For the temperature range of a typical single-effect application, carbon steel and copper are the preferred materials of construction. Over the extended life of a machine, significant corrosion can still occur and care must be taken to minimize the effects. The primary measures available are 1) pH control and 2) corrosion inhibitors.

Corrosion of steel (or copper) in the presence of an electrolyte, such as aqueous lithium bromide, is a multi-step oxidation-reduction reaction involving iron (or copper) ions leaving the solid surface and combining with oxygen at a distance from the surface. This leaves the solid surface of the base metal available for more attack. The oxidation potential of the solution is a strong function of the pH level in the acidic range. By controlling the solution to be only slightly basic, the hydroxyl radicals are in excess and this tends to cause oxide formation directly on the solid surface (passivation). Over time, alkalinity tends to increase as hydrogen gas is formed and it is preferable to keep pH close to neutral. pH control in a LiBr absorption machine can be achieved by adding small amounts of HBr (typically in aqueous form as hydrobromic acid) to achieve the desired pH. Since these components are already present within the solution, the addition of HBr does not significantly alter the solution properties.

Corrosion inhibitors provide a complementary reduction in corrosion rates. Various corrosion additives have been proposed and tested over the years including lithium chromate, lithium molybdate and lithium nitrate. These salts are added to the LiBr solution in amounts on the order of 1% by weight. Krueger et al. (1964) recommend 0.3% Li_2CrO_4 and 0.005% LiOH as the preferred inhibitor concentration. For molybdate, they recommend 0.1% Li_2MoO_4 and 0.2% LiOH. These inhibitors reduce corrosion rates, apparently by reacting with the metal surface and forming a relatively stable oxide coating (i.e., passivation). Lithium chromate was the corrosion inhibitor of choice for many years but its toxicity to plant and fish life in event of spillage has reduced its usage. Lithium molybdate, although somewhat less effective, has been found to provide adequate machine life in many applications.

High temperature applications, including some components in double-effect machines, require special materials to maintain long life. Copper-nickel alloys resist corrosion at high temperature better than copper. Tubes made from copper-nickel can be substituted for copper tubes without significant changes in design other than introducing a heat transfer penalty. However, copper-nickel is relatively expensive and is only used when copper does not provide adequate service life. Copper-nickel tubes are also used when the external environment is corrosive such as when ocean water is used as coolant. A major design issue is to minimize the electro-chemical potential between the tubes and the tube sheet. Stainless steels often exhibit pitting and corrosion cracking in the presence of high temperature LiBr.

Good compatibility is found with most rubber and polymer compounds. Seals in LiBr systems are not a particular problem since common rubber products are compatible. Leather is strongly attacked by LiBr solution. Use of leather garments (including shoes) is not recommended around LiBr systems since any splashing of the solution will destroy the leather rapidly.

5.4   Vacuum Requirements

Typical pressures in a LiBr absorption machine are subatmospheric. The pressures are determined by the vapor pressure characteristics of the working fluids. Since essentially pure water exists in the condenser and evaporator, the temperature of operation of these components defines the pressure. For an evaporator temperature of 5°C the corresponding vapor pressure of water is 0.872 kPa or approximately 0.009 atm. This extremely low pressure introduces several challenges to the designer including 1) large components due to the specific volume of the vapor, 2) requirement for hermetically sealed outer vessel, 3) sensitivity to trace amounts of internally generated gases and 4) hydrostatic head effects in evaporator design. These design issues are interrelated and are discussed in this section.

The pressure levels associated with LiBr absorption are not particularly low; however, the sensitivity of the technology to leaks is very high. Absorption machines are sensitive to leaks both because air in the machine hurts performance and because of corrosion considerations.

The pressure terminology used in vacuum technique is often confusing because practitioners sometimes measure the difference between the pressure in their system and the pressure of the atmosphere. The alternative method is to simply specify the absolute pressure. These two methods are compared in Figure 5.3 where a number of common pressure units are also shown. In this book, absolute pressure units are used throughout. But even when absolute pressure units are used, the different unit systems in common use can still be confusing. For example, practitioners in the field commonly use "mm Hg" (mm of mercury) and "in Hg" (inches of mercury) for vacuum gages. These are, of course, really measures of head and must be converted to a pressure by multiplying the head by the density of mercury and the acceleration of gravity.

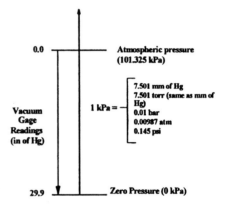

Figure 5.3 Vacuum terminology and units

The low pressures in an absorption system are initially attained by pumping out the air and other gases from the vapor space. Gases dissolved in the liquids in the machine must also be removed. This removal occurs naturally if enough time is allowed for the gases to diffuse out of the liquid into the evacuated vapor space. The process is significantly accelerated if the liquid is agitated while the vapor space is being evacuated. The agitation can be done mechanically, for example, by circulating the liquid with the pump, or thermally by boiling the fluids.

As the vacuum pump extracts from the vapor space, both unwanted gases and some water vapor are removed. The mass of the water removed is usually not significant from the standpoint of the absorption cycle operation since the specific volume of the vapor is so large. However, this vapor removal has some secondary effects. As the water vapor evaporates from the surface of the liquid pools within the machine, it sweeps the vapor space and assists in removing unwanted gases from the system. Although the system pressure never falls below the vapor pressure of the liquids in the system, the purging effect of the evaporating vapor is highly effective in reducing the partial pressure of the gases in the system to the necessary levels.

Another secondary issue associated with water vapor removal is the effect of this vapor

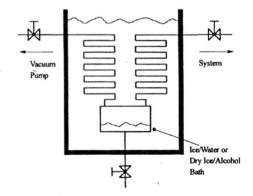

Figure 5.4 Cold trap design schematic

on the vacuum pump. For the pressure levels of interest in absorption technology, modern mechanical vacuum pumps designed for refrigeration practice are completely adequate. The main requirement is the ability to remove vapor at the vapor pressure of water. For oil lubricated pumps, the water vapor tends to condense in the exhaust stages of the pump and to end up in the oil sump. The water tends to reduce the lubrication effectiveness and will harm the pump if the condition is allowed to persist. Careful observation of the oil is necessary and when discoloration occurs, the oil must be replaced. To reduce or eliminate water condensation in the pump oil, a cold trap should be employed to trap the water before it gets to the pump. The cold trap can have a simple design, such as that shown in Figure 5.4, with the key feature being sufficient low temperature surface area on which to "trap" the water. The water must condense in such a way that it does not block the gas flow path. The water in the sump must be maintained at a temperature below its dew point.

Example 5.1 Pressure Units

Consider an absorption machine operating at the pressure levels 10 and 1 kPa. Convert these to all the other commonly used units indicated in Figure 5.3.

Known: Pressure levels

Find: Representation in other unit systems

Assumptions: Unit conversions given in Figure 5.3. The following equation converts a head reading (1 in Hg) into a pressure reading.

$$1 \; in \; Hg \;\; = \;\; h \, \rho \, g$$

$$= \;\; 1 \; in \; x \; \frac{0.0254 \; m}{in} \; x \; 13595 \; \frac{kg}{m^3} \; x \; 9.807 \; \frac{m}{sec^2}$$

$$= \;\; 3.3865 \; kPa$$

	P = 1 kPa	P = 10 kPa
mm Hg	7.501	75.01
torr	7.501	75.01
bar	0.01	0.1
atm	0.00987	0.0987
psi	0.145	1.45
in Hg (vacuum)	29.63	26.97

Analysis: The conversion to vacuum gage units requires the following analysis. One inch of mercury can be converted to the kPa equivalent as follows noting that the density of mercury is 13595 kg/m^3. This conversion is applied to the difference between the actual pressure and the standard atmospheric pressure (101.325 kPa). The unit conversions result in the preceding table.

Observations: The largest reading possible on a vacuum gage is 29.921 in Hg corresponding to zero pressure. In practice, vacuum gages often measure the difference between the system pressure and the local barometric pressure. A standard atmosphere was assumed in the calculation.

5.4.1 Component Size

The lowest pressure in an absorption machine is in the evaporator and absorber. Typical pressures are on the order of 1 kPa. At this pressure, the specific volume of saturated steam is 129.2 m^3/kg. This large volume leads to high velocities for vapor transfer between components. Large velocities generally imply large pressure drops. However, LiBr absorption machines are quite sensitive to pressure drop between the evaporator and the absorber. At a pressure of 1 kPa, the slope of the vapor pressure curve of water is 14°C/kPa indicating that small pressure drops can have a large influence on temperature. Furthermore, due to the logarithmic nature of the vapor pressure curve, the sensitivity of temperature to pressure drop increases with each increment of pressure drop. Pressure drop between the evaporator and absorber aggravates the design problems associated with crystallization since it forces the solution to achieve a higher concentration in order to maintain a lower pressure at the same temperature.

Thus, the designer must balance these two conflicting design requirements. The design solution is to provide very large vapor transport cross section between the evaporator and the absorber. In practice, these two components are usually housed in a single shell providing a minimum of viscous losses. A pressure difference is still required to drive the vapor from the surface of the evaporating liquid to the surface of the absorbing liquid. By minimizing tube bundle and wall viscous effects the pressure drop can be minimized.

The high velocities also provide design challenges within each of the components. The vapor velocity between tubes must be small enough so that viscous losses within the tube bundle are kept to acceptable values. This is a common problem in tube bundle design but it is more significant in LiBr absorption technology because of the sensitivity of the cycle operation to the pressure drop. For both the evaporator and the absorber, the designer has very little pressure drop to spend in getting the vapor out of (or into) the tube bundle. In this case, another design aspect normally drives the bundle design away from pressure drop problems. The need for small temperature differences, for example, between the evaporating refrigerant and the chilled water flowing in the tubes, requires a large heat exchanger area and low heat fluxes. Low heat flux leads to lower vapor velocities but also leads to large component size. The evaporator and absorber vessel of a typical LiBr absorption machine is quite large due to these various ramifications associated with the low vapor pressure of water.

5.4.2 Effect of Non-Absorbable Gases

Air leaks into an absorption machine cause unacceptable corrosion problems due to the oxygen. Therefore, a LiBr absorption machine must be essentially hermetic in design. Non-absorbable gases other than air can also cause poor performance.

The corrosion chemistry at work in an absorption machine causes a low level, continuous production of hydrogen gas. The rate of production depends on several factors including the condition of the machine, usage profile and solution chemistry. Hydrogen is essentially inert, non-absorbable in the temperature range of interest and has a very low solubility in both liquid water and aqueous LiBr solution. Thus the hydrogen produced accumulates in the vapor space of an absorption machine. The hydrogen has a tendency to migrate from the high pressure side to the low pressure side due to the influence of pressure on solubility (i.e., the solubility of hydrogen in both liquid water and liquid H₂O/LiBr solution is slightly higher at higher pressure). The primary effect of inert gases in such a machine is to reduce the performance of both the condenser and the absorber. Since the absorber represents the most critical component design, the discussion is based on the absorber.

Non-absorbable gases in an absorbable vapor tend to get swept toward the vapor-liquid interface by the bulk motion of the vapor. However, because the gas does not absorb appreciably into the liquid, the gas tends to accumulate on the vapor side of the interface. At steady state, the gas has a relatively high concentration at the interface and it diffuses back into the bulk vapor due to the concentration gradient. This diffusion process implies that the refrigerant must diffuse toward the interface. As such, the gas tends to blanket the surface and significantly reduces the rate of absorption. The effect of non-absorbable gas is most pronounced when the gas flow is laminar and not sweeping the absorbing surface. In a typical absorption machine, the vapor flows are turbulent. However, the effect of non-absorbables can still be significant (Minkowycz et al., 1966; Sparrow and Minkowycz, 1967).

Reductions in the performance of the absorber due to hydrogen buildup show up as reductions in capacity, COP, and more difficulty avoiding crystallization. Thus, the designer must provide a mechanism to purge the hydrogen from the system (Murray, 1993). A direct method is to simply evacuate the vapor space periodically with a vacuum pump. Various other methods have been proposed to continuously purge a machine. These include palladium cells which form a semi-permeable membrane in the system and ejector pumps which use the existing solution pump to collect the gas. A schematic of an ejector pump system is given in Figure 5.5. In

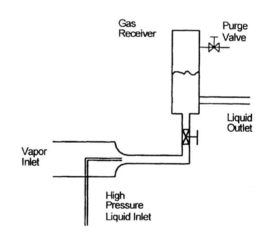

Figure 5.5   Ejector purge system

most designs, the machine operator must periodically purge the gas collection vessel with a vacuum pump. This is preferred over pumping on the entire vapor space of the machine since the gas is concentrated and removed quickly and easily.

5.4.3 Hydrostatic Head Effects in Evaporator Design

Due to the low operating pressure of the evaporator in a LiBr absorption machine, hydrostatic effects can be quite important in evaporator design. At a liquid surface temperature of 5°C, the vapor pressure of pure water is 0.873 kPa. For a nominal 1 g gravitational field, a column of water exerts a pressure of 0.09807 kPa/cm of height. Based on the vapor pressure curve for water, this implies that a 1-cm head of liquid water above the boiling surface would cause the saturation temperature at the boiling surface to increase to 6.5°C. A 10-cm head would cause the saturation temperature at the boiling surface to reach 16.3°C. These effects make a pool boiler design impractical for the evaporator in a LiBr absorption machine because the evaporator temperature is highly sensitive to hydrostatic head. Instead, a sprayed design is usually used where a film of refrigerant is sprayed onto horizontal tubes. The maximum thickness of the film on the tubes is determined by the viscosity of the liquid and generally is on the order of 1 to 2 mm.

5.5   Octyl Alcohol

The design of the absorber component in a LiBr absorption machine drives the design of the entire machine because it is the largest component. It is large for two reasons: 1) the large specific volume of water vapor at the absorber pressure requires large flow areas to maintain acceptable pressure drops and 2) the heat and mass transfer coefficients in the absorber are lower than the corresponding coefficients in the other components. The absorber transfer coefficients are low due to the relatively ineffective mass transfer process on the liquid side. Various techniques can be used to augment the mass transfer by inducing mixing in the liquid. One effective technique involves the addition of a surfactant additive, called octyl alcohol (2-ethyl-1-hexanol), which acts to induce Marangoni convection in the liquid. Use of octyl alcohol can increase transfer performance in absorber design by as much as two times (i.e., 200% performance improvement) but system effects (e.g., pressure drop and non-absorbable gas effects) usually limit the effectiveness of the additive in bundles to much more modest gains.

Marangoni convection is the name given to free surface flows driven by differences in surface tension. Surface tension driven flows can be easily observed in the kitchen by adding a drop of detergent to dirty dishwater. Prior to adding the detergent drop, the oil is distributed fairly uniformly over the surface and the surface is static. When the detergent drop is added, the surface tension is reduced locally. Careful observation of the surface shows a radial flow away from the drop. The surface tension forces at a distance from the drop are no longer in balance and cause the flow. The surface flow must be accompanied by a convection cell within the liquid to satisfy mass continuity. Thus, Marangoni convection mixes the surface liquid into the bulk liquid.

The processes in a LiBr absorber component are coupled heat and mass transfer processes. As the water vapor is absorbed into the liquid the latent heat associated with the phase change is released along with the heat of mixing. This energy release occurs at the vapor-liquid interface and causes the interface to be the location of highest temperature in the system. To allow the process to continue, the energy must diffuse through the liquid toward the cooling medium and the excess water at the surface must diffuse into the bulk of the liquid. These two diffusion processes are coupled in the sense that stopping either process causes the system variables to adjust in such a way as to stop the other process. For the overall transfer to take place, both processes have to proceed in tandem.

In the case of the absorber, the mass transfer process controls the coupled transfer. Thus, design efforts to improve overall absorber performance are focused on the mass transfer process. Octyl alcohol has been found to improve absorber performance and is routinely used in commercial machines.

5.6   Normal Maintenance and Expected Life

Normal maintenance for LiBr/water technology includes: 1) periodically purging non-absorbable gases, 2) periodic addition of octyl alcohol, 3) periodic addition of corrosion inhibitor, and 4) periodic addition of pH buffer. The appropriate period for performing these tasks depends on a number of variables, including size of the machine and the purging system, and is generally specified by the manufacturer of the machine. The basic procedures are quite simple and can be performed by trained technicians. The necessary chemicals are relatively inexpensive and readily available from suppliers.

Along with the maintenance procedures described above, an assessment of cycle performance against norms should be made on a regular basis to help diagnose any potential problems. The ultimate failure mode of a LiBr absorption machine is usually corrosion induced. For long life, attention must be paid to avoid introduction of air into the machine and to ensure that the corrosion inhibition regime is strictly followed.

Based on many years of experience, LiBr absorption machines have been proven to have a life expectancy of approximately 20 years. After 20 years significant corrosion can be observed in the steel surfaces and leakage may occur, particularly around the tubes in the tube sheets. At some point, a decision must be made to replace the machine instead of fighting a losing battle. The life of the machine is limited by corrosion. Thus, for long life, great attention must be paid to corrosion avoidance. Once a machine begins to leak, the presence of oxygen can greatly accelerate the degradation and thus leaks must be avoided. Introduction of oxygen during special maintenance procedures is another destructive event that must be avoided or minimized. If the vacuum must be broken for any reason, it is important to fill the vapor space with nitrogen or other inert gas to avoid introduction of oxygen. Long-term shutdown of charged machines should involve pressurizing the vapor space with nitrogen to a pressure above atmospheric so that any leakage that may occur does not introduce oxygen.

5.7   Controls

Considerable work has been done by absorption machine manufacturers to perfect controls that can provide trouble free operation for the user. With the exception of crystallization issues, absorption machines are inherently stable and self-starting. The user must turn on the pumps, including the solution pump and possibly an evaporator recirculation pump, and the cycle will run. The solution concentration will adjust to the imposed temperatures according to its thermodynamic properties. Thus, in general, no active controls are necessary. However, because of the problems associated with crystallization, a number of controls are generally provided to deal with this issue.

Unchecked inward leakage of air causes the absorber effectiveness to deteriorate. Then, conventional controls that sense chilled water temperature rising call for increased heat input to the desorber. This, in turn, tends to concentrate the salt solution and can cause crystallization.

One approach to avoiding crystallization is to ensure that the solution mass fraction never goes above some limiting value (e.g., 0.65 mass fraction LiBr). As long as the machine stays in the expected temperature range of operation, this restriction is enough to guarantee no crystals. The mass fraction of LiBr can be measured by sampling and performing a density measurement or by titration but an automatic method is sought which does not rely on sampling. A method of inferring the concentration is used by some manufacturers based on a known initial charge of solution with a known LiBr mass fraction. The refrigerant level in the evaporator is monitored and when it is too high, it is inferred that the solution is too highly concentrated. This scheme is relatively simple to automate and can be implemented using widely available level-sensing transducers.

HOMEWORK PROBLEMS

5.1   Determine the mass fraction at which solid precipitate will form (i.e. crystallization) at a temperature of 50°C.

5.2   Determine the composition of the precipitate from Problem 5.1.

5.3   Determine the mass fraction of each of the hydrates of lithium bromide.

5.4   Determine the hydrostatic pressure exerted by a column of 0.25 m of pure water and aqueous lithium bromide of 0.5 mass fraction LiBr.

Chapter 6

SINGLE-EFFECT WATER/LITHIUM BROMIDE SYSTEMS

The objective of Chapter 6 is to describe the operating characteristics of single-effect water/lithium bromide systems. That objective is addressed by examination of several examples with alternative design features and characteristics. The examples are presented as solutions of computer models of the respective cycle configurations. The computer models associated with the examples are included on the diskette which accompanies the book. The software is introduced in Appendix E and a description of the process of creating a computer model of a new cycle can be found in Appendix C.

6.1   Single-Effect Water/Lithium Bromide Chiller Operating Conditions

A single-effect, absorption cycle using water/lithium bromide as the working fluid is perhaps the simplest manifestation of absorption technology. A schematic of such a cycle is provided as Figure 6.1. The major components are labeled and the state points in the connecting lines are assigned state point numbers. The schematic shows the energy transfers external to the cycle as arrows in the direction of transfer with variable names representing the four heat transfers and one work term. The schematic is drawn as if it were superimposed on a Dühring chart of the working fluid properties as indicated by the coordinates shown in the lower left-hand corner. The relative position of the components with phase change in the schematic indicates the relative temperature and pressure of the working fluid inside those components.

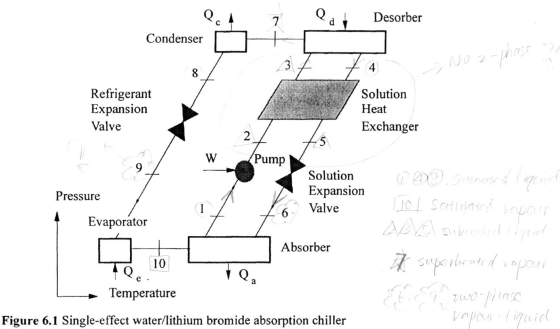

Figure 6.1 Single-effect water/lithium bromide absorption chiller

The exception to this is the subcooled and superheated states which cannot be accurately represented on a Dühring chart which displays only saturated states.

A set of operating conditions for the cycle in Figure 6.1 is listed in Table 6.1. This set of conditions was generated by imposing mass and energy balances on the components and the entire cycle as discussed in Chapter 4. As an introduction to the technology, these mass and energy balances are reviewed next.

Table 6.1 Operating conditions for a single-effect water/lithium bromide machine

i	h(i) (J/g)	m(i) (kg/sec)	P(i) (kPa)	Q(i) (Fraction)	T(i) (°C)	X(i) (% LiBr)
1	85.8225	0.0500	0.679	0.000	32.9	56.7
2	85.8266	0.0500	7.347		32.9	56.7
3	147.0	0.0500	7.347		63.2	56.7
4	221.2	0.0455	7.347	0.000	89.4	62.4
5	153.9	0.0455	7.347		53.3	62.4
6	153.9	0.0455	0.679	0.006	44.7	62.4
7	2644.6	0.0045	7.347		76.8	0.0
8	167.2	0.0045	7.347	0.000	39.9	0.0
9	167.2	0.0045	0.679	0.064	1.5	0.0
10	2503.4	0.0045	0.679	1.000	1.5	0.0

COP	= 0.720	P_{low}	= 0.679 kPa	Q_e	=	10.574 kW
ϵ_{shx}	= 0.640	Q_a	= 14.039 kW	Q_{shx}	=	3.06 kW
f	= 11.047	Q_c	= 11.213 kW	W	=	0.000206 kW
P_{high}	= 7.347 kPa	Q_d	= 14.678 kW			

6.1.1   Mass Flow Analysis

At steady state, the net mass flow into each of the components must be zero. Furthermore, since it is assumed that no chemical reactions occur between the water and the lithium bromide, the net mass flow of each of these species into any component must also be zero. Since there are two species (i.e., water and lithium bromide), there are only two independent mass balances (as discussed in Chapter 4). As an example, consider the mass balance on the desorber which can be written on an overall mass basis as

$$m_3 = m_4 + m_7 \qquad\qquad 6.1$$

A balance on the lithium bromide species, assuming that the vapor leaving the desorber has zero salt content, yields

$$m_3 x_3 = m_4 x_4 \qquad\qquad 6.2$$

The water mass balance can be obtained by subtracting Equation 6.2 from Equation 6.1 to obtain

$$m_3 (1 - x_3) = m_4 (1 - x_4) + m_7 \qquad\qquad 6.3$$

The fact that only two of the three mass balances are independent is emphasized by the fact that the third was obtained from the first two. However, it is also possible to write the water mass balance directly from the schematic as was done for the overall and lithium bromide mass balances. It is left as an exercise for the reader to prove that both procedures yield the same equation. Similar analyses can be performed on the other component data presented in Table 6.1. The conclusion is that mass continuity is satisfied for all components. It should be noted that the absolute values of the mass flow rates appearing in Table 6.1 were selected for illustration purposes only. In general, the mass flow rates scale with the capacity of the machine and will be different for each application.

Another mass flow parameter that is sometimes useful is the solution circulation ratio, f, defined in Equation 4.15. For the state points in Figure 6.1, f can be expressed as

$$f = \frac{m_3}{m_7} = \frac{x_4}{x_4 - x_3} = 11.05 \qquad\qquad 6.4$$

which shows that the liquid flow rate through the pump is 11.05 times the vapor flow rate leaving the desorber. This is a typical value and is useful to remember when dealing with design issues.

6.1.2   Thermodynamic States within the Cycle

The thermodynamic state of each of the points within the cycle must be understood to properly understand the cycle. A summary of the state point descriptions is listed in Table 6.2. As listed in the table, three of the points are saturated liquid (1, 4, and 8), one is saturated vapor (10), three are subcooled liquid (2, 3, and 5), one is superheated vapor (7), and two are two-phase vapor-liquid states for a total of 10 state points.

The vapor quality is assumed for four state points. Those are the three saturated liquid states and the saturated vapor state. These assumptions are made for convenience in modeling. In a real machine, the conditions at these points would not be exactly saturated. In general, the transfer processes within the components require a finite driving potential between the vapor and liquid phases. A saturated outlet condition would imply a zero potential difference at the outlet. This does not occur in practice. However, it has been found that this assumption does not introduce a large error and it is typical of a first order model of a cycle. In a real machine, the liquid streams would be expected to be subcooled and the vapor stream would be superheated. These states can also be modeled, but since additional data are then needed, they introduce more complication.

The state at the vapor outlet from the desorber (point 7) is specified as superheated water vapor (steam) based on the perspective that the stream is pure water. However, it is also possible to view the steam as the vapor component of a two-phase system where the solution in the desorber is the liquid phase. From this binary mixture perspective, the vapor is saturated. These two perspectives are both correct and they are both useful depending on the type of analysis being performed. This point is emphasized here since it is possible to generalize the assumption about the outlet state of the working fluid from each of the components. For this introductory model, saturated solution conditions are assumed at the outlet of each of the four major components (desorber, absorber, condenser and evaporator).

The outlet states from the expansion valves are determined by applying an energy balance to the valve assuming an adiabatic expansion as discussed in Section 4.8. It should be noted that the state point data for these states (points 6 and 9) listed in Table 6.1 represent the overall two-phase state. Thus, the enthalpy and mass flow rate values listed are for the overall two-phase flow at that point. A better understanding of the state at these points is obtained by computing the vapor quality to determine the amount of vapor that flashes as the expansion occurs. For the conditions listed and using the methods discussed in Section 4.8, the quality values obtained are listed in the table. At point 9, approximately 6.4% of the mass flow flashes into steam. Due to the substantial changes in volume that occur at such a low pressure, the flash gas significantly impacts the design of a refrigerant expansion device for this application. The amount of vapor that flashes at point 6 is much smaller for this example (only 0.6%) as a result of the significant subcooling that occurs in the solution heat exchanger. A change in the performance of the solution heat exchanger can cause more or less flash gas at point 6. As at point 9, the flash gas has a very high specific volume and causes the velocity of the two-phase stream at 6 to be significantly greater than the velocity at point 5.

The temperature drop that occurs across each of the expansion valves occurs because the vapor has a higher internal energy than the liquid. Thus, some energy must be extracted from the liquid to drive the phase change. The process attains its own equilibrium at a temperature below the starting temperature. The magnitude of the temperature drop correlates with the amount of vapor which flashes.

Table 6.2 Thermodynamic state point summary

Point	State	Notes
1	Saturated liquid solution	Vapor quality set to 0 as assumption
2	Subcooled liquid solution	State calculated from pump model
3	Subcooled liquid solution	State calculated from solution heat exchanger model
4	Saturated liquid solution	Vapor quality set to 0 as assumption
5	Subcooled liquid solution	State calculated from solution heat exchanger model
6	Vapor-liquid solution state	Vapor flashes as liquid passes through expansion valve
7	Superheated water vapor	Assumed to have zero salt content
8	Saturated liquid water	Vapor quality set to 0 as assumption
9	Vapor-liquid water state	Vapor flashes as liquid passes through expansion valve
10	Saturated water vapor	Vapor quality set to 1.0 as assumption

6.1.3   Energy Balance Analysis

Energy balances were imposed to obtain the solution listed in Table 6.1. In this section, the heat transfer values are manually reproduced by involving the energy balances on each of the components. This process of checking the model output is an extremely important step in ensuring that the model output is correct and meaningful.

An energy balance on the evaporator can be written as

$$Q_e = m_{10} h_{10} - m_9 h_9 = 10.57 \ kW \tag{6.5}$$

Similarly for the condenser

$$Q_c = m_7 h_7 - m_8 h_8 = 11.21 \ kW \tag{6.6}$$

These equations could be simplified slightly by using the fact that the mass flow rate is the same at the inlet and outlet. In fact the mass flow rate is identical throughout the entire refrigerant circuit as can be seen by examining the mass flow rate values in Table 6.1. An energy balance on the desorber can be written as

$$Q_d = m_7 h_7 + m_4 h_4 - m_3 h_3 = 14.68 \ kW \tag{6.7}$$

and similarly for the absorber as

$$Q_a = m_{10} h_{10} + m_6 h_6 - m_1 h_1 = 14.04 \ kW \tag{6.8}$$

An energy balance on the hot side of the solution heat exchanger can be written as

$$Q_{shx-h} = m_4 h_4 - m_5 h_5 = 3.06 \ kW \tag{6.9}$$

Alternatively, one can write the cold side balance as

$$Q_{shx-c} = m_3 h_3 - m_2 h_2 = 3.06 \ kW \tag{6.10}$$

The overall energy balance on the solution heat exchanger is satisfied if $Q_{shx-h} = Q_{shx-c}$ which is seen to be the case here.

An energy balance on the pump can be written similarly as

$$W = m_2 h_2 - m_1 h_1 = 0.000206 \ kW \tag{6.11}$$

The values listed for the enthalpy at points 1 and 2 in Table 6.1 have increased precision to display the small difference in enthalpy which occurs in the pumping process from points 1 to 2. This difference in enthalpy was obtained by assuming an isentropic pump model as discussed in Section 4.7. The pump work is quite small as compared with the heat transfer rates associated with the other components. Thus, it can be concluded that from a thermodynamic standpoint, the pump work is negligible for a single-effect water/lithium bromide cycle. Even if the pump were only 10% efficient, the pump work would still be less than 0.01% of the heat input in the desorber. One could also note that the availability (or exergy) of a work input is greater than that of a heat input. However, pump work still appears negligible when such an analysis is performed. Although the thermodynamic conclusion is that the pump can be ignored, practical experience shows that the pump is a critical component that must be carefully engineered. The major solution pump design issues include: 1) pump seals to avoid air leakage, 2) pump cost, and 3) sufficient net positive suction head to avoid cavitation in the suction line.

Once the heat transfer in each of the components has been determined, the coefficient of performance can be calculated as

$$COP_R = \frac{Q_e}{Q_d} = 0.720 \qquad\qquad 6.12$$

This value is fairly typical of single-effect water/lithium bromide chillers. Another view of the cycle performance is obtained by calculating the ratio of the rejected heat to the heat input. This ratio is the coefficient of performance for heating defined as

$$COP_h = \frac{Q_a + Q_c}{Q_d} = \frac{Q_d + Q_e}{Q_d} = 1 + COP_R = 1.720 \qquad\qquad 6.13$$

6.1.4   Discussion of the Operating Conditions

The highest pressure in a single-effect water/lithium bromide machine is typically less than 10 kPa absolute pressure. Thus, the entire machine operates well below atmospheric pressure. This characteristic requires hermetic design to avoid air in-leakage. The low pressure also constrains component design because viscous pressure drops must be minimized in all components.

Dühring Plot Representation

The cycle solution is plotted on a water/lithium bromide Dühring plot in Figure 6.2. The saturated states are plotted at their state point locations (points 1, 4, 6-9 and 10). The subcooled states are plotted at their respective temperatures and mass fractions. The pressure coordinate is not meaningful in this representation of the subcooled states.

The Dühring plot representation of a cycle solution is a very important step in visualizing the data. It is strongly recommended that such a plot be made as a tool to the understanding of any and all cycle solutions. This is particularly important for more complicated cycles but applies to single-effect cycles as well. A number of pitfalls can be avoided by checking any solution instead of assuming that it is correct. One such pitfall is the approach of the operating envelope to the crystallization line. Others include maximum and minimum temperatures, reasonable heat rejection temperatures, reasonable pressures and reasonable mass fractions. The values that are reasonable depend on the particular application. The ability to view all of the data in one plot allows the designer to rapidly perform a number of reasonableness checks on the data. If the operating conditions violate design limits, the model inputs must be changed, the cycle model run again and the plot re-created.

Temperatures

The desorber temperature glide covers the range 77.0 to 89.4°C. (Note that this range does not include the temperature changes associated with heating stream 3 up to saturation because the energy associated with sensible heating is relatively small in this case.) This is the temperature range over which the heat input must occur. The heat source temperature must be higher than the indicated internal temperatures. The evaporator temperature is 1.5°C which is quite close to the freezing point of water. Thus, one observation on this operating condition is that the evaporator may be in danger of freezing if the operating condition wanders.

Heat rejection occurs in both the absorber and condenser. The condenser saturation temperature is 39.9°C and the absorber solution glide is 44.7 to 32.9°C. For a typical cooling tower design temperature of 30°C, the pinch point in the absorber looks quite small. A heat transfer analysis is needed to determine if this design is workable. Such an analysis is done in Section 6.2.

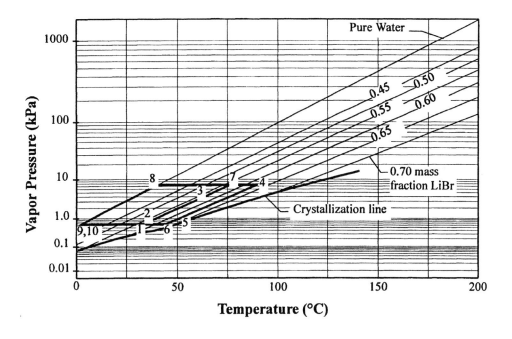

Figure 6.2 Dühring plot for cycle solution in Table 6.1

Mass Fractions

The lithium bromide mass fraction in the solution circuit changes from 0.567 in the pumped leg to 0.625 in the return leg to the absorber. As is typically the case, the outlet state from the solution heat exchanger into the absorber is the closest approach to the crystallization line. This is clearly seen in Figure 6.2 where the calculated operating conditions avoid crystallization but without much margin of safety.

Solution Heat Exchanger

The effectiveness of the solution heat exchanger is defined as in Equation 4.61 as

$$\varepsilon_{SHX} = \frac{T_4 - T_5}{T_4 - T_2} = 0.64 \qquad\qquad 6.14$$

The value of the effectiveness was an input to the model in this case. It is informative to examine the influence of the solution heat exchanger on cycle performance by running the model for a range of effectiveness values. This was done and the result is plotted in Figure 6.3. All other parameters of the cycle were held constant at the values tabulated in Table 6.1. As the effectiveness is varied, the outlet states on both sides of the solution heat exchanger change which then influences the heat transfer requirements in the absorber and the desorber. The COP is influenced through the effect on the desorber heat transfer rate.

The solution heat exchanger effectiveness can be viewed as a measure of the heat exchanger size. As the effectiveness is increased, the COP increases. The sensitivity of cycle COP to the heat exchanger effectiveness is quite high. With no solution heat exchanger, the cycle produces

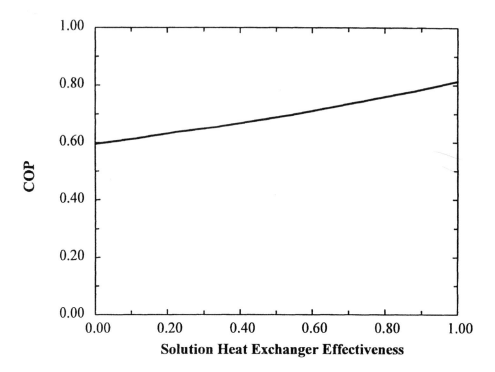

Figure 6.3 Cooling COP versus solution heat exchanger effectiveness for a single-effect chiller

a COP of only 0.6 because the desorber heat requirement is significantly higher in the absence of the internal heat exchange between the legs of the solution circuit. The absorber load is also substantially impacted by the effectiveness of the solution heat exchanger.

The conclusions here regarding the solution heat exchanger are generally valid but represent only a partial analysis. The performance of the solution heat exchanger is strongly tied to the solution flow rate and the cycle operating conditions. This subject is returned to in the following section where heat transfer models are added to the cycle model.

6.2   Single-Effect with Heat Transfer Models

In this section, the cycle introduced in Section 6.1 is viewed from a larger perspective so as to directly include the effects of the heat exchange processes between the cycle and its surroundings. A schematic of the cycle including fluid loops at all four corners is provided as Figure 6.4. All of the internal state point designations are identical to those introduced in Figure 6.1. Eight additional state points are added to represent the inlet and outlet state of each of the heat exchange fluid streams.

A set of operating conditions for the cycle in Figure 6.4 is listed in Table 6.3. It should be noted that the operating conditions internal to the absorption machine are identical to those listed in Table 6.1. This was arranged by running the more complete model first to determine the pressures, temperatures and mass fractions and then inserting those values to the more

Table 6.3 Operating conditions for a single-effect water/lithium bromide cycle with heat exchangers

i	h (i) (J/g)	m (i) (kg/sec)	P (i) (kPa)	Q (i) (Fraction)	T (i) (°C)	X (i) (% LiBr)
1	85.8	0.0500	0.679	0.000	32.9	56.7
2	85.8	0.0500	7.347		32.9	56.7
3	147.0	0.0500	7.347		63.2	56.7
4	221.2	0.0455	7.347	0.000	89.4	62.4
5	153.9	0.0455	7.347		53.3	62.4
6	153.9	0.0455	0.679	0.006	44.7	62.4
7	2644.6	0.0045	7.347		76.8	0.0
8	167.2	0.0045	7.347	0.000	39.9	0.0
9	167.2	0.0045	0.679	0.064	1.5	0.0
10	2503.4	0.0045	0.679	1.000	1.5	0.0
11	418.9	1.0000			100.0	
12	404.2				96.5	
13	104.8	0.2800			25.0	
14	154.9				37.0	
15	104.8	0.2800			25.0	
16	144.8				34.6	
17	42.0	0.4000			10.0	
18	15.6				3.7	

COP	= 0.720	P_{high} =	7.347 kPa	Q_{shx} =	3.06 kW	
ϵ_{shx}	= 0.640	P_{low} =	0.679 kPa	Ua_a =	1.800 kW/K	
$\Delta T_{lm,a}$	= 7.80 K	W =	0.000206 kW	Ua_c =	1.200 kW/K	
$\Delta T_{lm,c}$	= 9.34 K	Q_a =	14.039 kW	Ua_e =	2.250 kW/K	
$\Delta T_{lm,e}$	= 4.699 K	Q_c =	11.213 kW	Ua_d =	1.000 kW/K	
$\Delta T_{lm,d}$	= 14.678 K	Q_d =	14.678 kW	Ua_s =	0.132 kW/K	
$\Delta T_{lm,s}$	= 23.147 K	Q_e =	10.574 kW			

limited model to avoid the additional complexity of the heat exchangers in the introductory discussion.

For a more complete understanding, the heat exchanger models need to be included in an absorption cycle model since they represent the most important irreversibility in a practical machine. Furthermore, the inclusion of the heat exchanger models allows for a more realistic set of inputs to a simulation model. The insight obtained from running the simulation models with a range of variables for these realistic inputs is much greater than the insight obtainable from the approach in Section 6.1 (although significantly more computation is required). This is brought out in the discussion that follows.

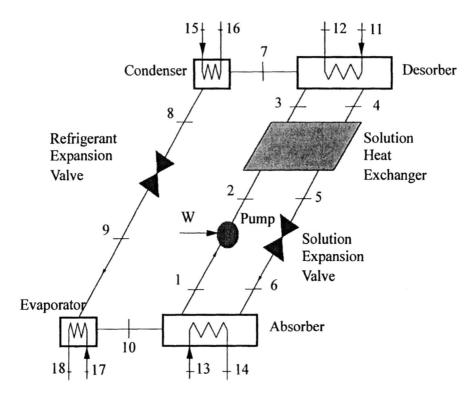

Figure 6.4 Single-effect water/lithium bromide absorption chiller with external heat transfer models

6.2.1   Heat Exchanger Models

The coupled heat and mass transfer processes on the four corners of the cycle are modeled using a UA formulation as discussed in Section 4.10. The UA values were input as indicated in Table 6.4 where all the model inputs are summarized.

Consider the evaporator first. The UA formulation is written as

$$Q_e \;=\; UA_e \, \Delta T_{lm,e} \;=\; 10.57 \; kW \qquad\qquad 6.15$$

where

$$\Delta T_{lm,e} \;=\; \frac{(T_{17} - T_{10}) - (T_{18} - T_9)}{\ln \dfrac{(T_{17} - T_{10})}{(T_{18} - T_9)}} \;=\; 4.70 \,^{\circ}C \qquad\qquad 6.16$$

This formulation is best visualized on a heat exchanger diagram such as that provided as Figure 6.5. The log mean temperature difference represents the average transfer potential between the hot and cold sides of the evaporator. As an aside, it can be seen by observation of Figure 6.5 that the heat exchanger is not particularly well matched since the heat transfer loop experiences a glide of 6.3°C while the evaporating fluid experiences zero glide. The design could achieve a better temperature match by increasing the flow rate of the heat transfer fluid (this is left as an exercise).

Table 6.4 Baseline inputs defining single-effect operating conditions in Table 6.3

Input Name	Value	Input Name	Value
ϵ_{SHX}	0.64	T_{13} (°C)	25
m_1 (kg/sec)	0.05	m_{13} (kg/sec)	0.28
UA_a (kW/K)	1.8	T_{15} (°C)	25
UA_c (kW/K)	1.2	m_{15} (kg/sec)	0.25
UA_d (kW/K)	1	T_{11} (°C)	100
UA_e (kW/K)	2.25	m_{11} (kg/sec)	1
		T_{17} (°C)	10
		m_{17} (kg/sec)	0.4

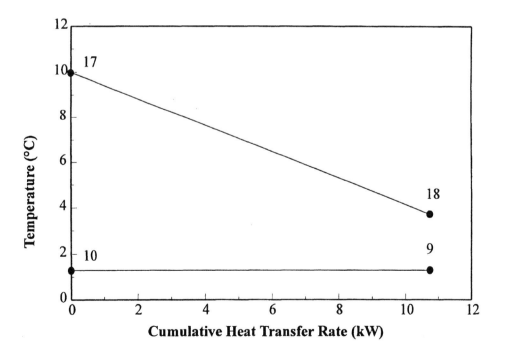

Figure 6.5 Heat exchanger diagram for evaporator

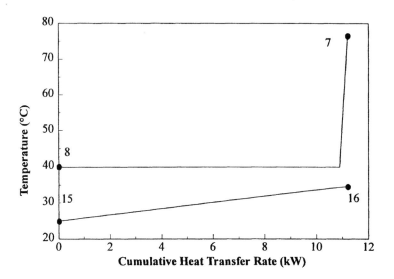

Figure 6.6 Heat exchanger diagram for condenser

A similar analysis on the condenser yields

$$Q_c = UA_c \Delta T_{lm,c} = 11.21 \ kW \tag{6.17}$$

with

$$\Delta T_{lm,c} = \frac{(T_{15} - T_8) - (T_{16} - T_8)}{\ln \dfrac{T_{15} - T_8}{T_{16} - T_8}} = 9.34\,°C \tag{6.18}$$

The condenser heat exchanger diagram is shown in Figure 6.6. The condenser log mean temperature difference defined in Equation 6.18 ignores the temperature effect of the de-superheating section (T_7-T_8). This model has been found to work very effectively and it has the advantage of simplicity for the current discussion. The condenser exhibits a better match than the evaporator as seen by the fact that the two sides of the heat exchanger are more closely parallel on this coordinate system.

A similar analysis for the desorber yields

$$Q_d = UA_d \Delta T_{lm,d} = 14.68 \ kW \tag{6.19}$$

where

$$\Delta T_{lm,d} = \frac{(T_{11} - T_4) - (T_{12} - T_7)}{\ln \dfrac{(T_{11} - T_4)}{(T_{12} - T_7)}} = 14.68\,°C \tag{6.20}$$

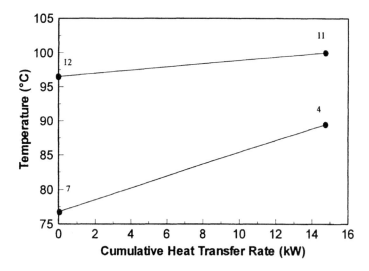

Figure 6.7 Heat exchanger diagram for desorber

The heat exchanger diagram for the desorber is plotted in Figure 6.7. In generating Figure 6.7, the temperature effects of the subcooled inlet (point 3) were ignored. In place of the subcooled state, the low temperature end of the cold side of the heat exchanger was plotted at the saturation temperature corresponding to the known pressure and the mass fraction at point 3. That temperature is the same as at point 7. The desorber is not particularly well matched, showing a pinch at the high temperature end of 10.6°C. The quality of the match can be seen by comparing the pinch against the log mean temperature difference. The closer they are, the better the match.

For the absorber, the heat transfer analysis yields

$$Q_a = UA_a \Delta T_{lm,a} = 14.04 \ kW \qquad \text{6.21}$$

with

$$\Delta T_{lm,a} = \frac{(T_6 - T_{14}) - (T_1 - T_{13})}{\ln \dfrac{T_6 - T_{14}}{T_1 - T_{13}}} = 7.80°C \qquad \text{6.22}$$

The absorber heat exchanger diagram is plotted in Figure 6.8. The temperature match exhibited by the absorber is the best among all the components examined here.

6.2.2   Cycle Performance

In operating an absorption machine, a change in any input variable causes changes in all the other dependent variables. When an input changes, the entire cycle reacts to reach a new equilibrium operating condition. The dynamic nature of this equilibration process must be taken into account when interpreting operating data. The model presented in Section 6.1 does not exhibit this type of complex behavior but the heat transfer model presented in this section does more closely reflect the actual cycle behavior at additional computational expense.

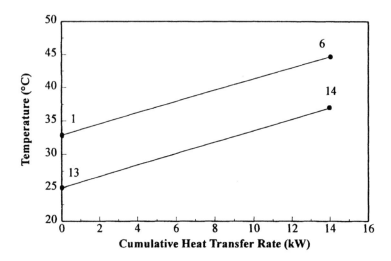

Figure 6.8 Heat exchanger diagram for absorber

6.2.3   Desorber Inlet Temperature Variations

A simple example of the interdependence of all operating variables can be obtained by varying the desorber water loop inlet temperature while holding all other inputs constant at the baseline values defined in Table 6.4. The cycle performance under these conditions is summarized in Figure 6.9 where coefficient of performance and capacity are plotted versus the desorber heat transfer fluid inlet temperature.

The coefficient of performance varies from a low value of 0.625 and exhibits a maximum of 0.75. The capacity varies approximately linearly starting from a low value of 2 kW up to 13.5 kW. The temperature range shown here was chosen because it spans the practical range for a single-effect application. The fact that the COP exhibits a maximum indicates that there are several competing changes occurring as the temperature is increased. The capacity plot helps to explain one of the primary effects. As the temperature increases, the capacity also increases. This increased duty appears in all of the heat exchangers in the system as shown in Figure 6.10. The increased duty results in increased heat transfer irreversibility and decreased COP.

The interesting question then is why is the COP curve relatively flat? This question was already addressed in Section 2.7 based on the results of the zero-order model. The answer comes from the fact that an absorption machine is fundamentally a three-temperature device. The performance of such a device can be largely determined by examining how the temperatures internal to the cycle change at each of the three temperature levels. The relevance of the three-temperature model is obscured somewhat by the fact that a real machine experiences a range of temperatures at each of the three temperature levels. Thus, the appropriate temperature to choose to make a meaningful comparison is not immediately obvious. However, some insight can be obtained by examining the trends exhibited by any of the temperatures. For example, Figure 6.11 shows a plot of T_1, T_4 and T_{10} versus the changes in the inlet temperature of the heating stream (T_{11}). It is observed that the internal cycle temperatures follow a peculiar trend. The rejection temperature stays approximately constant,

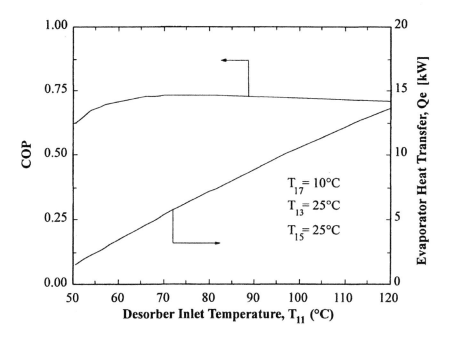

Figure 6.9 Effect of desorber inlet temperature on COP and capacity for a single-effect water/lithium bromide absorption chiller

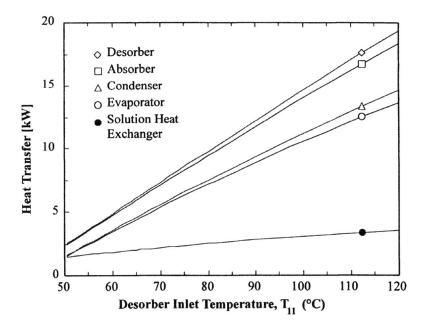

Figure 6.10 Effect of desorber inlet temperature on heat transfer rates for a single-effect water/lithium bromide absorption chiller

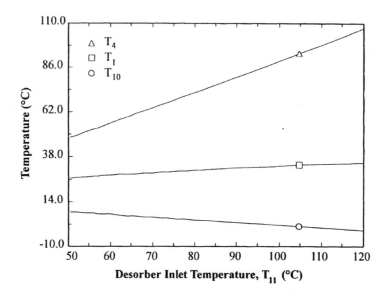

Figure 6.11 Effect of desorber inlet temperature on temperature for a single-effect water/lithium bromide absorption chiller

the desorber outlet temperature increases with the source temperature and the evaporator temperature actually decreases.

These effects can also be reproduced using the zero-order model of Section 2.7. The trends result from the constraints imposed by the working fluid properties. In particular, the slope of the vapor pressure curve for water on the Dühring plot is approximately the same as the slope of the vapor pressure curves for the lithium bromide solution. This implies that the temperature difference between the evaporator and the absorber is approximately the same as that between the absorber and desorber. In the zero-order model formulation, this characteristic is represented by Equation 2.22.

The performance of the cycle is strongly influenced by changes in the temperatures at all three levels. Since both the high temperature and the low temperature in the cycle are changing significantly, the net effect on COP is not obvious. The COP would be expected to increase with increasing desorber temperature but to decrease with decreasing evaporator temperature. In this particular example, the temperature trends are strong enough to cause an overall increase in COP at low values of heat input temperature but as the heat exchanger duties increase, the heat transfer irreversibilities eventually overwhelm the effect and cause COP to decrease slightly.

Although not shown here, the pressures and mass fractions within the cycle change significantly as the heat source temperature changes. From a thermodynamic standpoint, the performance is influenced most directly by the temperatures. Thus, the discussion here focuses on the temperatures. However, for a complete understanding of the changes, it is recommended that plots of all the variables be produced to fully understand a particular trend.

From the perspective of a user of the technology, the shape of the COP curve exhibited in Figure 6.9 is very important. Unlike one might expect from intuition, there is no COP benefit to be obtained by firing a single-effect machine at a higher temperature. This conclusion must be made carefully. It applies to single-effect technology. Double effect (see Chapter 7) and

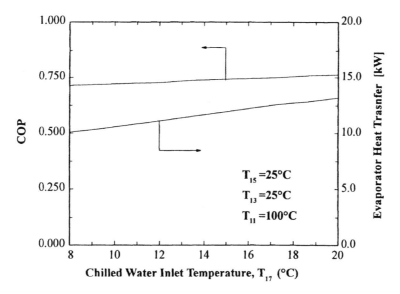

Figure 6.12 Effect of chilled water inlet temperature on COP and capacity for a single-effect water/lithium bromide absorption chiller

other cycle variations can provide higher performance when higher temperature heat sources are available. In fact, double-effect technology exhibits a similar flat trend in COP as a function of heat input temperature but the heat input temperature and COP values are higher. Thus, this characteristic shape to the COP curve can be viewed as a fundamental characteristic of absorption technology. There is a step change in COP in going from single- to double-effect technology but within each technology the COP is nearly constant as any of the imposed temperatures are changed. It should be noted that some advanced cycles incorporate designs that overcome this limitation by essentially becoming variable effect cycles (an example is the GAX cycle discussed in Chapter 11).

6.2.4   Evaporator Temperature Variations

The effect of the variation of the inlet chilled water temperature (T_{17}) is shown in Figure 6.12. In this case, both the COP and the capacity vary very little over the range of inlet temperature considered. A full understanding of these trends again requires a detailed examination of the other variables in the cycle. As the inlet temperature is increased, the pressure in the evaporator increases in such a way that the heat transfer driving potential increases only slightly. The effect of the pressure on the absorber is to decrease the mass fractions slightly. Overall, the evaporator inlet temperature variation causes temperature effects that balance the increased capacity in such a way that the COP actually increases slightly. The sensitivity of capacity to changes in the chilled water temperature is a function of the design of the machine. Different values for the heat exchanger sizes (the UA values) yield some variation in that sensitivity.

6.2.5   Rejection Temperature Variations

The cooling streams to the condenser and absorber can be connected in parallel or series flow arrangements. The cases discussed above involved parallel flow where the inlet temperature of both components are identical (25°C in the above discussion). To complete the examination of the effect of inlet temperature variations the inlet coolant temperature was varied and the results are plotted in Figure 6.13. Once again, the COP is seen to vary only slightly as the combined effects of temperature and capacity changes effectively cancel one another. The cycle capacity is seen to be more sensitive to coolant temperature than it is to chilled water temperature. Once again, these sensitivities depend somewhat on the design conditions. Series flow piping of the condenser and absorber is preferred by operators since a single pump can be utilized and control problems inherent in a parallel design are avoided. Two different series flow arrangements are possible depending on which component is first in the flow path. The cases are compared in Table 6.5 including parallel flow and both series flow arrangements. For the same total cooling water flow rate, the series flow with condenser first has higher COP and capacity than the other series flow case or the parallel flow case. However, if the same flow rate per component is provided, as in the second parallel flow case, then parallel flow looks preferable. This is intuitive since the effective rejection temperature is lower for the parallel flow case with the higher total cooling water flow rate. The tradeoffs between series and parallel flow are indicated in the table. Other factors may influence the design and operating decisions. For example, the series flow arrangement where the absorber is first is the better of the two series designs as far as avoiding crystallization.

Table 6.5 Condenser and absorber coolant piping comparisons based on inputs in Table 6.4

Configuration	COP	Capacity (kW)
Parallel, with mass flow rate in each component equal to 50% of series flow value	0.696	8.161
Parallel, with mass flow rate in each component equal to 100% of series flow value	0.720	10.574
Series, condenser first ($m_{13} = 0.28$ kg/sec)	0.712	8.913
Series, absorber first ($m_{13} = 0.28$ kg/sec)	0.690	8.418

6.2.6   Solution Flow Rate Variation

The effect of solution flow rate on cycle operating conditions is plotted in Figure 6.14. The coefficient of performance varies from 0.790 to 0.649 over the range of solution flow rates considered. The key to understanding this effect is to examine the load on the solution heat exchanger which is also plotted in Figure 6.14. As the solution flow rate increases, the load on the solution heat exchanger increases considerably since there is more energy available in the solution stream leaving the desorber and more energy needed in the stream leaving the absorber. This increased load implies increased losses (irreversibility). This explains why the COP decreases with increasing solution flow rate.

The explanation for the capacity maximum exhibited in Figure 6.14 is more subtle. As was found before, changes in the solution flow rate cause changes throughout the entire cycle. In particular, the evaporator temperature exhibits a minimum temperature at the capacity maximum. This behavior is due to the temperature changes that occur in the absorber. At the capacity maximum, the temperature profiles in the absorber are well matched in the sense that the temperature difference throughout the device is approximately constant. For flow rates that are greater than or less than the optimum value, the temperature profiles become skewed so that

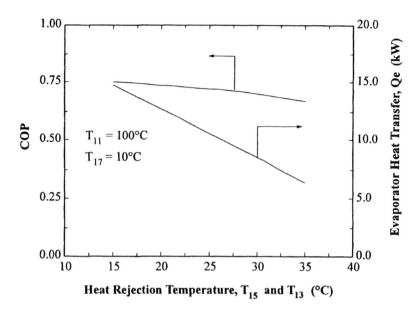

Figure 6.13 Effect of cooling water inlet temperature on COP and capacity for a single-effect water/lithium bromide absorption chiller

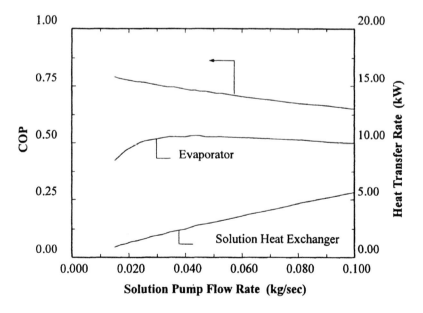

Figure 6.14 Effect of solution pump flow rate on COP, capacity and solution heat exchanger heat transfer rate for a single-effect water/lithium bromide absorption chiller

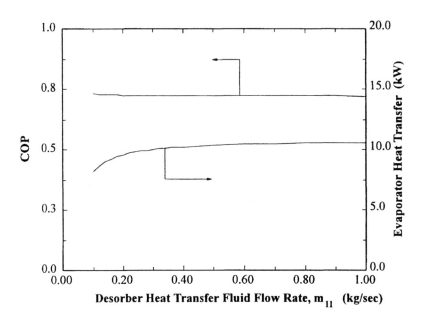

Figure 6.15 Effect of desorber heat transfer fluid rate on COP and capacity for a single-effect water/lithium bromide absorption chiller

there is a pronounced pinch at one end of the heat exchanger. The well-matched design yields lower losses, resulting in a lower absorber pressure at the optimum. This lower pressure then causes the lower evaporator temperature which causes the increased evaporator capacity.

Matching temperature profiles on both sides of a heat exchanger is an important consideration in optimum design of thermal systems, including absorption systems. The pronounced temperature glides occurring in the absorber and desorber in water/lithium bromide technology provide an opportunity or a design constraint, depending on your perspective. Clever system design should incorporate temperature matching in both components.

6.2.7   Heat Transfer Loop Flow Rate Variation

An example of the influence of the heat transfer loop flow rates on performance can be seen in Figure 6.15 where the COP and capacity are plotted versus desorber heat transfer fluid flow rate. The curves show very little sensitivity and a similar trend was also exhibited by the other three heat transfer loop flow rates. Since the curves are not particularly interesting, they are not included here. It should be noted that the current model does not include any influence of flow rate on heat transfer coefficient. In many cases the controlling side of the heat exchanger may be the heat transfer fluid and, in such cases, changes in heat transfer coefficient with flow rate may be important. Such a feature could be simply added to the present models to compute the heat transfer coefficient from correlations.

6.2.8   Evaporator-Absorber Pressure Drop Variation

All of the results discussed so far were based on zero pressure drop between the evaporator and absorber. In reality, a pressure difference must exist between the surface of the evaporating liquid and the surface of the absorbent to drive the flow. Pressure drop must be minimized by design, primarily to avoid crystallization. The influence of pressure drop on system performance is plotted in Figure 6.16. The capacity effects are primarily due to changes in the evaporator temperature which rises significantly as the pressure drop increases.

6.2.9   Heat Exchanger Size Variations

The influence of heat exchanger size on performance is summarized in Figures 6.17 to 6.21. In each figure, the COP and capacity are plotted versus the UA of that particular heat exchanger. The abscissa scales were chosen to span the UA values used in the baseline case. In all cases, the same general trends are observed. That is, the COP is relatively insensitive to changes in UA except for very small values of UA where the COP decreases as UA decreases. Capacity is more sensitive to UA with the exception of the solution heat exchanger UA which does not have much influence on capacity. The baseline values of UA that were chosen were somewhat arbitrary. It can be seen from these plots that the baseline values are well beyond the knee of the capacity curve with the desorber UA exhibiting the greatest effect on capacity. Based on these curves it appears that the largest influence on capacity of adding one additional unit of heat exchange area would be in the desorber. This gives a clue as to the likelihood that the baseline distribution of heat exchange area is not optimum.

6.2.10   Summary of Single-Effect Operating Conditions

In the present section, a single-effect water/lithium bromide cycle was examined from several angles. The interpretation and understanding of the technology is complicated by the cyclic nature which means that changes in any cycle inputs influence all the dependent variables in the system. The approach taken here is to use a computer model to examine the effect of changes in various parameters. The results presented are a cross section of the many variables in the system. This presentation has been designed to bring out the key aspects without plotting each variable. However, as a part of the learning process, it is recommended that the student exercises the models provided and reproduces the presented results as well as examines other variables in the system to get a complete understanding of the interdependence of the absorption system variables.

6.3   Single-Effect Water/Lithium Bromide Heat Transformer (Type II Heat Pump)

A Type II heat pump, sometimes called a temperature booster or heat transformer, is the second basic variation of a three-temperature device as described in Sections 1.2 and 2.3. A typical application for which a Type II heat pump is considered is a case where waste heat is available and where heat is needed at a somewhat higher temperature than that of the waste heat. The Type II cycle can convert a portion of the waste heat energy into a heat transfer out of the cycle at the highest temperature.

Type II cycles based on water/lithium bromide have been built and tested at numerous research facilities and at a limited number of demonstration sites around the world. The cycle works well and it is a proven energy saving technology appropriate for industrial facilities with substantial thermal energy flows. Applications have been limited, apparently due to a combination of factors including a limited number of potential customers which match the description given above and a lack of economic incentive to save energy (i.e., an environment of relatively cheap fuel).

Application challenges have been experienced at some of the demonstration sites. The primary challenge seems to be corrosion. A Type II cycle naturally operates at a higher temperature than a single-effect Type I cycle due to the fact that the Type II cycle transfers heat out at the high temperature end. Thus, assuming a nominal heat transfer driving potential of 20°C in the high temperature device, the Type II cycle will operate 40°C above the highest

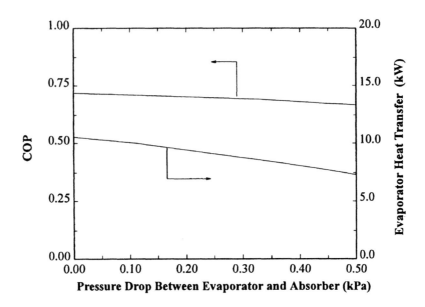

Figure 6.16 Effect of pressure drop between evaporator and absorber on COP and capacity for a single-effect water/lithium bromide absorption chiller

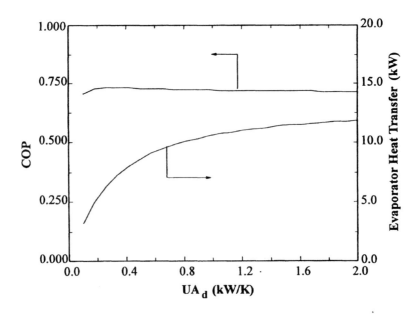

Figure 6.17 Effect of desorber heat exchanger on COP and capacity for a single-effect water/lithium bromide absorption chiller

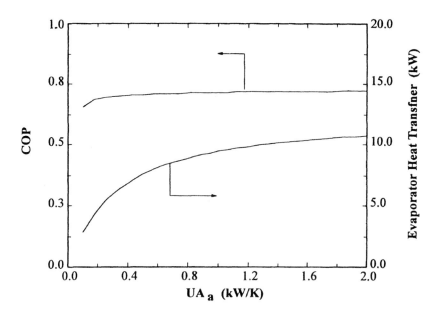

Figure 6.18 Effect of absorber heat exchanger on COP and capacity for a single-effect water/lithium bromide absorption chiller

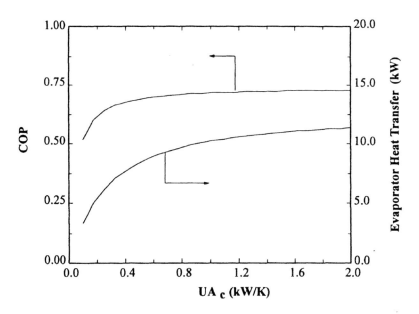

Figure 6.19 Effect of condenser heat exchanger on COP and capacity for a single-effect water/lithium bromide absorption chiller

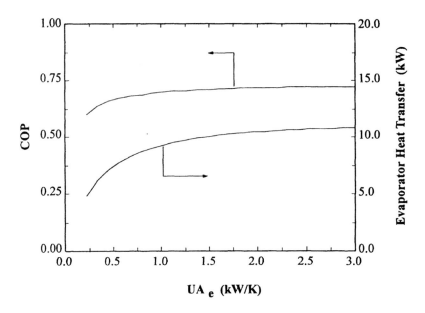

Figure 6.20 Effect of evaporator heat exchanger on COP and capacity for a single-effect water/lithium bromide absorption chiller

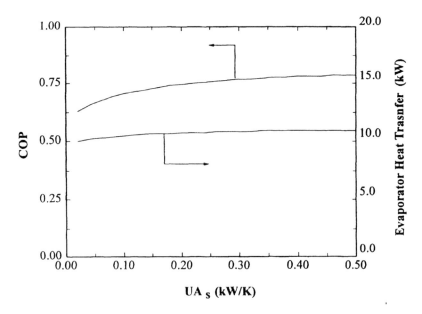

Figure 6.21 Effect of solution heat exchanger on COP and capacity for a single-effect water/lithium bromide absorption chiller

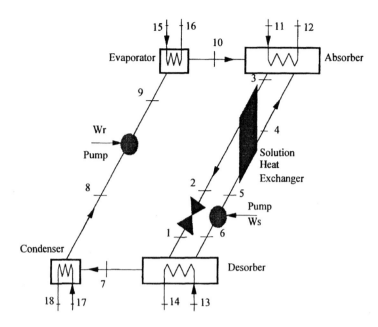

Figure 6.22 Single-effect water/lithium bromide temperature booster heat pump (Type II absorption cycle) with external heat transfer models

temperature in the Type I cycle. However, the design temperatures for a Type II cycle are within the same range as the typical temperatures routinely experienced in a double-effect machine. Double-effect technology, which forms the focus of Chapter 7, is a proven long-life technology where corrosion challenges have been effectively met. Thus, Type II technology should be viewed as an available technology. The design of a Type II machine is a straightforward extension of chiller technology for an experienced water/lithium bromide engineering team.

The operating characteristics of a Type II cycle are best understood in terms of an example. A baseline set of operating conditions is given in Table 6.6 based on the cycle schematic in Figure 6.22. There are major differences and major similarities between the cycles pictured in Figures 6.4 and 6.22 The presentation here is designed to highlight both for a complete understanding of the cycle. A major difference is that all the internal mass flows are in the opposite direction between the Type I and Type II cycles. In the Type II cycle the solution is pumped up through the right-most leg of the solution circuit and it flows down through the left leg. The flow in the refrigerant loop is from bottom to top and a pump is required in the line to drive the flow from a low pressure to a high pressure.

A summary of the internal state points is provided in Table 6.7 to clarify a number of key assumptions used in the model. In particular, the condition of the fluid exiting each of the components is assumed to be saturated and thus the quality is set to either 0 or 1 for saturated liquid and vapor, respectively. To further clarify the model, the input variables are specified in Table 6.8. The inputs were chosen only to illustrate one potential operating condition and are emphasized here only to help the reader understand the model results presented.

In the Type II cycle, the absorber and evaporator are together at nearly the same pressure, but now that pressure is the high pressure in the system. This fact is emphasized in Figure 6.22 which is a Dühring-type schematic such that components at the top of the figure operate at a higher pressure. The desorber and condenser operate at the low pressure in the system. The

Table 6.6. Single-effect Type II operating conditions

i	h(i) (J/g)	m(i) (kg/sec)	P(i) (kPa)	Q(i) (Fraction)	T(i) (°C)	x(i) (% LiBr)
1	277.40	1.078	16.05	0.016	102.8	59.4
2	277.40	1.078	104.08		124.7	59.4
3	332.80	1.078	104.08	0.000	153.0	59.4
4	329.93	1.000	104.08		144.8	64.0
5	270.2051	1.000	104.08		112.0	64.0
6	270.1540	1.000	16.05	0.000	112.0	64.0
7	2688.3	0.078	16.05		100.5	
8	231.8400	0.078	16.05	0.000	55.4	
9	231.9291	0.078	104.08		55.4	
10	2676.9	0.078	104.08	1.000	100.8	
11		4.000			135.0	
12					145.7	
13		4.000			120.0	
14					109.2	
15		4.000			120.0	
16					108.6	
17		4.000			30.0	
18					41.4	

COP	= 0.484	T_{out}	= 163.689 °C	Q_{shx}	=	59.72 kW
ϵ_{shx}	= 0.800	x_{out}	= 63.140 %	Ua_a	=	10.000 kW/K
$\Delta T_{lm,a}$	= 18.002 °C	h_{out}	= 361.742 kJ/kg	Ua_c	=	10.000 kW/K
$\Delta T_{lm,c}$	= 19.166 °C	P_h	= 104.076 kPa	Ua_d	=	25.000 kW/K
$\Delta T_{lm,d}$	= 7.234 °C	P_l	= 16.054 kPa	Ua_e	=	15.000 kW/K
$\Delta T_{lm,e}$	= 12.718 °C	Q_{abs}	= 180.0 kW	Ua_s	=	5.811 kW/K
$\Delta T_{lm,s}$	= 10.278 °C	Q_{cond}	= 191.7 kW	W_R	=	0.007 kW
m_{in}	= 0.014 kg/s	Q_{evap}	= 190.8 kW	W_S	=	0.051 kW
m_{out}	= 1.014 kg/s	Q_{des}	= 180.9 kW			

pressure levels for the Type II cycle are considerably different than the Type I. For the Type II cycle operating conditions in Table 6.6, the high pressure is 104.1 kPa and the low pressure is 16.1 kPa. Thus, the high pressure end can operate above atmospheric pressure while the low pressure end is still in vacuum.

The basic operation of the components of the Type II cycle is familiar from earlier discussion of these devices. For example, the solution circulation ratio can be calculated as

$$f = \frac{m_6}{m_7} = 12.82 \qquad\qquad 6.23$$

The value obtained here is slightly higher than that obtained for a Type I cycle since the definition is now in terms of the low flow rate side of the solution circuit. Although much of the technology is similar to Type I considerations, it may still be useful to briefly describe the

Table 6.7 Thermodynamic state point summary

Point	State	Notes
1	Vapor-liquid solution state	Vapor flashes as liquid passes through expansion valve
2	Subcooled liquid solution	State calculated from solution heat exchanger model
3	Saturated liquid solution	Vapor quality set to 0 as assumption
4	Subcooled liquid solution	State calculated from solution heat exchanger model (Note: $T_4 < T_3$)
5	Subcooled liquid solution	State calculated from solution pump model
6	Saturated liquid solution	Vapor quality set to 0 as assumption
7	Superheated water vapor	Assumed to have zero salt content
8	Saturated liquid water	Vapor quality set to 0 as assumption
9	Subcooled liquid water	State calculated from refrigerant pump model
10	Saturated water vapor	Vapor quality set to 1.0 as assumption

Table 6.8 Baseline inputs defining Type II operating conditions in Table 6.6

Input Name	Value	Input Name	Value
ϵ_{SHX}	0.8	T_{13} (°C)	120
m_6 (kg/sec)	1.0	m_{13} (kg/sec)	4.0
UA_a (kW/K)	10.0	T_{15} (°C)	120
UA_c (kW/K)	10.0	m_{15} (kg/sec)	4.0
UA_d (kW/K)	25.0	T_{11} (°C)	135
UA_e (kW/K)	15.0	m_{11} (kg/sec)	4.0
		T_{17} (°C)	30
		m_{17} (kg/sec)	4.0

operation of the cycle. A Dühring plot of the cycle state points is provided as Figure 6.23. This plot and the schematic should be referred to regularly during study of the cycle.

At low pressure, water vapor is driven out of the solution in the desorber by heat transfer into the component. This heat transfer would typically be waste heat. The desorbed vapor flows to the condenser where it changes phase and heat rejection occurs, as required by the Second Law. The liquid refrigerant is pumped up to the high pressure and sent to the evaporator. Waste heat is applied to the evaporator to effect a liquid to vapor-phase change. The vapor then flows to the absorber where it recombines with the salt solution to complete the cycle. The absorption process is accompanied by an energy release that is the product of the cycle. By transferring energy out at the high temperature of the cycle, the Type II cycle has a role in upgrading the temperature of a waste heat stream to a temperature level that is more useful. This functional description of the cycle has glossed over several additional differences between the Type I and II cycles. These differences include the operation of the solution heat exchanger and the absorber. These differences are best described with reference to the operating conditions in Table 6.6.

The solution heat exchanger transfers heat from the hot leg to the cold leg in both Type I and Type II cycles. However, in the Type II cycle the hot leg is the left leg of the solution circuit as drawn in Figure 6.22. This observation implies that state point 4 must have a temperature less than or equal to the temperature of point 3. This situation is emphasized on the schematic by drawing the solution heat exchanger box in a different manner than that drawn for the Type I solution circuit. The basic methods of modeling and designing the solution heat exchanger are not different but the fact that the solution enters the absorber subcooled at state 4 has a major impact on cycle performance since that solution must be heated to the saturation temperature in the absorber before any energy can be transferred out as the end product. The solution heat exchanger has a more critical impact on cycle performance in a Type II cycle than it does in a Type I cycle.

In steady-state operation, the highest temperature in the absorber is much higher than the temperature at point 4. The exothermic nature of the absorption of water into lithium bromide plays a key role in obtaining this high temperature in the absorber. Since high temperature heat is the product of the cycle, it is very important to understand and to be able to predict this temperature. The actual highest temperature achieved in a real Type II absorber will depend on both heat transfer and thermodynamic design details. A limiting thermodynamic value is the adiabatic saturation temperature defined as the equilibrium temperature obtained by mixing the solution stream at state 4 with the vapor at state 10. The resulting temperature depends primarily on the level of subcooling in stream 4 as follows.

The adiabatic saturation analysis is based on the schematic in Figure 6.24. The solution flow rate and state at point 4 are assumed known. The vapor inlet state is assumed to be identical to state 10 in Figure 6.22. The inlet vapor flow rate, outlet solution flow rate and outlet solution mass fraction are all unknowns which are determined from mass and energy balances.

The overall mass balance is

$$m_4 + m_{in} = m_{out} \qquad\qquad 6.24$$

The lithium bromide mass balance is

$$m_4 x_4 = m_{out} x_{out} \qquad\qquad 6.25$$

The energy balance on the saturation process is

$$m_4 h_4 + m_{in} h_{10} = m_{out} h_{out} \qquad\qquad 6.26$$

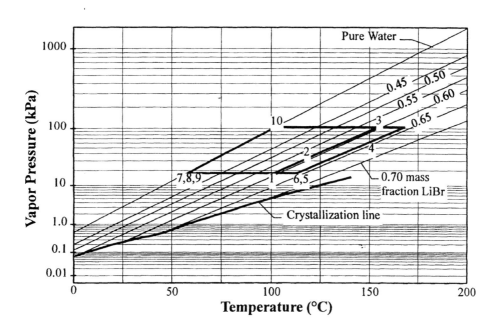

Figure 6.23 Dühring plot for the cycle solution in Table 6.6

These three equations are solved for the three unknowns m_{in}, m_{out}, and x_{out}. Since the outlet state is assumed saturated, the pressure (which is assumed to be known) and the mass fraction are sufficient to define the state, and the enthalpy and temperature can be determined from these two. Due to the substantial subcooling of the inlet solution stream, considerable vapor must be absorbed to reach saturation and the mass fraction of the saturated stream is considerably different than the mass fraction at state 4. Thus, the saturation temperature corresponding to the concentration at point 4 would not provide an accurate prediction of the highest temperature in the absorber. Numerical results of this calculation are provided in Table 6.6 as a part of the cycle solution. The adiabatic saturation process requires approximately 18% of the total vapor flowing into the absorber. The latent heat of this vapor, added to the heat of mixing released during the absorption, raises the temperature of the solution to a saturation temperature of 163.7°C from an inlet temperature of 144.8°C.

Due to the fact that the pressure differences are greater in the Type II cycle, the pump work requirements are larger than that found for the Type I cycle. In this case, the sum of the pump work for both the solution and the refrigerant pumps is approximately 0.01% of the total thermal input required in the evaporator and desorber. Thus, the pump work requirement is seen to be still quite small in relation to the thermal energy input requirements.

In contrast to the Type I cycle, the Type II analysis shows that the heat transfer rates in the evaporator and condenser are larger than the rates in the desorber and absorber. This difference comes about because of changes in the input streams to the absorber and desorber. In the case of the absorber, the heat output is reduced due to the subcooling of state 4. In the case of the desorber, the heat input is reduced due to the fact that a significant fraction of the vapor required at 7 flashes as the solution passes through the expansion valve.

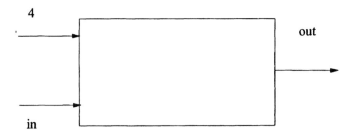

Figure 6.24 Adiabatic absorption temperature calculation schematic

Solution Heat Exchanger Size Effects

A key choice in the design of a Type II cycle is the size of the solution heat exchanger. The influence of the solution heat exchanger on performance is plotted in Figure 6.25 where COP and capacity are plotted versus solution heat exchanger effectiveness. The effectiveness is varied over the full range of possible values. It is observed that the COP increases significantly as effectiveness increases. With no solution heat exchanger the COP is 0.41 and with a very large solution heat exchanger the COP is 0.5, a 22% increase. The capacity increase is approximately 32% over the same range. The UA of the solution heat exchanger for an effectiveness of 0.8 was calculated as a part of the solution given in Table 6.6 and the value is $UA_s = 5.8$ kW/K. Thus, even though the UA value required is relatively small compared to the other heat exchangers, the performance benefits are significant.

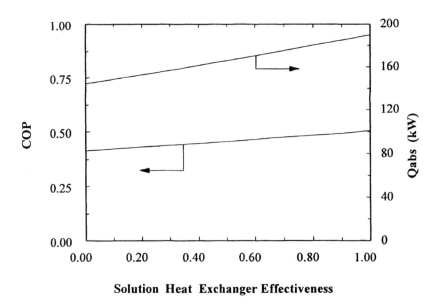

Figure 6.25 Effect of solution heat exchanger effectiveness on COP and capacity of the Type II cycle

6.4   Discussion of Available Single-Effect Systems

Single-effect, Type I, water/lithium bromide systems are manufactured and sold throughout the world. The technology is widespread including manufacturing plants in India and China as well as throughout Western Europe, the Americas, and the Pacific Rim. The technology was pioneered in the U.S. in the 1940s and has been widely available since the 1950s. Japanese manufacturing started in the 1960s and absorption machines (including single and double effect) reportedly supply close to 50% of the space cooling load in Japan today with at least eight major manufacturers in the market. Four companies supply single-effect machines in the U.S. today (AGCC, 1995).

Single-effect machines are typically indirect fired with heat input provided by steam or hot water. Applications are usually related to waste heat recovery. Typically the heat input temperatures must be greater than 75°C. Available sizes range from 18 to 5840 kW (5 to 1660 Tons of refrigeration) as the evaporator heat transfer rate. The COP of these systems is approximately 0.7.

HOMEWORK PROBLEMS

6.1   Using the internal model which was used to generate Table 6.1, vary the solution mass flow rate through the solution pump to generate plots of COP and capacity versus solution flow rate.

6.2   Using the heat transfer model which was used to generate Table 6.2, vary the solution mass flow rate through the solution pump to generate plots of COP and capacity versus solution flow rate.

6.3   Compare the results of Problems 6.1 and 6.2. Note that the approach taken in Section 6.1 is quite limited and often results in confusing predictions if care is not taken in the interpretation. The approach taken in Section 6.2 is more realistic because the internal variables all adjust to changes in the operating conditions.

6.4   Using the model which generated Table 6.3, perform the following experiment. Reset the initial guesses to the default values provided by EES. Then, rerun the model to see what will happen. Do the same with the model used to generate Table 6.1.

6.5   Convert the model used to generate Table 6.3 such that the solution heat exchangers are modeled using a UA approach instead of the effectiveness approach used. Using a fixed UA value, reproduce the results found in Figures 6.9 to 6.21 to determine whether the solution heat exchanger modeling approach changes the nature of the results.

6.6   Convert the model used to generate Table 6.3 into a model for a machine with double the capacity of the existing version. Note: It should be possible to obtain a solution with identical state points but double the capacity by carefully choosing the appropriate parameters to change.

6.7   Using the model which generated Table 6.3, change the absorber UA value to 20% of its existing value and rerun the model. Be sure to check the approach of the solution to the crystallization line.

6.8   Repeat Problem 6.4 for the heat transformer model used to generate Table 6.6.

6.9   Using the model which generated Table 6.6, vary the parameters to determine the effects of the following variables on performance.
   a.   Absorber heat transfer fluid inlet temperature from 100 to 145°C
   b.   Desorber and evaporator inlet temperatures from 100 to 140°C
   c.   Condenser inlet temperature from 20 to 40°C
   d.   Solution circuit flow rate from 0.5 to 2 kg/sec.
   e.   Heat transfer fluid flow rates in each of the four heat transfer loops. Vary over the range of 50 to 150% of existing flow rate.
   f.   UA values for each of the heat exchangers in the machine. Vary over the range of 50 to 150% of existing values.

6.10   Determine the effect of solution heat exchanger effectiveness on the highest temperature obtainable in the absorber of a Type II machine.

Chapter 7

DOUBLE-EFFECT WATER AND LITHIUM BROMIDE TECHNOLOGY

As discussed in Chapter 6, one of the limitations of single-effect absorption cycles is that they cannot take advantage of the higher availability of high temperature heat sources to achieve higher COP. Although the COP of a reversible cycle is quite sensitive to heat input temperature, the COP of a real absorption machine is essentially constant due to the irreversible effects associated with heat transfer. Thus, the cooling COP of a single-effect water/lithium bromide machine is around 0.7, essentially independent of the heat input temperature. To achieve higher cycle performance, it is necessary to design a cycle that can take advantage of the higher availability (or exergy) associated with a higher temperature heat input. Double-effect technology represents one such cycle variation. The present chapter is devoted to describing the operating characteristics and the performance potential of double-effect, water/lithium bromide technology. Other high-performance cycles are also possible and an overview of these is presented in Chapter 8.

7.1   Double-Effect Water/Lithium Bromide Cycles

Due to the relatively low COP associated with single-effect technology, it is difficult for single-effect machines to compete economically with conventional vapor compression systems except in low temperature waste heat applications where the input energy is free. Double-effect technology, with COP in the range of 1.0 to 1.2, is much more competitive. Gas-fired double-effect water/lithium bromide technology is a mature technology that competes for the gas cooling market segment. Competing gas-fired technologies include gas engine-driven vapor compression systems and desiccant systems. Double-effect machines, using water/lithium bromide as working fluid, are produced by a large number of manufacturers world-wide. Each manufacturer uses a different design depending on its view of the market economics. Instead of focusing on the design details of one machine, the discussion here of design tradeoffs is based on generic double-effect machines.

A schematic of a double-effect machine is provided as Figure 7.1 in the Dühring plot format. In Figure 7.1 the external heat transfer interactions are represented by arrows. Heat is transferred into the cycle in both the high desorber and the evaporator. Heat is transferred out from the cycle in the absorber and low condenser. The double-effect cycle includes two solution heat exchangers that have a similar role in the solution circuits as was described for the single-effect cycle. A new feature of the double effect is the internal heat exchange between the high condenser and the low desorber. This internal heat exchange is achieved in practice by incorporating these two components into a single transfer device. One side of the exchanger is the high condenser and the other side is the low desorber. This combined component is represented by the dotted border in Figure 7.1.

The low desorber and low condenser of the double effect operate at approximately the same conditions as the desorber and condenser of a single-effect machine. The operating temperatures and pressures of the high pressure end of the double effect can be inferred from Figure 7.1. The heat input in the double effect occurs at a much higher temperature than in the single effect. The COP of the double-effect technology is greater than single effect because it

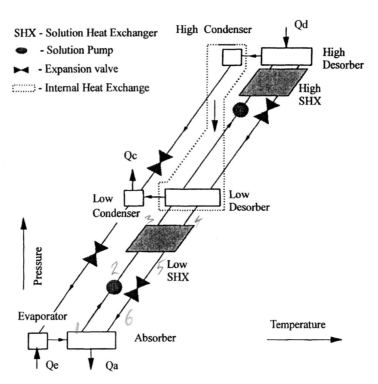

Figure 7.1 Double-effect water/lithium bromide chiller Dühring chart schematic (parallel flow)

is able to utilize the increased availability of the higher temperature input heat. As compared to a single-effect machine, the double-effect machine takes heat in at a higher temperature but it rejects heat at about the same temperature and it provides refrigeration at about the same temperature. Thus, the double effect can still be viewed as a three temperature device as introduced in Section 2.2. However, because the cascade of energy through the machine is accompanied by irreversible effects, the actual COP falls short of the reversible value obtained in Section 2.2. The reversible analysis is still useful since the relative COP changes associated with increasing the heat input temperature for a real cycle roughly follows the reversible analysis.

Example 7.1 Reversible COP of Double-Effect Machine

Determine the reversible performance of a three-temperature chiller operating between the temperatures 150, 30 and 12°C which represent a typical double-effect operating condition.

Solution: Direct application of Equation 2.11 yields

$$COP_{rev} = \frac{T_0}{T_2}\frac{(T_2 - T_1)}{(T_1 - T_0)} = 4.49$$

Observations: The reversible analysis requires the use of absolute temperatures. This calculation indicates that the thermodynamic limit of COP for this set of temperatures is quite high. Actual double-effect machines that operate over this temperature range achieve a COP of approximately 1.0 to 1.2. Thus, great care must be taken in using the reversible COP since it is far from realistic.

Example 7.2   Zero-Order Model of a Double-Effect Machine

Using a zero-order model, determine the capacity and COP of a double-effect machine operating between the temperatures given in Example 7.1 and with UA values at the high, heat rejection and evaporator temperatures given as UA_h = 1 kW/K, UA_c = 3 kW/K and UA_e = 2 kW/K.

Solution: The zero-order model for the double-effect uses Equations 2.17-2.21 as written. Equation 2.22 must be modified to better model the double effect. One possibility is

$$T_{hi} - T_{ci} = 2(T_{ci} - T_{ei})$$

Based on these conditions and equations, the model yields a COP of 1.35 and a capacity of 24.1 kW. The calculated internal temperatures are T_{hi} = 405.3 K, T_{ci} = 317.1 K and T_{ei} = 273.1 K.

Observations: The difference between the reversible COP calculated in Example 7.1 and the more realistic COP calculated from the zero-order model is heat transfer irreversibility at all three temperature levels. The thermal resistance causes the internal temperatures to be different than the external ones and this difference is characteristic of a real cycle. The model also gives a zero-order prediction of the capacity. This capacity prediction is quite idealized but it is still useful. In Section 7.3 a more realistic model (UA model) is described. When that model is used to calculate the performance of a similarly specified cycle (i.e., with the same UA distribution) it yields a capacity of only 15 kW and a COP of 1.1 (demonstration of this is left to the reader). Thus, both the COP and capacity predictions from the zero-order model are too high, but it has the advantage of the ease with which it can be set up and solved and it correctly predicts the trends in COP and capacity as source and sink temperatures are changed.

A simplified but quite accurate view of a double-effect machine can be obtained by viewing it as a three-pressure device. As was done in modeling the single-effect machines in Chapter 6, it is assumed here that pressure drops occur only in the throttling valves and pressure increases occur only in pumps. This approximation is quite accurate because other design constraints require that the machine be designed to minimize pressure drops. The key design constraint here is identical to what was found for single-effect technology. There it was found that the pressure drops in the machine must be minimized due to the low absolute pressure that is associated with water/lithium bromide and pressure drop tends to aggravate most of the other design issues including crystallization. With reference to Figure 7.1, the three pressure levels are represented by the pairs of components connected by horizontal lines. The high condenser and high desorber operate at the high pressure, the low condenser and low desorber operate at the intermediate pressure and the evaporator and absorber operate at the lowest pressure in the cycle. The model discussed in Section 7.3 is based on such a three-pressure model of the double effect. The pressure levels are quantified in that section.

The term "double effect" refers to the fact that the heat input at the high temperature is used twice within the cycle to generate vapor. Vapor is generated in the high desorber as the heat is input. This vapor then flows to the high condenser where it changes phase by rejecting heat. This heat is at a sufficiently high temperature that it can be used to drive vapor out of the solution in the low desorber. Thus, the heat is used twice and the term double effect reflects this. Another term that is useful in describing the cycle pictured in Figure 7.1 is "two stage". An absorption machine can always be decomposed into a series of basic stages which are just single-effect machines. Such a building block approach is quite useful when designing new cycles (Alefeld and Radermacher, 1994). By inspection, the double-effect cycle in Figure 7.1 can be seen to consist of two single-effect stages stacked one on top of the other. Thus, the cycle in Figure 7.1 can also be described as a two-stage system. It is noted that these two terms are not interchangeable since they have different meanings. It is possible to have a two-stage cycle that is triple effect (such as the cycle discussed in Chapter 10). Thus, the term stage is reserved for describing the physical configuration of the cycle and effect is reserved for describing the performance level of the resulting cycle. The number of effects is an approximation to the increase in COP that can be expected. In reality, a double-effect machine typically shows an increase in COP of less than two times due to heat transfer losses associated with the cascade of energy through the system. A superposition method for predicting the COP of multi-stage cycles can be found in Alefeld and Radermacher (1994).

7.2   Solution Circuit Plumbing Options

One of the major design choices in double-effect technology is the choice of how to connect the solution circuits. The basic options are parallel or series flow. The cycle in Figure 7.1 was drawn assuming parallel flow and assuming that the solution mass fraction change across each of the desorbers is identical. The actual piping configuration used depends on the manufacturer. One possible parallel flow piping diagram is provided as Figure 7.2. Parallel flow offers thermodynamic and heat transfer benefits over series flow but achieving these relatively small benefits requires more control complexity.

Series flow schematics are included as Figures 7.3 and 7.4. The two cases are differentiated by the plumbing of the solution leaving the absorber. In the configuration in Figure 7.3, the solution is sent first to the high desorber and then to the low desorber. In the configuration in Figure 7.4, these are reversed. In both cases, the internal heat exchange process between the high condenser and the low desorber constrains the temperatures. The high condenser must have a high enough temperature so as to provide a heat transfer driving potential to drive heat into the low desorber. Thus, the different series flow arrangements lead to different temperatures in the various components, and, as expected, the performance of the various configurations differs. A summary of the performance of the three configurations is presented in Table 7.1.

Table 7.1 Comparison of parallel and series flow for double-effect water/lithium bromide cycles

Configuration	COP	Capacity (kW)
Parallel	1.325	354.4
Series, high desorber first	1.244	371.1
Series, low desorber first	1.238	370.2

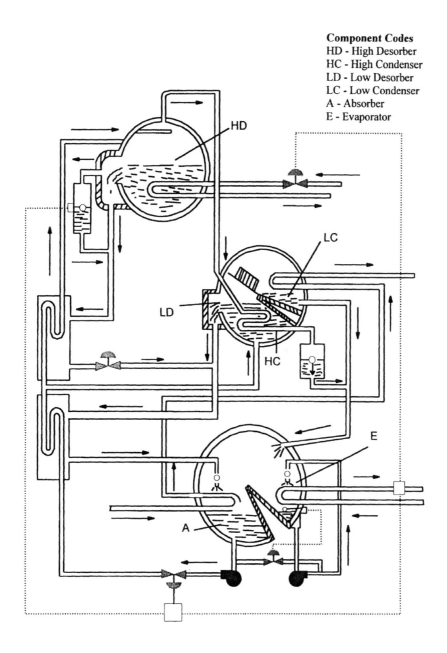

Component Codes
HD - High Desorber
HC - High Condenser
LD - Low Desorber
LC - Low Condenser
A - Absorber
E - Evaporator

Figure 7.2 Double-effect water/lithium bromide chiller piping chart schematic (parallel flow)

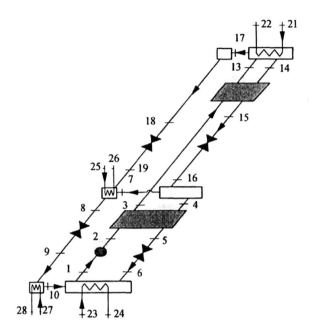

Figure 7.3 Series flow double-effect water/lithium bromide chiller Dühring chart schematic (solution to high temperature)

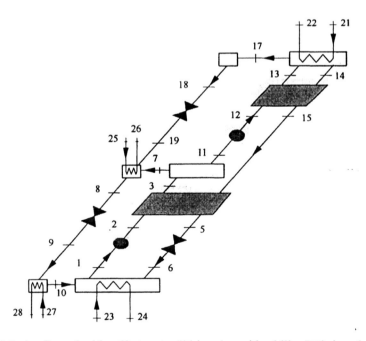

Figure 7.4 Series flow double-effect water/lithium bromide chiller Dühring chart schematic (solution to low desorber first)

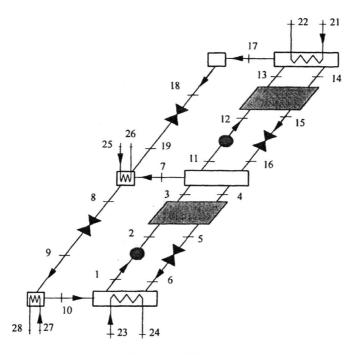

Figure 7.5 Parallel flow double-effect water/lithium bromide chiller Dühring chart schematic showing state points

It can be seen that the difference in COP favors the parallel flow configuration. However, the capacity favors the series flow configurations. The results in Table 7.1 are all based on the same heat exchanger sizes and the same external fluid loop inlet conditions as discussed in detail in Section 7.3 (Grossman, 1990).

7.3   Operating Conditions of Double Effect Machines

7.3.1   Parallel Flow Double Effect
The parallel flow configuration of Figure 7.1 is redrawn in Figure 7.5 showing the state points between components and showing the external heat transfers as heat transfer loops. The assumptions about the internal state points are quite similar to those discussed in Chapter 6 with reference to the single-effect technology. However, it may still be useful to summarize them here. This is done in Table 7.2.

The cycle was modeled by imposing mass and energy balances on all components and by requiring the heat exchangers to follow UA or effectiveness models. The methods used are quite similar to those described in Chapter 6. A better idea of the type of model can be obtained by examining the model inputs which are summarized in Table 7.3.

The solution flow rate leaving the absorber is specified (m_1). In general, one would expect to also specify the flow rate in the upper solution circuit (Pump 2). However, in this case the energy balance between the high condenser and the low desorber is used to calculate the upper circuit flow rate. There is only one upper circuit solution flow rate that exactly balances the heat requirements of these two components (assuming zero jacket losses).

Table 7.2 State point summary for parallel flow double-effect cycle

Point	State	Notes
1	Saturated liquid solution	Vapor quality set to 0 as assumption
2	Subcooled liquid solution	State determined from pump model
3	Subcooled liquid solution (typically)	State determined from solution heat exchanger model
4	Saturated liquid solution	Vapor quality set to 0 as assumption
5	Subcooled liquid solution	State determined from solution heat exchanger model
6	Two-phase solution state (typically)	Vapor quality determined from valve model
7	Superheated water vapor	Assumed to have zero salt content
8	Saturated liquid water	Vapor quality set to 0 as assumption
9	Vapor-liquid water state	Vapor quality determined from valve node 1
10	Saturated water vapor	Vapor quality set to 1.0 as assumption
11	Saturated liquid solution	Vapor quality set to 0 as assumption
12	Subcooled liquid solution	State determined from pump model
13	Subcooled liquid solution (typically)	State determined from solution heat exchanger model
14	Saturated liquid solution	Vapor quality set to 0 as assumption
15	Subcooled liquid solution	State determined from solution heat exchanger model
16	Two-phase solution state (typically)	Vapor quality determined from valve model
17	Superheated water vapor	Assumed to have zero salt content
18	Saturated liquid water	Vapor quality set to 0 as assumption
19	Vapor-liquid water state	Vapor quality determined from valve node 1

Table 7.3 Inputs for the parallel flow double-effect cycle model

Parameter	Value	Parameter	Value
m_1 (kg/sec)	1.0	T_{25} (°C)	25
m_{21} (kg/sec)	8.0	T_{27} (°C)	12
m_{23} (kg/sec)	12.0	UA_d (kW/K)	25
m_{25} (kg/sec)	14.0	UA_c (kW/K)	65
m_{27} (kg/sec)	20.0	UA_e (kW/K)	40
T_{21} (°C)	150	UA_a (kW/K)	50
T_{23} (°C)	25	UA_{cd} (kW/K)	10
		ϵ_{SHX1}	0.5
		ϵ_{SHX2}	0.5

The mass flow rate of heat transfer fluid in the four external heat transfer loops is specified as well as the inlet temperature to each loop. The UA values are input to represent the size of each of the external heat exchangers. A UA model was also used for the internal heat exchange between the high condenser and the low desorber (designated Ua_{cd}). Effectiveness models were used for the solution heat exchangers. The effectiveness model was used for convenience since such a model tends to be more robust and allowed parametric studies of the cycle with less sensitivity to the initial guess values.

The operating conditions calculated for the baseline case defined by the inputs in Table 7.3 are listed in Table 7.4. Point 20 is missing from the list because this point was not used in this particular point numbering scheme. It is noted that the COP and capacity match the values reported in Table 7.1. All of the results in Table 7.1 were made with the same baseline inputs as listed in Table 7.3. It is also noted that the heat transfer fluid was assumed to be water with a fixed $c_p = 4.2$ J/g-K for all four of the external loops.

The double-effect machine can be visualized by displaying the operating conditions in a graphical format such as the Dühring chart in Figure 7.6. The machine operates at three pressure levels which are 0.9, 4.2 and 64.2 kPa for this baseline case. Of course these pressures depend on the other operating conditions. A change in any of the inputs will affect all three pressures. The pressures shown here are typical. The high end pressure approaches atmospheric pressure but remains subatmospheric.

A key aspect of the operating conditions is that the temperature of the high condenser must be greater than that of the low desorber. This is seen to be the case here. The high end temperature in the machine is approximately 145°C. This is considerably greater than that encountered in single-effect machines. The high temperature has consequent corrosion implications. Corrosion problems are more severe in double-effect technology than they are in single-effect technology due to the higher temperatures.

The high mass fraction in the machine must be maintained below the point at which crystallization will occur. The margin shown by this case is typical of that encountered in practice. The mass fraction change in each of the solution circuits is the same for this model.

Table 7.4 Baseline operating conditions for a parallel flow double-effect water/lithium bromide machine

i	h (i) (J/g)	m (i) (kg/sec)	P (i) (kPa)	Q (i) (Fraction)	T (i) (°C)	X (i) (% LiBr)
1	65.5911	1.000	0.880	0.000	29.85	52.765
2	65.5932	1.000	4.171		29.85	52.765
3	102.7	1.000	4.171		47.33	52.765
4	195.0	0.852	4.171	0.000	76.39	61.967
5	151.4	0.852	4.171		53.12	61.967
6	151.4	0.852	0.880	0.003	47.91	61.967
7	2608.7	0.067	4.171		57.47	0.000
8	124.5	0.148	4.171	0.000	29.72	0.000
9	124.5	0.148	0.880	0.041	5.13	0.000
10	2511.0	0.148	0.880	1.000	5.13	0.000
11	124.284	0.550	4.171	0.000	57.47	52.765
12	124.322	0.550	64.231		57.49	52.765
13	194.1	0.550	64.231		90.18	52.765
14	323.3	0.469	64.231	0.000	144.84	61.967
15	241.4	0.469	64.231		101.16	61.967
16	241.4	0.469	4.171	0.015	78.60	61.967
17	2726.2	0.082	64.231		122.80	0.000
18	367.2	0.082	64.231	0.000	87.73	0.000
19	367.2	0.082	4.171	0.100	29.72	0.000
21		8.000			150.00	
22					142.04	
23		12.000			25.00	
24					33.65	
25		14.000			25.00	
26					28.16	
27		20.000			12.00	
28					7.78	

COP	= 1.325	$\Delta T_{lm,s2}$	= 48.961 K	SI	= 2.000
cp	= 4.20 J/g-K	P_h	= 64.231 kPa	UA_a	= 50.000 kW/K
ϵ_{hx}	= 0.500	P_l	= 0.880 kPa	UA_c	= 65.000 kW/K
hl_{16}	= 203.915 J/g	P_m	= 4.171 kPa	UA_{cd}	= 10.000 kW/K
hl_6	= 142.880 J/g	$W_{p,1}$	= 0.002 kW	UA_d	= 25.000 kW/K
hv_{16}	= 2647.604 J/g	$W_{p,2}$	= 0.022 kW	UA_e	= 80.000 kW/K
hv_6	= 2589.969 J/g	Q_a	= 436.179 kW	UA_{s1}	= 1.424 kW/K
$\Delta T_{lm,a}$	= 8.724 K	Q_{cd}	= 192.776 kW	UA_{s2}	= 0.784 kW/K
$\Delta T_{lm,c}$	= 2.857 K	Q_c	= 185.702 kW	v_1	= 0.645 cm³/g
$\Delta T_{lm,cd}$	= 19.278 K	Q_e	= 354.366 kW	v_{11}	= 0.651 cm³/g
$\Delta T_{lm,d}$	= 10.700 K	Q_d	= 267.492 kW	Xl_{16}	= 62.932 %
$\Delta T_{lm,e}$	= 4.430 K	$Q_{s,1}$	= 37.116 kW	Xl_6	= 62.183 %
$\Delta T_{lm,s1}$	= 26.061 K	$Q_{s,2}$	= 38.387 kW		

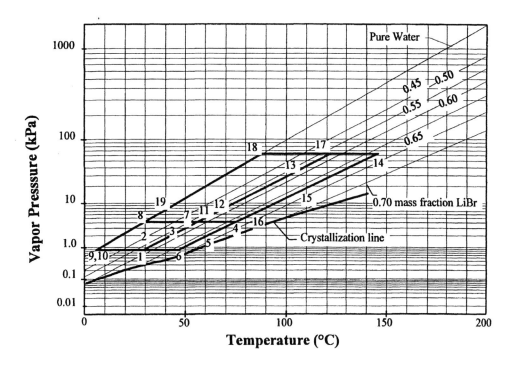

Figure 7.6 Dühring state plot for baseline double-effect water/lithium bromide example from Section 7.3

This requirement was imposed as an assumption for convenience in modeling. In a real machine, the mass fraction change across a given component will be a complex function of the operating characteristics and conditions. Thus, in general, it is not possible to design such that the mass fraction difference is the same. In practice, some additional mixing irreversibility is encountered when the solution returning from the high desorber is mixed with the solution returning from the low desorber before being sent on to the absorber.

Another observation about the mass fraction is that the change across the solution circuits displayed here is quite large (approximately 0.09 change in mass fraction) compared to standard practice. A more typical design would use a larger mass flow rate for a machine of this capacity so that the mass fraction change is around 0.05. This case was also run by changing the solution flow rate input from 1.0 to 2.0 kg/sec. The result is quite informative about the technology and it is left as an exercise for the reader to run that case to illustrate the following observations. When the solution flow rate is increased to 2.0 kg/sec, the capacity increases only slightly to 361.3 kW (2% increase). However, the COP decreases to 1.19 which is a 10% reduction. The explanation for this is complex since many things are changing. The temperature matching in the external heat exchangers actually improves so that is not the source of the COP decrease. The COP decrease can be traced, finally, to substantially increased heat transfer duty in the solution heat exchangers associated with the increased flow rate. The change is even more pronounced if fixed UA models are used for the solution heat exchangers (fixed effectiveness models were used here for convenience).

This effect was also found in single-effect technology. The cycle performance is quite sensitive to the solution flow rate due to irreversibilities in the solution heat exchangers. In practice, such a wide mass fraction change is not usually practical due to the limited space between the available cooling temperature and the crystallization line. In the baseline case, a low cooling temperature was used which allowed the cycle solution to work. If this design were put in operation, a hot ambient condition would be expected to cause operational problems due to crystallization.

This operating characteristic indicates one possible control scheme to optimize performance. If a variable speed solution pump were provided, it would be possible to allow a wider solution band when the cooling water temperature is lower. The performance advantage that this option would provide depends somewhat on the size of the solution heat exchangers. The benefit is less when the solution heat exchangers are more effective.

7.3.2   Series Flow Double Effect

The series flow option is preferred in some cases since the solution path is simpler. Maintaining the correct flow split in parallel flow is problematic without active controls. Thus, particularly for small systems, series flow is attractive as a design compromise. The best series flow option seems to be the one shown in Figure 7.3 where the solution is sent first to the high desorber. A model of this cycle, using the same inputs and same basic assumption as were used to model the parallel flow configuration, was written for the series flow case. The operating conditions for the series flow baseline case are given in Table 7.5.

The state point numbers in Table 7.5 refer to Figure 7.3. Point numbers 11, 12 and 20 were omitted intentionally in an attempt to maintain some consistency with the solution to the parallel flow example. As was already noted in the discussion of Table 7.1, the COP of the series flow configuration is lower than the parallel flow but the capacity is higher. A number of things change between the two cycle solutions obscuring the cause of the COP difference. Careful study of the solutions reveals, however, that the key difference is the increased heat transfer load on the high solution heat exchanger. The relatively larger solution flow rate in the upper solution circuit causes a small mass fraction difference but also implies more heat transfer load on the solution heat exchanger. This higher load implies higher irreversibility and lower thermodynamic performance.

The increased capacity of the series flow configuration is apparently the result of a better temperature match in the high desorber and the internal heat transfer between the high condenser and the low desorber. A better temperature match in the high desorber would be relatively simple to achieve in the parallel flow case by simply decreasing the heat transfer fluid flow rate. However, since this causes the average temperature of the heat transfer fluid to decrease, it actually decreases the capacity of the parallel flow machine. The point of this is that the comparisons between the two design choices are not very straightforward. Design optimization for each plumbing option causes different choices for the relative sizes of the components and for the optimum choice of heat transfer loop flow rates and all other inputs. Thus, it is somewhat simplistic to compare the configurations on the basis of an arbitrary set of design parameters. When the optimization is done, it would be expected that the differences between the two technologies would narrow somewhat but that the overall conclusions about the preferred choice would not change.

Table 7.5 Operating conditions for a series flow double-effect water/lithium bromide machine

i	h (i) (J/g)	m (i) (kg/sec)	P (i) (kPa)	Q (i) (Fraction)	T (i) (°C)	X (i) (% LiBr)
1	66.3	1.000	0.876	0.000	30.00	52.900
2	66.3	1.000	4.353		30.01	52.900
3	108.1	1.000	4.353		49.70	52.900
4	203.0	0.844	4.353	0.000	78.84	62.656
5	157.7	0.844	4.353		54.42	62.656
6	157.7	0.844	0.876	0.003	49.23	62.656
7	2610.7	0.079	4.353		58.55	0.000
8	127.6	0.156	4.353	0.000	30.47	0.000
9	127.6	0.156	0.876	0.043	5.06	0.000
10	2510.8	0.156	0.876	1.000	5.06	0.000
13	190.6	1.000	77.520		88.46	52.900
14	300.7	0.923	77.520	0.000	138.59	57.325
15	211.3	0.923	77.520		94.14	57.325
16	211.3	0.923	4.353	0.018	69.46	
17	2736.5	0.077	77.520		128.66	0.000
18	388.0	0.077	77.520	0.000	92.69	0.000
19	388.0	0.077	4.353	0.107	30.47	0.000
21		8.000			150.00	
22					141.12	
23		12.000			25.00	
24					34.08	
25		14.000			25.00	
26					28.66	
27		20.000			12.00	
28					7.58	

COP	=	1.244	$\Delta T_{lm,s1}$	= 26.712 K	$Q_{s,2}$	=	82.501 kW
cp	=	4.200 J/g-K	$\Delta T_{lm,s2}$	= 47.231 K	SI	=	2.000
ϵ_{shx}	=	0.500	P_h	= 77.520 kPa	UA_a	=	50.000 kW/K
hl_{16}	=	165.80 J/g	P_l	= 0.876 kPa	UA_c	=	65.000 kW/K
hl_6	=	149.341 J/g	P_m	= 4.353 kPa	UA_{cd}	=	10.000 kW/K
hv_{16}	=	2630.4 J/g	$W_{p,1}$	= 0.002 kW	UA_d	=	25.000 kW/K
hv_6	=	2592.4 J/g	Q_a	= 457.823 kW	UA_e	=	85.000 kW/K
$\Delta T_{lm,a}$	=	9.156 K	Q_{cd}	= 181.303 kW	UA_{s1}	=	1.564 kW/K
$\Delta T_{lm,c}$	=	3.308 K	Q_c	= 215.039 kW	UA_{s2}	=	1.747 kW/K
$\Delta T_{lm,cd}$	=	18.130 K	Q_e	= 371.083 kW	v_1	=	0.644 cm^3/g
$\Delta T_{lm,d}$	=	11.929 K	Q_d	= 298.223 kW	Xl_{16}	=	58.404 %
$\Delta T_{lm,e}$	=	4.366 K	$Q_{s,1}$	= 41.771 kW	Xl_6	=	62.871 %

7.4   Systems on the Market

Five manufacturers actively market double-effect water/lithium bromide machines in the U.S. A summary of the systems offered can be found in AGCC (1995) which includes system schematics and overall performance levels and operating conditions. Differences between the operating conditions claimed by the manufacturers and the performance predictions presented in this chapter are the result of design choices in the sizing of the machine components. The primary tradeoff here is economic. The main obstacle to market penetration of double-effect absorption systems is still the first cost. Alternative technologies with lower first cost are readily available, in most cases. Thus, to make their absorption products as attractive as possible, manufacturers naturally reduce the heat exchanger sizes to reduce cost. This has the effect of reducing COP.

HOMEWORK PROBLEMS

7.1   Use the zero-order model of Example 7.2 and maximize the capacity of the double-effect machine under the conditions of fixed total UA to find the optimum distribution of UA around the machine. Use a total UA value of 6 kW/K and the temperatures in Example 7.2.

7.2   Use the model which generated Table 7.4 to investigate changes in operating conditions for a parallel flow double effect machine. For the following parts, hold all inputs constant except for the one being varied in that particular problem. In each case, determine the COP and capacity as a function of that particular variable.
   a)   Vary solution circuit flow rate
   b)   Vary solution heat exchanger effectiveness in each circuit
   c)   Vary the UA in each of the five heat exchangers where UA models are used.
   d)   Vary the heat transfer fluid inlet temperature in each of the four external heat exchangers
   e)   Vary the heat transfer fluid flow rate in each of the four external heat exchangers

7.3   Convert the solution heat exchanger models to a UA formulation where the UA values are input. Run the model to determine the effect of this change by repeating some of the studies in Problem 7.2 and comparing the results. Appropriate UA values can be obtained by running the effectiveness model first and allowing it to calculate the UA values for those components.

7.4   Experiment with the parallel flow model to determine whether series piping of the absorber and condenser is feasible. The model run in Table 7.4 is for a parallel flow arrangement where the inlet temperatures to each component are identical (and the flow rates are different). In the series flow arrangement, the flow rate to each is the same and the outlet temperature from one component is the inlet to the next. Calculate performance for both cases, differentiated by which component comes first. Note that by adding an additional constraining equation to the model set, one less input is required.

7.5   Examine the effect on performance of the temperature difference between the high and low temperature stages. This can be done by varying the UA between the high temperature condenser and the low temperature desorber. Note the effects on COP, capacity and the highest temperature in the cycle.

7.6   Estimate the thermodynamic losses in the refrigerant expansion valves by performing a calculation to determine the work which could be obtained by expanding the vapor through a reversible, adiabatic turbine. Is the isentropic work which would be produced sufficient to drive the solution pumps?

7.7   Prepare heat exchanger plots for each of the two solution heat exchangers. By inspection, determine whether these components exhibit temperature matching.

7.8   Replace the heat transfer fluid in the high-temperature desorber with steam input at constant temperature. Describe the changes needed in the model to make this work.

Chapter 8

ADVANCED WATER/LITHIUM BROMIDE CYCLES

The cycles discussed in Chapters 6 and 7 represent absorption chillers that are currently on the market as commercial products from various manufacturers. The focus of the present chapter is on advanced cycles that have been conceived as the solution to some particular thermal management challenge. The cycles covered include the half-effect, triple-effect, resorption cycle and the absorption power cycle. Each technology has a particular niche application which makes it desirable. However, in all cases, some barrier exists which limits the usage of the technology and which differentiates these technologies as experimental or development stage concepts. In the treatment that follows, these barriers are discussed along with the potential applications to give an overview of the potential of each concept.

8.1   Half-Effect Cycle

The half-effect cycle is used when the temperature of the available heat source is less than the minimum necessary to fire a single-effect cycle. The determination of the minimum temperature is not completely straightforward since it depends on the other two temperatures in the system. An idea of the minimum temperature needed to fire a single effect was obtained in Chapter 2 by examining the zero-order model in Example 2.1. If the chilled water temperature is 10°C and the heat rejection temperature is 32°C, a 5°C heat transfer driving potential at each level yields a minimum heat input temperature of 74°C. A similar analysis with zero heat transfer driving potential yields a minimum heat input temperature of 54°C. When the heat source temperature is low, there is a practical limit below which single-effect design becomes over-constrained by the characteristics of the working fluid as represented in Equation 2.22. For such low temperature heat sources, the utilization options are limited. One option is the half-effect cycle discussed here.

A Dühring schematic of the half-effect machine is provided as Figure 8.1. This figure shows the general relationship between the operating pressures and temperatures of the primary components. The half-effect machine is a three-pressure level machine. The high and low pressure levels function in ways familiar from single-effect practice. The intermediate pressure level is the new feature. At the intermediate pressure level, the low desorber delivers refrigerant vapor to the high absorber. The high solution circuit transports the refrigerant up to the high desorber where it is boiled out of solution a second time. The refrigerant then traverses the condenser, evaporator and low absorber as usual.

The unique feature of the half-effect machine is that the required heat input temperature is lower than that for a single-effect with the same chilled water and heat rejection temperatures. Unfortunately, there is a thermodynamic penalty that must be paid to allow the cycle to be fired at a lower temperature. The cooling COP of a half-effect machine is typically half of that for a single-effect machine. Thus, for water/lithium bromide the half-effect machine would be expected to have a COP of approximately 0.35. This level of performance has been verified in our laboratory (CAC, 1985).

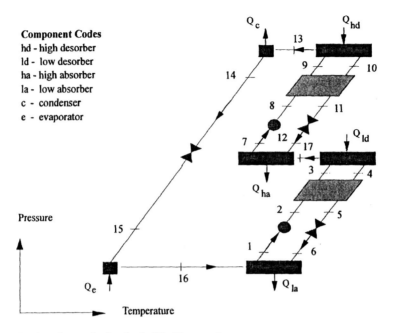

Figure 8.1 Cycle schematic for the half-effect cycle

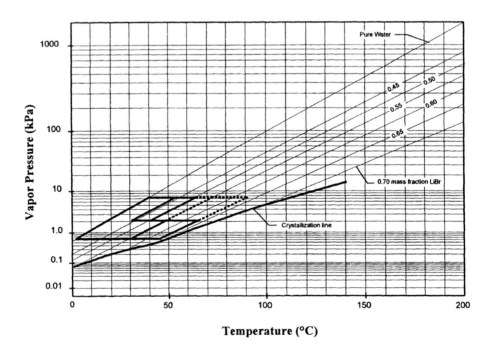

Figure 8.2 Dühring plot for the half-effect cycle

Table 8.1 Operating conditions for the half-effect cycle

i	h (i) (J/g)	m (i) (kg/sec)	P (i) (kPa)	Q (i) (Fraction)	T (i) (°C)	x (i) (% LiBr)
1	74.7	3.000	0.657	0.000	30.0	55.4
2	74.7	3.000	1.932		30.0	55.4
3	91.4	3.000	1.932		38.1	55.4
4	182.7	2.614	1.932	0.000	65.0	63.6
5	163.5	2.614	1.932		54.5	63.6
6	163.5	2.614	0.657	0.003	46.6	63.6
7	57.1	2.500	1.932	0.000	30.0	42.7
8	57.1	2.500	7.368		30.0	42.7
9	76.5	2.500	7.368		38.2	42.7
10	137.3	2.114	7.368	0.000	65.0	50.5
11	114.4	2.114	7.368		54.5	50.5
12	114.4	2.114	1.932	0.012	40.3	50.5
13	2602.9	0.386	7.368		54.8	0.0
14	167.4	0.386	7.368	0.000	40.0	0.0
15	167.4	0.386	0.657	0.065	1.0	0.0
16	2502.3	0.386	0.657	1.000	1.0	0.0
17	2591.7	0.386	1.932		48.1	0.0

COP	=	0.391	Q_c = 939.9 kW	Q_{shxl} = 50.1 kW
ϵ_{shx}	=	0.300	Q_e = 901.1 kW	T_{abs} = 30.0 °C
P_h	=	7.368 kPa	Q_{ha} = 1099.2 kW	T_{con} = 40.0 °C
P_l	=	0.657 kPa	Q_{hd} = 1090.1 kW	T_{evap} = 1.0 °C
P_m	=	1.932 kPa	Q_{la} = 1103.5 kW	T_{des} = 65.0 °C
W_{ph}	=	0.0097 kW	Q_{ld} = 1203.7 kW	
W_{pl}	=	0.0024 kW	Q_{shxh} = 48.5 kW	

A model of the half-effect cycle was written to predict the performance and the details of the internal operating conditions. The results for a typical operating condition are given in Table 8.1. A plot of these conditions on a Dühring chart is included as Figure 8.2. The cycle is arranged such that the heat rejection at all three pressure levels occurs over approximately the same temperature range. Also, the heat input in the two desorbers occurs over approximately the same temperature range.

The dotted line in Figure 8.2 represents the extension of the lower solution circuit as if the cycle were a single-effect cycle. This shows the significant difference in heat input temperature of the two cycles. By incorporating the intermediate pressure level, the half-effect cycle can realistically accept heat at 70 to 80°C while the single-effect cycle requires more like 100 to 110°C. And this is while both cycles provide refrigeration at the same temperature.

The model provided was written without heat exchanger models to maintain simplicity. Heat exchanger models could be added to this model with minimum effort by using the existing internal state point solution as a building block. For example, using the known heat transfer rate in any one of the heat exchangers, the temperature glide in a companion heat transfer fluid loop could be easily calculated from an energy balance. Then the log mean temperature would be known. Finally, the UA required to match the postulated coefficients could be calculated. This procedure would yield a consistent set of parameters. Once the appropriate UA is determined, the internal cycle temperature inputs could be replaced by the UA input and the model would be expected to converge without problems (it is a good idea to ensure that the initial guesses of all variables are updated after the internal solution is found and before the input changeover is made). This conversion of the internal model to a model with heat exchangers is straightforward if done on a step-by-step basis realizing that the solution procedure is quite sensitive to initial guesses.

By examination of the solution in Table 8.1, it is noted that the solution mass fraction values in the upper circuit are somewhat low as compared to traditional single-effect design practice. One challenge associated with this is that the property correlations used must be valid over the full range of conditions inside the cycle. In this case, the ASHRAE property data equations were extended to cover the full range of interest before the solution was attempted. The details of these correlation equations can be found in the model code.

It is noted that the half-effect cycle rejects approximately 50% more heat than the single effect. This is consistent with the fact that the COP of the half effect is 0.392 for the operating conditions in Table 8.1 as compared with a typical single-effect COP of 0.7. This increased heat rejection is the thermodynamic penalty associated with using low-grade input energy.

Various versions of the half-effect cycle have been built and tested in laboratory or one-of-a-kind installations. A water/lithium bromide machine with a nominal size of 175 kW was built at Battelle Memorial Institute in 1982. This machine was funded by U.S. Department of Energy for use in low temperature heat recovery from government facilities (CAC, 1985). The machine was thoroughly tested and found to perform with a nominal COP of 0.35 as designed. More recently, an ammonia/water half-effect cycle was built and operated in Alaska for several seasons (Erickson, 1995). These machines run well but they have not captured any market share due to the poor economics. Although they can capture and use waste heat, the first cost is sufficiently high as to be prohibitive to many users. The costs are high partly due to the fact that no design experience exists for such machines. The extrapolation of the designs from well-known absorption technology is reasonably straightforward but it still requires additional design effort.

8.2   Triple-Effect Cycle

Triple-effect technology is currently under active development by several of the leading absorption equipment manufacturers. The promise of triple effect is to raise the gas-fired cooling COP to the range of 1.4 to 1.5 with only a modest increase in first cost. Since these systems have not reached the market as yet, the true potential of triple-effect absorption technology has not been well defined. The discussion in this section focuses on a version of the technology that appears to be a simple extension of double effect. The main challenges associated with this concept, and with triple-effect concepts in general, are corrosion and materials. Triple effect inherently implies higher temperatures. The thermodynamic basis of the higher COP values comes from the increased availability of the high temperature heat input. The higher temperatures cause significant increases in the corrosion rates for traditional materials of construction. Thus, most triple-effect concepts revolve around the solution of the high temperature corrosion challenge.

8.2.1   Four-Pressure Triple-Effect Using Water/Lithium Bromide

A simple extension of the double-effect concept can be made to arrive at the four-pressure triple-effect machine shown schematically in Figure 8.3. This cycle includes two internal heat exchange processes between a condenser and a desorber as shown by the dotted boxes in the

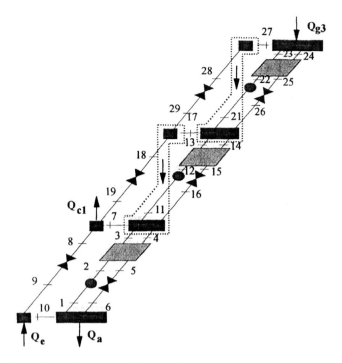

Figure 8.3 Cycle schematic for the triple-effect cycle

figure. Thus, each unit of heat is used in three different desorbers to generate vapor and hence the name triple effect. This particular cycle is a three-stage machine. A number of other three-stage, triple-effect configurations are also possible in theory. However, for water/lithium bromide, the cycle shown in Figure 8.3 shows the greatest promise.

By examination of the figure and in comparison with double-effect design, it can be seen that the triple-effect concept requires an additional desorber and solution heat exchanger at the highest temperature level. Also, the high temperature condenser must be integrated into the highest temperature desorber of the double-effect components. The relative sizes of the heat exchangers change somewhat because the heat transfer rates change. As expected with additional components, the number of design choices is increased and this brings along increased difficulty in arriving at an optimized design.

The approach in this section is to present modeling results for the cycle pictured in Figure 8.3 using water/lithium bromide under the assumption that the corrosion challenges are solvable. The cycle pictured uses a parallel-flow solution circuit. An internal model is used that does not include heat exchanger models for the four external heat transfer interactions. Instead, a set of nominal temperatures were input in those four components to set the operating conditions. This approach has the benefit of eliminating some complexity from the model which was not needed for this discussion. A UA formulation for the heat exchange processes could be simply added to the model based on the known internal solution.

Another complicating detail in the modeling of triple-effect cycles is that the property correlations readily available do not encompass the necessary temperature range. The property routines used for water/lithium bromide here, and throughout this book, are based on the work of McNeely (1979). These correlations are valid only up to 175°C but the temperatures in the high temperature components of the triple effect go above this value. An examination of the correlations indicates that in the temperature range of interest, the McNeely correlations

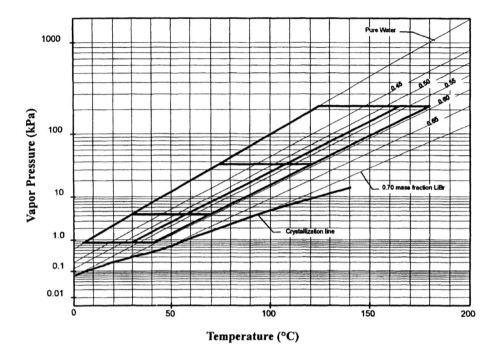

.Figure 8.4 Dühring plot for the triple-effect cycle

extrapolate smoothly and appear to be well-behaved even at temperatures above the range of accuracy. Thus, the approach here is to simply utilize the extrapolated form of the property routines. This approach is likely to introduce some error and the absolute values of the predicted performance parameters need to be treated as less accurate than model predictions based on accurate property data. However, it appears that the model does predict the trends in performance and thus the model has some utility in understanding the potential of triple-effect technology.

The inputs for the triple-effect model are summarized in Table 8.2. Only a single pump flow rate is input because the other two solution pump flow rates are calculated from energy balances on the condenser-desorber internal heat exchange processes. For this modeling approach, internal temperatures at four locations are imposed to eliminate the needed heat exchanger models for the external heat exchange processes. Effectiveness models were used for the solution heat exchangers and all three utilized the same value of effectiveness. For the condenser-desorber internal heat exchange processes, a pinch point model was imposed with a pinch temperature difference as indicated in Table 8.2. This set of inputs results in the cycle solution in Table 8.3 and the Dühring plot in Figure 8.4. Note that point 20 is omitted to obtain a more orderly numbering scheme.

Several observations can be made regarding the cycle solution. The pinch temperature difference used in the internal heat exchange processes is quite tight at 5°C. To achieve this small temperature difference, large heat exchange areas would be needed. It is noted that the highest temperature in the cycle is sensitive to the pinch temperature difference used. An increase of 1°C in pinch temperature difference increases the heat input temperature by 2°C (with everything else held constant). The high end pressure is also of note in triple-effect

Table **8.2** Inputs for the triple-effect cycle model

Parameter	Value	Parameter	Value
m_1 (kg/sec)	1.0	T_1 (°C)	30
T_{24} (°C)	180	T_{10} (°C)	5
T_8 (°C)	30	ϵ_{SHX}	0.5
		ΔT (°C)	5

cycles. From Figure 8.4 it is seen that the highest pressure in the cycle is above atmospheric pressure (238 kPa). The relatively high pressure results in somewhat different design conditions for the high pressure components as compared to the next lower pressure level. In particular, the lower specific volume of the vapor implies lower boiling heat transfer coefficients but it also implies some size reductions in the condenser geometry.

The vapor fractions listed include 10 points for which the vapor fraction was specified as an assumption. In general, these represent the assumption the fluid leaving the component is saturated. These entries can be identified in Table 8.3 as the entries in the vapor fraction column that do not have a decimal point.

For the same heat input, the capacity of the triple effect is greater than double effect. This is because more vapor is generated for each unit of heat input. That vapor then finds its way to the evaporator to provide refrigeration capacity. The capacities of the components in this cycle calculation are arbitrary. The solution mass flow rate that was input essentially defines the capacities and the input value was arbitrarily set to 1 kg/sec. The cooling COP calculated for this cycle is 1.645. This is a cycle COP that does not account for additional losses that would be expected in a gas burner. Assuming a burner efficiency of 90%, the gas-fired cooling COP for this cycle would be 1.48. This is believed to represent the practical upper limit of performance potential of triple-effect configurations. Design optimization may be able to improve that value slightly but it is unlikely that triple effect will do significantly better in practice. Practical design tradeoffs may result in reductions in that value. Based on the percentage improvement obtained in going from single effect to double effect, the cooling COP of triple effect will be lower. Using the superposition principles of Alefeld and Radermacher (1994) the double-effect machine should achieve a COP of 1.24 and the triple effect should achieve 1.61. In practice, double-effect machines are built with additional design compromises that do not allow them to achieve their full performance potential. If triple-effect machines achieve the same performance level as current double-effect machines, a realistic performance projection for gas-fired triple effect is a cooling COP of 1.3 (based on assuming double-effect technology has COP of 1.0). The actual gas-fired cooling COP for triple effect will be somewhere between 1.3 and 1.5.

The design tradeoffs that result in reduced performance are typically economic ones. First cost of such machines is an important issue in the market and first cost can be reduced by reducing the size of heat exchangers and giving up a few points on COP. Compactness is also an economic issue. If the machine is made small, it will sell better but the high velocities which result tend to cause entrainment and carryover between components. Liquid carryover can be a major source of COP reduction in a real machine. Mist eliminators can minimize this problem but the ultimate solution is to increase the flow areas so that the vapor velocities are low enough to avoid the problem altogether. Such issues need to be addressed in triple-effect design if the highest performance is to be obtained.

Table 8.3 Operating conditions for the triple-effect cycle

i	h (i) (J/g)	m (i) (kg/sec)	P (i) (kPa)	Q (i) (Fraction)	T (i) (°C)	x (i) (% LiBr)
1	66.4	1.000	0.87	0	30.0	52.94
2	66.4	1.000	4.24		30.0	52.94
3	101.4	1.000	4.24		46.5	52.94
4	166.8	0.902	4.24	0	69.5	58.66
5	128.0	0.902	4.24		49.7	58.66
6	128.0	0.902	0.87	0.0060	41.4	58.66
7	2609.9	0.027	4.24		58.1	
8	125.6	0.098	4.24	0	30.0	
9	125.6	0.098	0.87	0.0420	5.0	
10	2510.7	0.098	0.87	1	5.0	
11	126.0	0.724	4.24	0	58.1	52.94
12	126.0	0.724	37.66		58.1	52.94
13	181.8	0.724	37.66		84.3	52.94
14	268.0	0.653	37.66	0	120.8	58.66
15	206.2	0.653	37.66		89.5	58.66
16	206.2	0.653	4.24	0.0133	71.2	58.66
17	2700.6	0.028	37.66		108.2	
18	311.5	0.071	37.66	0	74.5	
19	311.5	0.071	4.24	0.0764	30.0	
21	232.8	0.437	37.66	0	108.2	52.94
22	232.9	0.437	237.84		108.3	52.94
23	296.8	0.437	237.84		138.1	52.94
24	384.9	0.394	237.84	0	180.0	58.66
25	314.0	0.394	237.84		144.1	58.66
26	314.0	0.394	37.66	0.0493	76.1	58.66
27	2799.4	0.043	237.84		166.0	
28	528.2	0.043	237.84	0	125.8	
29	528.2	0.043	37.66	0.0933	74.5	

COP	=	1.645	W_2	=	0.0157 kW	Q_c	= 232.7 kW
ΔT_1	=	5.000 °C	W_3	=	0.0577 kW	Q_{shx1}	= 35.0 kW
ΔT_2	=	5.000 °C	Q_{abs}	=	294.059 kW	Q_{shx2}	= 40.4 kW
ϵ_{shx}	=	0.500	Q_{c1}	=	80.1 kW	Q_{shx3}	= 27.9 kW
P_h	=	37.657 kPa	Q_{c2}	=	76.0 kW	T_c	= 30.000 °C
P_{h2}	=	237.843 kPa	Q_{c3}	=	96.8 kW	T_{evap}	= 5.000 °C
P_l	=	0.872 kPa	Q_{d1}	=	76.0 kW	T_{des}	= 180.000 °C
P_m	=	4.238 kPa	Q_{d2}	=	96.8 kW		
W_1	=	0.0022 kW	Q_{d3}	=	141.4 kW		

8.3   Resorption Cycle

Another cycle variation, which has the potential of expanding the design options of water/lithium bromide technology, is the resorption cycle. A simple single-effect resorption cycle is shown schematically in Figure 8.5. The cycle employs two solution circuits instead of only one. The condenser/expansion valve/evaporator section of a conventional single-effect cycle are replaced, in the resorption cycle, by a solution circuit consisting of an absorber, solution heat exchanger, pump and desorber. The absorber takes the role of the condenser and rejects heat whereas the desorber takes the role of the evaporator. It should be noted that the new solution circuit flows counterclockwise similar to the solution circuit in a Type II cycle. The overall result of this arrangement is a cycle which is no longer tied to the thermodynamic properties of the pure refrigerant. By allowing the average mass fraction of the new solution circuit to be a design variable, the resorption cycle provides an additional degree of freedom in cycle design. This additional design flexibility is needed in certain advanced cycles where the resorption cycle can be used as a building block. The cycle solution provided here is an example of how to analyze such a configuration.

One aspect of the resorption cycle that complicates the modeling slightly is that the mass fraction of the aqueous lithium bromide in the new solution circuit tends to be much lower than that encountered in a conventional cycle. The complication arises due to the fact that the property correlations used are valid only in a limited mass fraction range between approximately $0.4 < x < 0.7$. For mass fractions below 0.4, the polynomial curve fits are very poorly behaved (as can be demonstrated by plotting the extrapolation). Thus, a property correlation is needed that covers the entire range of mass fraction. The work of McNeely (1979) includes property data over the whole range but correlations over only the limited range usually found in practice. These data were fitted to a simple function and the result is tabulated in Appendix A. The compound correlation consisting of the McNeely correlation plus our own correlation of the McNeely data in the low mass fraction range was then programmed to enable

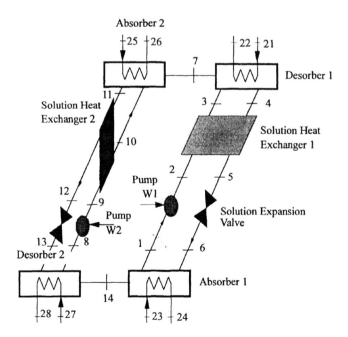

Figure 8.5 Cycle schematic for the resorption cycle

simulation of the resorption cycle.

The model which was created is a UA type model which utilizes effectiveness models for the solution heat exchanger. The inputs to the model are summarized in Table 8.4. The mass flow rates in both solution pumps are input as well as the mass flow rates in all four external heat transfer loops. The UA values in each of the four heat exchangers is input as well as the solution heat exchanger effectiveness (the same value was used in both solution heat exchangers). The inlet temperature to each of the heat transfer loops is also input. The remaining input is the mass fraction difference between the two legs of the solution circuit that supplies the high temperature desorber. The need for such an input to close the system of equations requires a separate discussion.

By imposing the mass fraction difference across one of the solution circuits (it does not matter particularly which one), the cycle mass balances see one more constraint. The need for this constraint to close the system stems from a fundamental difference between the conventional cycle and the resorption cycle. In the conventional cycle, the additional constraint is imposed by the fixed composition of the refrigerant stream. When that constraint is not present, as in the resorption cycle, it must be replaced by some other constraint. From a modeling perspective, it would be equally valid to choose the value of the mass fraction in one of the legs of either solution circuit. In practice, the resorption cycle is less stable than the conventional cycle and the instability can be traced to the same origin as the need for the additional input. Because none of the mass fractions in the resorption cycle are fixed by the working fluid properties, they can all wander. Small differences in the flow rate of the pure refrigerant between the solution circuits tend to cause the average mass fraction of the two circuits to approach each other. Active controls are required to make the resorption cycle operate stably. However, these controls can be quite simple and might involve only a single float-controlled valve. The magnitude of the control problem is indicated by the need for one input in the model. As soon as the control system fixes one additional variable, the system will operate in a controlled manner.

A cycle solution for the inputs in Table 8.4 is given as Table 8.5. A Dühring plot of the cycle solution is given as Figure 8.6. Several features are worthy of note. The heat transfer loop connected to the low temperature desorber operates over the temperature range 18.2 to 20°C. Thus, although the cycle provides refrigeration, it does not appear to be very practical for conventional applications. The low temperature could be lowered by allowing the low pressure in the cycle to drop but then the absorber crystallization problem would be aggravated. The cycle solution used in this illustration demonstrates the concept. In practice, the resorption cycle would probably be utilized as a component of a more complex, multi-stage cycle.

The cooling COP of the resorption cycle is 0.573. This value is lower than that found for the conventional single-effect cycle due primarily to increased irreversibilities in the new solution circuit. In particular, streams 7 and 10 are far from equilibrium when they are brought together in the absorber. Another factor is the increased flow losses in the expansion valve. In the new solution circuit, the amount of vapor that flashes is considerably greater and this is indicative of higher irreversibility. This results from the fundamental difference in the thermal configuration of a solution heat exchanger in a counterclockwise flowing solution circuit. Even though the effectiveness values of the two solution heat exchangers are identical, the amount of subcooling in stream 12 is much less than the amount of subcooling in stream 5. This is typical of the difference between such oppositely flowing solution circuits.

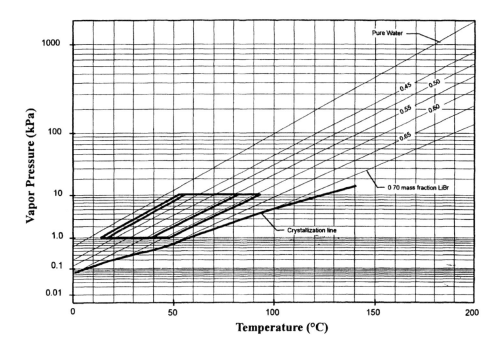

Figure 8.6 Dühring plot for the resorption cycle

Table 8.4 Inputs for the resorption cycle model

Parameter	Value	Parameter	Value
m_1 (kg/sec)	0.05	T_{21} (°C)	120
m_8 (kg/sec)	0.05	T_{23} (°C)	25
UA_{d1} (kW/K)	0.47	T_{25} (°C)	25
UA_{d2} (kW/K)	1.626	T_{27} (°C)	20
UA_{a1} (kW/K)	0.7	m_{21} (kg/sec)	1.0
UA_{a2} (kW/K)	0.28	m_{23} (kg/sec)	1.0
ϵ_{SHX}	0.64	m_{25} (kg/sec)	1.0
$\Delta x_{4\text{-}3}$ (mass fraction)	0.05	m_{27} (kg/sec)	1.0

Table 8.5 Operating conditions for the resorption cycle

i	h (i) (J/g)	m (i) (kg/sec)	P (i) (kPa)	Q (i) (Fraction)	T (i) (°C)	x (i) (% LiBr)
1	98.8	0.0500	1.045	0	39.7	56.6
2	98.9	0.0500	10.490		39.7	56.6
3	160.9	0.0500	10.490		70.3	56.6
4	229.3	0.0459	10.490	0	95.6	61.6
5	161.9	0.0459	10.490		59.8	61.6
6	161.9	0.0459	1.045	0.006	50.4	61.6
7	2658.4	0.0041	10.490		84.3	0.0
8	22.5	0.0500	1.045	0	14.9	35.4
9	22.5	0.0500	10.490		14.9	35.4
10	88.9	0.0500	10.490		40.2	35.4
11	131.4	0.0541	10.490	0	54.4	32.8
12	70.0	0.0541	10.490		31.8	32.8
13	70.0	0.0541	1.045	0.020	13.9	32.8
14	2529.9	0.0041	1.045		14.9	0.0
21	503.5	1.0000			120.0	
22	490.2				116.9	
23	104.8	1.0000			25.0	
24	117.5				28.1	
25	104.8	1.0000			25.0	
26	112.9				26.9	
27	83.9	1.0000			20.0	
28	76.3				18.2	

COP	= 0.573	m_{in}	= 0.00083 kg/sec	Q_{d2}	=	7.610 kW
ϵ_{shx1}	= 0.640	m_{out}	= 0.0508 kg/sec	Q_{shx1}	=	3.100 kW
ϵ_{shx2}	= 0.640	P_{high}	= 10.490 kPa	Q_{shx2}	=	3.319 kW
$\Delta T_{lm,a1}$	= 18.234 K	P_{low}	= 1.045 kPa	T_{out}	=	55.7 °C
$\Delta T_{lm,a2}$	= 29.042 K	W_1	= 0.000293 kW	UA_{a1}	=	0.700 kW/K
$\Delta T_{lm,d1}$	= 28.266 K	W_2	= 0.000361 kW	UA_{a2}	=	0.280 kW/K
$\Delta T_{lm,d1}$	= 4.680 K	Q_{a1}	= 12.764 kW	UA_{d1}	=	0.470 kW/K
$\Delta T_{lm,shx1}$	= 22.645 K	Q_{a2}	= 8.132 kW	UA_{d2}	=	1.626 kW/K
$\Delta T_{lm,shx2}$	= 15.497 K	Q_{d1}	= 13.285 kW	x_{out}	=	34.859 % LiBr

8.4   Additional Water/Lithium Bromide Technologies

A number of novel configurations, employing water/lithium bromide, have appeared in the literature and several have been reduced to practice in hardware. These variations are quite interesting but do not represent significant market opportunities, at the present time. They are of interest more from the standpoint of understanding that they are possible. In many cases, something can be learned about chiller physics by examining these more exotic variations on the theme.

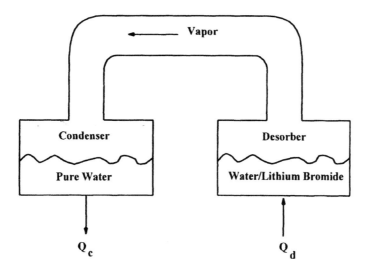

Figure 8.7 Energy storage system

8.4.1   Energy Storage

Energy storage is possible in a water/lithium bromide system in the form of a chemical potential difference. Consider a configuration such as that in Figure 8.7 where a desorber and condenser are connected by a vapor duct. This system could be started with no water in the condenser and all the working fluid mixed in the desorber. As energy is added to the system as heat transfer in the desorber, the pressure in the vapor space will rise until it is sufficiently high to begin condensing in the condenser. As water vapor leaves the desorber and the lithium bromide mass fraction of the remaining liquid increases, the temperature of the working fluid will rise as required by the properties of water/lithium bromide. Thus, the process pictured will eventually stop when the desorber working fluid temperature rises to the point where it equals the heat source temperature. At this point, the vapor duct valve can be shut to isolate the two components. Once the valve is shut, the system temperatures can be allowed to equilibrate with the environment. As the desorber temperature decreases, the pressure in the desorber section will decrease as well. The pressure difference between the condenser and desorber sections represents a potential to do work. This potential can be harnessed in a number of ways including: 1) refrigeration, 2) heating or 3) power generation. These options are discussed next.

The stored energy in the system, which is stored in the form of segregated masses with a chemical potential difference between them, can be used to produce refrigeration or heating by reversing the processes that were used to store the energy. To produce refrigeration, the valve between the two spaces must be opened and the section that was originally the desorber must be cooled so that it remains close to ambient temperature. By cooling the component, it will become an absorber for the higher pressure vapor from the pure water component. In addition, to enhance the rate of the process, it would be necessary to find a way to increase the vapor liquid surface area such as by a recirculating spray system in the absorber. As vapor is absorbed into the aqueous lithium bromide solution, the pure water in the other component will evaporate and the temperature of the component will drop as internal energy is utilized to achieve the phase change. The low temperature produced can then be used as a source of refrigeration.

The heating variation is quite similar to the refrigeration variation except that for heating the evaporator is kept close to ambient temperature and the absorber is allowed to rise in temperature. In both the refrigeration and heating modes the temperatures which can be achieved will change with time as the mass fraction of the solution in the absorber changes. Eventually, all the water in the evaporator will be used up and the process will stop.

An interesting variation on the theme is to utilize the pressure difference between the evaporator and the absorber to drive a machine to produce a work output. Such machines have been built for test purposes as discussed by Isshiki (1991). By utilizing the energy rejected in the absorber to heat the evaporator, the system can be self-contained thermally. The main difficulty is in competing with other energy storage technologies which have better economics. The concept is discussed here because it provides a fascinating variation on absorption technology.

8.4.2   Cogeneration

The term cogeneration is interpreted loosely here as covering all energy conversion technologies where two separate systems are integrated together by a cascade of thermal energy. Integration of absorption machines with other energy conversion equipment can lead to increased system performance. The only requirement for the integration of two technologies in a cascade of energy is that the temperature of the available heat from one system must match the temperature requirements of the heat from the mating system. Examples include 1) gas turbine inlet air pre-cooling using an absorption chiller powered by rejected heat, 2) absorption chiller fired by the engine cooling system, 3) regeneration of a desiccant using reject heat from an absorption chiller, and 4) absorption chiller fired by reject heat from a fuel cell. This list is by no means exhaustive but instead is meant to indicate the range of options that are possible for system integration.

In examples 1, 2, and 4 the absorption machine is a bottoming cycle whereas in example 3 the absorption machine is a topping cycle. This terminology simply refers to the temperature scale. In the cascade of energy from a high temperature to a low temperature, the various technologies are arranged according to their temperature level. The topping cycle is the one that receives the energy at the highest temperature while the bottoming cycle accepts heat rejected by the topping cycle.

Many opportunities exist for improving system efficiency using a combination of technologies including absorption. Few of these ideas have been given a full examination due to the economics of energy and due to the additional complexity inherent in such an integration of systems. As the energy economy changes, it seems likely that more attention will be paid to cogeneration concepts.

8.4.3   Solar Cooling

Water/lithium bromide and ammonia/water chillers have been used for solar cooling where the solar thermal arrays collect energy in a heat transfer fluid and it is pumped to the chiller. For non-concentrating collectors the efficiency is a strong function of the collection temperature. Thus, the best combination, generally, utilizes a single-effect chiller. Numerous solar-powered absorption systems have been installed in laboratories and in commercial settings where the economics of energy is skewed such that electricity is very expensive. However, these systems are not generally economical in the context of building design in the U.S. in 1995 and therefore the total number of such installations is relatively small.

The combination of solar thermal energy collection and absorption cooling is a natural match because the temperature levels of both technologies are complementary. It is expected that as solar energy applications expand, more opportunities will exist for solar/absorption combinations.

HOMEWORK PROBLEMS

8.1   Check the mass and energy balances on each of the components in the half-effect cycle described in Table 8.1. Note that for mass fractions below 0.45, low-mass fraction correlations must be used as in the model.

8.2   Run the half-effect model for a series of cases where the solution heat exchanger effectiveness is varied from 0 to 1.0 and plot the COP and capacity.

8.3   Convert the half-effect model such that the solution heat exchangers are modeled using a UA approach. Using the solution heat exchanger UA values calculated from the effectiveness solution, run the model to see if the results match the results in Table 8.1.

8.4   Convert the half-effect model into a UA model by adding heat transfer loops at each of the heat exchangers transferring heat external to the cycle. Run the model to examine the effects on performance and capacity of
    a)   Vary solution circuit flow rate in both solution circuits
    b)   Vary solution heat exchanger effectiveness in each circuit
    c)   Vary the UA in each of the heat exchangers where UA models are used.
    d)   Vary the heat transfer fluid inlet temperature in each of the six external heat exchangers
    e)   Vary the heat transfer fluid flow rate in each of the six external heat exchangers

8.5   Check the mass and energy balances for each of the components in the triple-effect cycle described in Table 8.3.

8.6   Vary the inputs to the triple-effect model to examine the effect on COP and capacity. The inputs are listed in Table 8.2.

8.7   Check the mass and energy balances for each of the components in the resorption cycle described in Table 8.5. Note that the low mass fraction states, i.e. those below 0.45, require extended property routines as programmed in the resorption cycle model.

8.8   Vary the inputs to the resorption cycle model to examine the effect on COP and capacity. The inputs are listed in Table 8.4.

8.9   Compute the energy storage density of a water/lithium bromide chiller designed to allow liquid refrigerant to accumulate in the evaporator during times of low cooling demand. Assuming that the solution is concentrated from 0.55 to 0.65 during the energy storage mode, how much solution volume would be required to produce 1 Ton of refrigeration (1 Ton equals 3.52 kW) for 1 hour? For this estimate, assume the cycle operating conditions are steady state and at typical values.

8.10   Using the extended properties for water/lithium bromide, as used in the resorption and half-effect cycle models, plot enthalpy mass fraction and pressure-temperature-mass fraction diagrams over the full range of mass fraction from 0 to crystallization.

Chapter 9

SINGLE-STAGE AMMONIA/WATER SYSTEMS

9.1 General Considerations

Single-stage ammonia water systems have been in use since the mid-1800s. They come in many different variations and implementations. Originally, the applications were refrigeration. Since the 1960s a considerable number of these units were built for residential air-conditioning and since the 1970s they are under consideration for residential and commercial heating and cooling, as well.

The basic single-stage absorption cycle for ammonia/water mixtures is on the first glance very similar to that for water/lithium bromide. However, there are several important details that are significantly different. These are all a consequence of the fluid properties.

9.1.1 Properties of Ammonia and Safety Concerns

Ammonia is a naturally occurring substance that is produced and used in large quantities (in the U.S. alone 20 million tons per year, IPCS, Ammonia Health and Safety Guide, Publ. World Health Org. Programme on Chemical Safety, Geneva, 1990) for agriculture as fertilizer and as the source material for fibers, plastics and explosives. Consequently it is shipped in large quantities by rail and ship. Ammonia is also used as a cleaning and descaling agent and in food additives.

Ammonia is a colorless gas of low density at room temperature with a pungent smell. It has a relative molecular mass of 17.03 and is lighter than air at atmospheric conditions. It can be stored and transported as a liquid under a pressure of 1 MPa at 25°C. The critical point of ammonia is at 132.3°C and 11.3 MPa. The critical density is 235 kg/m^3.

Since ammonia is highly soluble in water generating NH_4^+ and OH^- ions, it reacts very quickly with mucus membranes. However, it is not absorbed through the skin. It can be smelled by humans in concentrations of very few ppm. At about 50 ppm, the odor is almost unbearable. This is also the concentration range (25 ppm) to which long-term exposure is limited from an occupational health point of view, (IPCS, 1990, Ammonia Health and Safety Guide, Publ. World Health Org. Programme on Chemical Safety, Geneva). At high dosages ammonia exposure can be lethal. Ammonia is flammable and explosive in the range of 16 to 25 vol.% (IPCS, 1990, Ammonia Health and Safety Guide, Publ. World Health Org. Programme on Chemical Safety, Geneva) in air.

The strong odor of ammonia can be seen as an asset. It is self-alarming. Even very small leaks in systems are easily noticed and therefore a significant incentive exists for early repairs and consistent maintenance. One method of leak detection is to use wet indicator paper, which will quickly change its color once it is exposed to air with a few ppm of ammonia content. However, traditional leak detection devices such as soap (or bubble) solutions do not work since the ammonia is dissolved in the water without creating bubbles.

9.1.2 Water Content of the Refrigerant Vapor

The first important difference is the vapor pressure of ammonia/water mixtures compared to that of water/lithium bromide. Here, ammonia is the refrigerant. The normal boiling

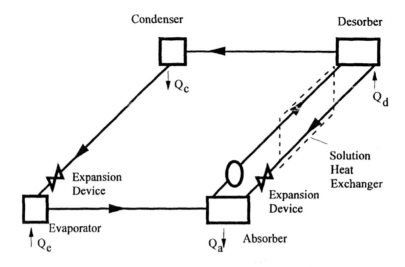

Figure 9.1 Single-stage ammonia/water cycle

point is -33.35°C and therefore the pressure in the desorber component at the temperatures normally encountered in air-conditioning and refrigeration applications is relatively high. The pressure is comparable to vapor compression systems that utilize R22 as the refrigerant. The high vapor pressure leads to rather small pipe diameters, and relatively compact heat exchangers as compared to water/LiBr.

A second important difference is that the absorbent (water) has a vapor pressure that is not negligible relative to that of ammonia. As a consequence, the vapor generated in the desorber contains a certain amount of water. The mass fraction depends on the mass fraction of the liquid mixture in the desorber, the temperature and the desorber design. Any water contained in the desorber vapor is detrimental to the performance of the system. As can be seen in Figure 9.1, which depicts a basic single-stage cycle, the water will pass with the vapor into the condenser and then into the evaporator where the water tends to accumulate if a pool boiler design is used. The vapor leaving the evaporator is rich in ammonia. Although it still contains some water, the ammonia mass fraction of this vapor is considerably higher than that of the vapor leaving the desorber. Thus, water will remain in the evaporator. This is demonstrated in Example 9.1. If no other measures are taken, the evaporator temperature has to be increased considerably to evaporate the remaining water-ammonia solution.

At constant evaporator temperature, an accumulation of water will lead to a decrease in evaporator pressure which in turn will affect the absorber conditions. Figure 9.2 shows a single-stage cycle superimposed on the pressure-temperature diagram. The lines of constant mass fraction (isosteres) are also shown for pure ammonia and the mixtures within the solution circuit. As the water accumulates in the evaporator, the pressure drops for a constant evaporator temperature which is fixed by the application. The new location for the evaporator on this diagram is indicated by the dashed lines. The absorber has either to be cooled to a lower temperature or the mass fraction of the solution has to change to a lower ammonia content. The latter is shown in Figure 9.2. Assuming that the desorber temperatures do not change, the condenser temperature must drop which is usually prohibited by the cooling water temperature

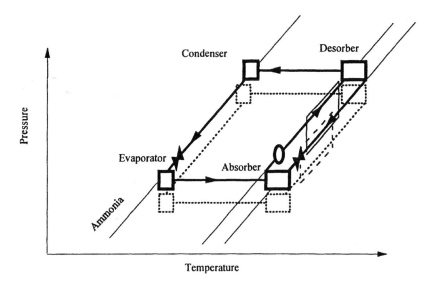

Figure 9.2 Single-stage ammonia/water cycle

available to the condenser. As a consequence, the high pressure level has to be raised and the desorber temperature will increase as well. As the water continues to accumulate in the evaporator, the operating conditions of the entire system drift and operation at design conditions is not possible.

One way of preventing excessive water accumulation is to drain it periodically into the absorber. However, this method represents a loss of efficiency in two ways. First, the water was evaporated in the desorber, requiring desorber heat input, but it does not evaporate in the evaporator; thus it does not provide to the cooling capacity. Second, it contains a considerable amount of ammonia that is retained as liquid, amplifying this effect.

Example 9.1 Evaporator Temperature Glide

In the desorber of an NH_3/H_2O absorption system the rich solution enters at a temperature of 150°C and a mass fraction of 0.3. The vapor produced in this desorber is in equilibrium with the incoming liquid. File: ex9_1.ees.

1) Determine the mass fraction of the vapor in equilibrium with this liquid.

2) Assume that saturated liquid of the same mass fraction as the vapor leaving the desorber enters the evaporator ($T_{evap} = 0°C$). What is the mass fraction of the vapor in equilibrium with this liquid?

3) Assume further that the entire liquid is evaporated at constant pressure. What is the size of the temperature glide?

Solution: Using the equation of state software described in Appendix E or the enthalpy-mass fraction diagram in Figure 3.11, the following property data are obtained:

Table 9.1 Data for Example 9.1

	T (°C)	P (bar)	x, Liquid	x, Vapor
Desorber	150	25.8	0.3	0.853
Evaporator inlet	0	3.7	0.853	1.0
Evaporator outlet	88	3.7	0.20	0.853

Discussion: The vapor leaving the desorber has an ammonia mass fraction of 0.853. This fluid passes through the condenser and evaporates in the evaporator. The mass fraction of the first vapor bubble generated in the evaporator is 1.0, essentially pure ammonia. However, as the evaporation process proceeds, the remaining liquid is gaining in water content. By the time the evaporation process is completed as a steady-state, steady flow process at constant pressure, the vapor leaving the evaporator has a mass fraction of 0.853, the same as the entering liquid, and the mass fraction of the last droplet evaporating is 0.20. The resulting temperature glide amounts to 88 K. This temperature glide is not suitable for conventional applications (see also Figure 9.4, plotted for a different set of operating conditions).

It should be noted that the temperature glide is highly non-linear. This is discussed further in Problem 9.1.

One way of reducing the water content of the vapor is to use rectification or dephlegmation. Alternatively, the condensate precooler presents an opportunity to utilize the large glide that exists in the evaporator. These measures are addressed in Section 9.2.

9.1.3   Material Considerations

Ammonia is a very good solvent for copper. Thus the use of any copper or copper-containing material is impossible. Experience in the laboratory indicates that even chromium-plated brass parts are susceptible to ammonia corrosion. The most common material for the construction of ammonia/water systems is steel or stainless steel. When steel is the material of choice, corrosion inhibitors are required for most applications. These are salts that are added in small quantities (1% by weight or so). They form a protective oxide layer on the metal surface so that there is no direct contact with the working fluid. The influence of these salts on the thermodynamic properties of the working fluid is usually neglected. Traditionally, the corrosion inhibitors are salts that contain heavy metals. These are being banned by the U.S. Environmental Protection Agency and manufacturers find themselves needing to develop new, environmentally acceptable replacements.

The thermal conductivity of steel is about one tenth of that of cooper. Thus the heat transfer resistance of the wall material may no longer be negligible.

Since ammonia has a relatively high vapor pressure, the machines are not as susceptible to performance degradation due to the presence of noncondensable gases as water/lithium bromide systems. However, it is nevertheless good practice to keep all noncondensables out of the system.

9.2   Performance Calculations

The following performance calculations are conducted for single-stage ammonia/water systems. More advanced cycles are discussed in subsequent chapters. These sections are easier

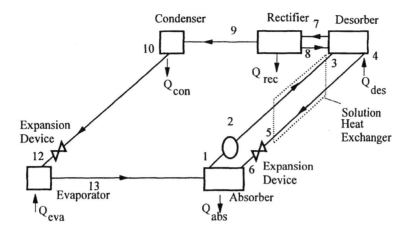

Figure 9.3 Single-stage ammonia/water cycle

to understand once the chapters on water/ lithium bromide systems are mastered.

9.2.1.   Single-Stage Ammonia/Water Systems

As a first example the cycle shown in Figure 9.3 is considered. It is essentially the same as the single-stage water/lithium bromide cycle, however, with one important difference. The rectifier is added. The performance calculation is shown in full detail in the following example.

Example 9.2 Single-Stage Ammonia/Water System

A single-stage ammonia/water absorption system is given as shown in Figure 9.3. The evaporator saturation temperature at the outlet is assumed to be -10°C with two-phase fluid leaving at a vapor quality of 0.975. The mass flow rate of solution through the solution pump is 1 kg/s. The temperature of the saturated liquid leaving the absorber and the condenser is 40°C. The difference in mass fraction of the two solution streams is given to be 0.10. The rectifier produces a vapor with a mass fraction of 0.999634 ammonia. It is assumed that the pump efficiency is 100% and that the effectiveness of the solution heat exchanger is 80%. File: ex9_2.ees.

Find:
The COP, all amounts of heat exchanged and the pump work.

Solution:
Using the procedures outlined in Chapter 4 or suitable software, the state points can be found for all conditions as shown in Table 9.2. Based on these values, and the equations specified in Chapter 4, the following detailed results are obtained. The solution circulation ratio is

Table 9.2 State points for the ammonia/water system according to Figure 9.3

State Points	h (J/g)	m (kg/sec)	P (kPa)	Vapor Quality	T (°C)	x (kg/kg)
1	-50.0	1.00	288.0	0.0	40.0	0.3971
2	-48.5	1.00	1555	n/a	40.1	0.3971
3	207.5	1.00	1555	n/a	97.1	0.3971
4	358.5	0.858	1555	0.0	124.0	0.2971
5	60.1	0.858	1555	n/a	57.4	0.2971
6	60.1	0.858	288.0	n/a	57.6	0.2971
7	1516	0.153	1555	1.00	101.7	0.9586
8	229.7	0.010	1555	0.0	101.7	0.3971
9	1294	0.142	1555	1.00	44.0	0.999634
10	190.1	0.142	1555	0.0	40.0	0.999634
11	used	later				
12	190.1	0.142	288.0	0.183	-10.2	0.999634
13	1224	0.142	288.0	0.975	-10.0	0.999634
14	used	later				

$$f = \frac{(x_9 - x_4)}{(x_3 - x_4)} = 7.025 \qquad 9.1$$

the pump power, W_p amounts to

$$W_p = (P_{high} - P_{low})\frac{v\,m_1}{\eta_p} = 1.5\ kW \qquad 9.2$$

with $v = 0.0012$ m³/kg, the specific volume of the rich solution, and a pump efficiency of 1.0. The enthalpy at the pump exit (point 2) is found by adding the pump work per unit of mass flowing through the pump to the enthalpy at point 1. Next, the performance of the solution heat exchanger is evaluated. Here an effectiveness model is used and the following equation holds.

$$\epsilon_{shx} = 0.8 = \frac{(T_4 - T_5)}{(T_4 - T_2)} \qquad 9.3$$

It is further required that the energy balance is fulfilled, i.e., that the amount of heat released by the hot stream is the same as the amount of heat received by the cold stream. From this condition, a second equation results. With two temperatures and enthalpies known, the liquid leaving the pump and the liquid leaving the desorber, which is assumed to be saturated, there are two equations for two unknowns. These are the temperatures of the streams leaving the solution heat exchanger.   Solving for those conditions it is found that the amount of heat exchanged in the solution heat exchanger amounts to

$$Q_{shx} = m_1(h_3 - h_2) = 256.0 \ kW \qquad\qquad 9.4$$

All the properties of all the state points are now known; all other quantities of heat exchanged can be calculated. The absorber heat amounts to (with $f = m_1/m_{13}$)

$$Q_{abs} = m_{13} \ [h_{13} - h_6 + f(h_6 - h_1)] = 275.8 \ kW \qquad\qquad 9.5$$

The heat released by the rectifier is found to be (with $r = m_9/m_7$)

$$Q_{rec} = m_7 \ [h_7 - h_8 + r(h_8 - h_9)] = 45.0 \ kW \qquad\qquad 9.6$$

It is assumed here that the heat rejected by the rectifier can be approximated by utilizing the inlet and desired outlet conditions without undertaking the detailed design calculations involving individual plates since this is not the focus for this example. However, the reader should bear in mind that for actual system design a more detailed calculation is necessary especially when the water content of the vapor is high. The reason is that the desired outlet conditions are not necessarily attainable. The desorber heat requirement amounts to

$$Q_{des} = m_7 \ [h_7 - h_4 + f(h_4 - h_3)] + Q_{rec} = 329.2 \ kW \qquad\qquad 9.7$$

The condenser heat amounts to

$$Q_{con} = m_9 \ (h_9 - h_{10}) = 157.2 \ kW \qquad\qquad 9.8$$

and the evaporator heat to

$$Q_{eva} = m_9 \ (h_{13} - h_{12}) = 147.2 \ kW \qquad\qquad 9.9$$

The overall energy balance for the entire system yields 0.0 as it should. Thus the COP can be evaluated as

$$COP = \frac{Q_{eva}}{Q_{des}} = 0.445 \qquad\qquad 9.10$$

Discussion:
1) The lines not used in Table 9.2 will be used later for additional example calculations.
2) The vapor mass fraction (point 9) specified in the problem statement is intended to represent "pure ammonia". However, in order to have the equation of state subroutines that are used here work properly, a value of slightly less than 1.0 has to be specified, in our case 0.999634.
3) The evaporator outlet is not single phase; there is still a small amount of liquid accompanying the vapor. This is a consequence of the fact that the vapor quality is 0.975. There exists a small temperature glide. From examination of the temperatures for points 12 and 13, it is seen that the glide is 0.16 K.
4) The temperature might be expected to drop slightly as the solution is expanded in the

expansion valve from point 5 to 6. However, in the results above, this temperature is seen to increase. This is a consequence of the Joule Thompson effect for an essentially incompressible liquid.

5) In the energy balance equations (i.e. Equations 9.4 to 9.9) the form used deviates from the form introduced in Chapter 4 where the equations were written on the basis of one kg of vapor circulating through the condenser and evaporator. In this case, since the flow rates are known, it is more direct to simply use the appropriate flow rate for each stream. These energy balances could be converted to the form in Chapter 4 by dividing through by m_7.

The performance of an actual absorption system that follows the above example may not be as good as calculated in Example 9.2. The effectiveness of the solution heat exchanger may not be as high as 0.8; there are also pressure drops in the heat exchangers, less effective rectification and a less efficient pump. As discussed in Problem 9.2, by accounting for these non-idealities, a somewhat lower performance will be found. Another source of performance degradation is the fact that the solution in the absorber has to be subcooled to a certain degree to provide a temperature driving potential for the absorption process. Similarly, the vapor in the desorber may be superheated. Both effects degrade the heat exchange process.

On the other hand, there are a number of ways of increasing the performance of a single-stage absorption system that are explored next.

9.2.2 Measures to Improve the Performance

There are several opportunities to considerably improve the performance of the absorption system of Section 9.2.1. All measures that are discussed next apply in principle to all absorption systems, single or multi-stage and independent of the fluids used. However, depending on the application and the challenges involved in introducing additional heat exchangers, only some are used for any given design.

All measures that lead to performance improvements have an important feature in common. They all are based on reusing quantities of heat within the cycle. This is often referred to as "internal heat exchange". One important example that has already been introduced previously is the solution heat exchanger.

9.2.2.1 Condensate Precooler

The condensate precooler subcools the refrigerant that leaves the condenser by pre-heating the vapor entering the absorber. It can be used in two ways. The less complex method is discussed first. The condensate precooler reduces the enthalpy of the condensate. Thus the enthalpy at the evaporator inlet is reduced. Since the enthalpy at the evaporator outlet is unchanged, the cooling capacity per kilogram of fluid is increased. This is the desired benefit. Possible disadvantages of this method are the following. The fact that the vapor entering the absorber is superheated leads to an increased absorber heat; thus the absorber may increase in size. The condensate precooler causes an additional pressure drop requiring either a higher evaporator pressure or a lower heat rejection temperature for the absorber or lower ammonia mass fractions of the solution streams, which in turn will lead to higher desorber temperatures. Nevertheless, in ammonia/water systems, this heat exchanger is usually quite beneficial.

The second method to utilize the condensate precooler is the following. When the vapor purity leaving the rectifier is not very high, then the temperature glide for the evaporation process is quite considerable as shown in Figure 9.4. Here the ammonia mass fraction is assumed to be 0.99. Even with a water content of only 1%, the temperature glide associated with the evaporation process becomes excessive, especially for high vapor qualities. This leads

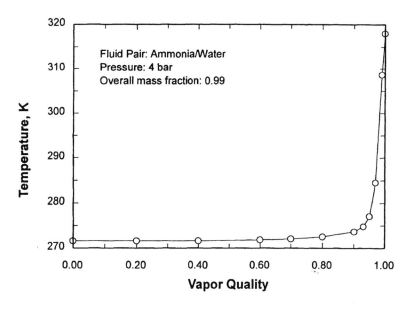

Figure 9.4 Temperature glide in evaporator

in pool boiler designs to accumulation of water in the evaporator unless precautions are taken. One option to remove the excess water from the evaporator is to use a blow-down system in which the accumulated water is drained into the absorber. Another option is to have water droplets carried over into the absorber by high velocity vapor. However, this option may require an undesirably high pressure drop. Also, both methods degrade performance because the water carries a considerable amount of (liquid) ammonia with it that is lost for refrigeration purposes.

The third and best option is to evaporate this water in the condensate precooler. In this way the enthalpy of the incoming condensate is reduced even further than by just preheating the vapor. It actually turns out that this method can also reduce the rectifier load and increase performance once again. Figure 9.5 shows the single-stage ammonia/water cycle with a condensate precooler. The liquid leaving the condenser is being subcooled, while the vapor leaving the evaporator is heated. Any water droplets contained in this vapor stream are evaporated.

The following set of examples shows the influence of the various options. First, the effect of the condensate precooler itself is examined by extending Example 9.2

Example 9.3 Condensate precooler

Reevaluate the performance of the absorption system of Example 9.2. All operating conditions remain the same, except the effectiveness of the condensate precooler which is increased from 0 (i.e., no condensate precooler) to 0.8. Table 9.3 is based on the same inputs as Table 9.2 (Example 9.2) except for a precooler effectiveness of 0.8. The two previously unused rows refer to the outlet states of the condensate precooler as shown in Figure 9.5. File: ex9_3.ees.

Discussion:
An inspection of Table 9.3 reveals the following observations:

1) The enthalpy of the condensate drops by approximately 120 kJ/kg, while the vapor stream picks up the same amount of energy. The temperature of the condensate is reduced by 25 K while the temperature of the vapor increases by 36.8 K. Thus, most of the transferred heat contributes to heating the vapor since there was only a small amount of liquid left at the evaporator outlet that requires evaporation. If the state at 13 was all vapor, then the vapor temperature would increase by approximately twice the amount by which the condensate temperature drops because the specific heat of the vapor is about half that of the liquid.

2) The only change in state points occurs at the condensate precooler (all others are not affected and are the same as in Example 9.2).

3) The COP increases by about 11.7% due to an increase in the evaporator capacity. As a consequence, the absorber heat increases by the same amount by which the evaporator capacity increases. The cold vapor entering from the evaporator is no longer available to contribute to absorber cooling.

It should be noted here that the condensate precooler always improves the COP in an absorption system, while in vapor compression systems its benefits are often compensated by performance penalties in the compressor. Further, in water/lithium bromide systems, the condensate precooler is not employed because it is a major design challenge to develop such a heat exchanger without incurring significant pressure drop penalties on the vapor side. Based on the state points in Table 9.3 the values in the last row are obtained for all amounts of heat exchanged and other variables of interest.

In the following example, the performance of the system is investigated when the condensate precooler is used to evaporate a significant amount of liquid phase.

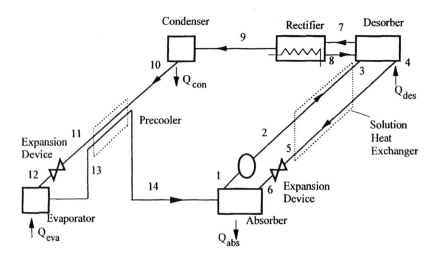

Figure 9.5 Single-stage ammonia/water cycle with precooler

Table **9.3** State points for the ammonia/water system of Example 9.3

State Points	h (J/g)	m (kg/sec)	P (kPa)	Vapor Quality	T (°C)	x (kg/kg)
1	-50.0	1.00	240.2	0.0	40.0	0.3679
2	-48.5	1.00	1555	n/a	40.1	0.3679
3	207.5	1.00	1555	n/a	102.3	0.3679
4	358.5	0.863	1555	0.0	131.0	0.2679
5	60.1	0.863	1555	n/a	58.9	0.2679
6	60.1	0.863	240.2	n/a	59.2	0.2679
7	1516	0.150	1555	1.00	108.0	0.9444
8	229.7	0.013	1555	0.0	108.0	0.3679
9	1294	0.137	1555	1.00	44.0	0.999634
10	190.1	0.137	1555	0.0	40.0	0.999634
11	70.0	0.137	1555	n/a	21.3	0.999634
12	70.0	0.137	240.2	0.128	-14.5	0.999634
13	1224	0.137	240.2	0.998	-10.0	0.999634
14	1345	0.137	240.2	n/a	30.0	0.999634

f = 7.317 W_p = 1.5 kW Q_{shx} = 277.8 kW Q_{sc} = 12.3 kW
Q_{rec} = 51.4 kW Q_{des} = 338.0 kW Q_{abs} = 296.3 kW Q_{con} = 150.9 kW
Q_{eva} = 159.1 kW   COP = 0.469

Example 9.4 Condensate Precooler with Liquid Evaporation

Reevaluate the performance of the absorption system from Example 9.3 holding all operating conditions the same except that the mass fraction of the ammonia vapor leaving the rectifier is set to 0.99 (Point 9). Table 9.4 shows the results. File: ex9_4.ees.

Based on state points in Table 9.4 the values in the final row are obtained for all amounts of heat exchanged and other variables of interest.

Discussion:

An inspection of Table 9.4 and a comparison of the summary results with Example 9.3 reveals the following observations:

1) The COP is decreased by 17% as compared to Example 9.3 due primarily to three effects: a) lower temperature at point 12, b) higher rectifier load due to higher water content in desorber outlet (point 7), and c) lower evaporator load due to lower mass flow rate of refrigerant.

Table 9.4 State points for the ammonia/water system according to Figure 9.5 (Example 9.4)

State Points	h (J/g)	m (kg/sec)	P (kPa)	Vapor Quality	T (°C)	x (kg/kg)
1	-9.2	1.00	132.8	0.0	40.0	0.2811
2	-7.6	1.00	1540	n/a	40.0	0.2811
3	336.3	1.00	1540	n/a	118.2	0.2811
4	538.6	0.876	1540	0.0	152.4	0.1811
5	146.2	0.876	1540	n/a	63.3	0.1811
6	146.2	0.876	132.8	0.005	61.2	0.1811
7	1674	0.148	1540	1.00	127.4	0.8736
8	379.7	0.024	1540	0.0	127.4	0.2811
9	1410	0.124	1540	1.00	76.5	0.99
10	184.1	0.124	1540	0.0	40.0	0.99
11	55.9	0.124	1540	n/a	13.5	0.99
12	55.9	0.124	132.8	0.140	-27.5	0.99
13	1234	0.124	132.8	0.975	-10.0	0.99
14	1362	0.124	132.8	n/a	25.4	0.99

$f = 8.089$ $W_p = 1.6 \text{ kW}$ $Q_{shx} = 343.9 \text{ kW}$ $Q_{sc} = 15.8 \text{ kW}$
$Q_{rec} = 64.0 \text{ kW}$ $Q_{des} = 374.1 \text{ kW}$ $Q_{abs} = 305.7 \text{ kW}$ $Q_{con} = 151.6 \text{ kW}$
$Q_{eva} = 145.6 \text{ kW}$ $\text{COP} = 0.388$

2) The three effects mentioned result from the assumptions and inputs used in the model. A somewhat unexpected result is that the rectifier heat load is greater when the vapor mass fraction is lower as in Example 9.4. This result can be traced to the rectifier input where the vapor leaving the desorber is found to have a much higher water content than that in Example 9.3. This is a prime example of the complicated effects which often result from the coupled nature of processes in an absorption cycle.

3) The much lower temperature at point 12 is a major difference between Examples 9.4 and 9.3. It is also noted that the low pressure in this simulation is much lower in Example 9.4. In a more realistic simulation, the evaporator temperature would be linked to the external temperatures through a heat exchanger that would feedback and tend to keep the evaporator pressure high by transferring more heat as the temperature difference increased. However, the internal model discussed here does not have that level of sophistication (thus care must be taken in interpreting the results)

A well-designed ammonia/water absorption system requires a careful analysis of the tradeoffs between the use of the rectifier and that of the condensate precooler. The above examples do not discuss additional effects that may contribute to additional savings but also to a complication of the design task as shown next.

9.2.2.2 Rectifier Integration

When designing the solution heat exchanger, it can be observed that the heat capacities of the two streams do not match perfectly. Usually, the stream leaving the absorber has the higher mass flow rate and therefore a higher capacitance rate than the stream leaving the desorber. This imbalance can be compensated by having the rich solution leaving the absorber circulate through the rectifier as coolant before entering the solution heat exchanger, as shown in Figure 9.6. Thus the rectifier heat is not rejected to a heat sink outside of the absorption system but rather utilized within the system. This internal heat exchange can be an important means to improve performance. The following example demonstrates this alternative.

9.5 Example Rectifier Heat Integration

Consider the absorption system of Example 9.4. Now the rectifier heat is used to preheat the rich solution before it enters the solution heat exchanger as in Figure 9.6. File: ex9_5.ees.

Discussion:
An inspection of Table 9.5 and a comparison of the summary results reveal the following observations:
The COP is increased by 5.4% as compared to Example 9.4. By using the rectifier heat internally, the heat load of the solution heat exchanger is decreased and the absorber heat rejection increased. This implies that a significant amount of "rectifier heat" is still rejected at the ambient temperature level. At the same time, the desorber heat decreased, leading to a COP improvement. In the case selected here, the improvement is marginal. There are other operating conditions under which the rectifier integration can lead to more dramatic effects.

It should be noted that the heat capacity of the rich solution is limited. Only that portion that cannot be satisfied by the poor solution can be utilized to cool the rectifier. As soon as the rectifier heat exceeds this quantity, then no additional benefits are gained. State point 16 represents the rich solution after it has passed through the rectifier.

Based on the state points in Table 9.5 the values in the final row for all amounts of heat exchanged and other variables are obtained.

9.2.2.3 Solution Recirculation

In some instances it is desirable to allow a large temperature glide in the absorber and the desorber. This has several consequences. The pump flow rate and pump power requirements are reduced. In addition, the absorber requires a smaller cooling water flow rate and it will heat the cooling water to higher temperatures which may be important in certain heat pump applications. It also can lead to a reduction in the size of the cooling tower if such a device is used. Further, the desorber outlet temperature of the poor solution increases as well. Thus the benefit in the absorber is paid for by an increased temperature of the heat supplied to the desorber. The imbalance in the solution heat exchanger is increased. Because of the reduced pump flow rate and assuming that the vapor flow rate through the condenser and evaporator

Table 9.5 State points for the ammonia/water system according to Figure 9.6 with rectifier heat integration (Example 9.5)

State Points	h (J/g)	m (kg/sec)	P (kPa)	Vapor Quality	T (°C)	x (kg/kg)
1	-9.2	1.000	132.8	0.0	40.0	0.2811
2	-7.6	1.000	1540	n/a	40.0	0.2811
3	356.2	1.000	1540	n/a	122.5	0.2811
4	538.6	0.876	1540	0.0	152.4	0.1811
5	196.5	0.876	1540	n/a	75.1	0.1811
6	196.5	0.876	132.8	0.025	64.3	0.1811
7	1674	0.148	1540	1.000	127.4	0.8736
8	378.6	0.024	1540	0.0	127.4	0.2811
9	1410	0.124	1540	1.000	76.5	0.99
10	184.1	0.124	1540	0.0	40.0	0.99
11	55.9	0.124	1540	n/a	13.5	0.99
12	55.9	0.124	132.8	0.140	-27.5	0.99
13	1234	0.124	132.8	0.975	-10.0	0.99
14	1362	0.124	132.8	n/a	25.4	0.99
16	56.4	1.000	1540	n/a	54.9	0.2811

f = 8.089	W_p = 1.6 kW	Q_{shx} = 299.8 kW	Q_{sc} = 15.8 kW
Q_{rec} = 64.0 kW	Q_{des} = 354.1 kW	Q_{abs} = 349.8 kW	Q_{con} = 151.6 kW
Q_{eva} = 145.6 kW	COP = 0.409		

remains constant, the difference between the two solution flow rates in the solution heat exchanger increases. This increases not only the differences in the heat capacities of the two streams but also the temperature mismatch. For example, the liquid leaving the desorber is considerably warmer than required for preheating the incoming rich solution.

One way of taking advantage of this situation is to use the heat content of the poor solution that exits the desorber to further generate refrigerant vapor. For this purpose the liquid line containing the poor solution is recirculated in counter-flow through the desorber as shown in Figure 9.7. The figure shows the overall schematic of an absorption system with the new feature displayed in thicker lines. Thus the solution is cooled to the lowest desorber temperature before it enters the solution heat exchanger. In this way the availability in that solution stream is utilized in the most efficient way. However, this method used by itself will not be beneficial. The solution heat exchanger is usually limited by the poor solution. Thus,

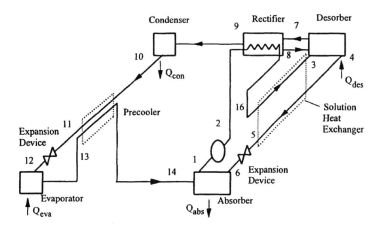

Figure 9.6 Single-stage ammonia/water cycle with integrated rectifier

reducing the heat content of the poor solution further will not help unless special operating conditions exist.

On the other hand, when the rich solution is first preheated in the absorber before entering the solution heat exchanger (or the rectifier if this is integrated), then the solution recirculation in the absorber can be quite advantageous. Thus, the principle of solution recirculation that was explained for the desorber above can also be applied to the absorber. Here the rich solution, once it passed through the pump, reenters the absorber in a separate pipe and flows in counter- current heat exchange with the absorbing solution as shown in Figure 9.8. In this way, the solution is preheated by reducing the absorber heat. The next example shows the effect the solution recirculation has on a single-stage system when it is applied in the absorber only.

Example 9.6 Solution Recirculation in the Absorber

Example 9.5 is used as the baseline and solution recirculation is added in the absorber (with point 15 defined in Figure 9.8 and all remaining points identical to Figure 9.6). To demonstrate the effect most clearly, it is assumed that the solution leaving the absorber solution recirculation heat exchanger, after it has picked up additional heat, has reached the highest temperature occurring in the absorber (the solution recirculation heat exchanger effectiveness is 100%). File: ex9_6.ees.

Discussion:
The COP is increased by another 8.6% as compared to Example 9.5. The heat load of the solution heat exchanger is decreased by about 20% and the absorber heat rejection is reduced by 7.9%. Further, the desorber heat requirement is reduced by 7.8% as well. The desorber heat load is reduced because the entering solution reaches a higher temperature when it exits the solution heat exchanger . The rich solution, while passing through the absorber, picks up about 88 kW of heat. - The heat duty of the solution heat exchanger is reduced by a similar

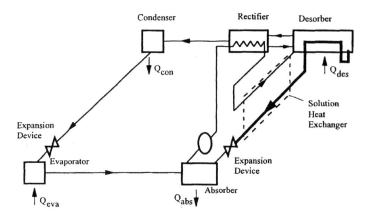

Figure 9.7 Solution recirculation in desorber (thick lines)

amount. The values do not match exactly, because other variables are affected as well. However, the absorber heat duty is reduced by 27.6 kW. This reduced duty results because the poor solution is now entering the absorber at a higher temperature, since it cannot be cooled as effectively because of the higher solution heat exchanger inlet temperature of the rich solution. While the total heat exchange area needed for the absorber increases, that of the solution heat exchanger decreases by a similar amount.

It should be noted that the rectifier performance is not affected by the fact that the coolant, the rich solution, enters at a higher temperature. The point 7 temperatures are identical in Tables 9.5 and 9.6. However, a comparison of the temperatures indicates also that the solution temperature leaving the rectifier, point 16, may not increase beyond the limit set by point 7. Otherwise the vapor purity could not be maintained. State point 16 represents the rich solution after it has passed through the rectifier. Based on the state points in Table 9.6 the values for all amounts of heat exchanged and other variables are obtained and are shown in the last row of Table 9.6.

There are several options of re-using various amounts of heat internally within the system. This internal heat exchange is important for efficiency improvements when conditions are right. However, not all of these measures show benefits at all times. Solution recirculation, for example, is the more effective with larger temperature glides. It would not be very beneficial in a typical LiBr/water system. Also, attempting to utilize the rectifier heat internally and the recirculation at the same time may not be productive. It turns out that at times the solution recirculation in the absorber competes for the same heat sink as the rectifier heat. In many circumstances only one of the two measures can be employed.

Solution recirculation as it is shown in Figures 9.7 and 9.8 implies that three-path heat exchangers are employed. These devices are difficult to manufacture. An alternative is to utilize two-path heat exchangers in parallel to replace, for example, the absorber. While this

Table 9.6 State points for the ammonia/water system according to Figure 9.8 using rectifier integration and solution recirculation within the absorber (Example 9.6)

State Points	h (J/g)	m (kg/sec)	P (kPa)	Vapor Quality	T (°C)	x (kg/kg)
1	-9.2	1.000	132.8	0.0	40.0	0.2811
2	-7.6	1.000	1540	n/a	40.0	0.2811
3	383.8	1.000	1540	0.002	127.7	0.2811
4	538.6	0.876	1540	0.0	152.4	0.1811
5	265.3	0.876	1540	n/a	91.2	0.1811
6	265.3	0.876	132.8	0.052	68.2	0.1811
7	1674	0.148	1540	1.000	127.4	0.8736
8	378.9	0.024	1540	0.0	127.4	0.2811
9	1410	0.124	1540	1.000	76.5	0.99
10	184.1	0.124	1540	0.0	40.0	0.99
11	55.9	0.124	1540	n/a	13.5	0.99
12	55.9	0.124	132.8	0.140	-27.5	0.99
13	1234	0.124	132.8	0.975	-10.0	0.99
14	1362	0.124	132.8	n/a	25.4	0.99
15	80.3	1.000	1540	n/a	60.5	0.2811
16	144.3	1.000	1540	n/a	75.3	0.2811

f = 8.089		W_p = 1.6 kW		Q_{shx} = 239.5 kW		Q_{sc} = 15.8 kW	
Q_{rec} = 64.0 kW		Q_{des} = 326.5 kW		Q_{abs} = 322.2 kW		Q_{con} = 151.6 kW	
Q_{eva} = 145.6 kW		COP = 0.444					

alternative is easier to implement from a hardware point of view, it poses considerable control challenges. The two-phase flow entering the absorber, as well as the incoming vapor stream, has to be split and supplied to both heat exchangers in just the required flow rates, requiring sophisticated flow controls.

9.2.2.4   Solution Cooled Absorber
An alternative is to utilize a so-called solution cooled absorber. Here the absorber is also split into two-path heat exchangers, but they are arranged in series as shown in Figure 9.9. The original absorber is essentially unmodified. It rejects its heat to an external sink not shown in Figure 9.9. However, the high temperature end of the temperature glide is now accommodated

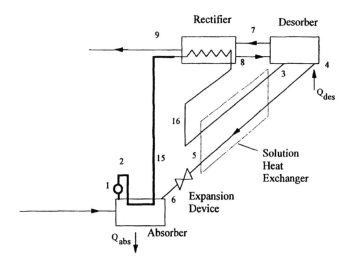

Figure 9.8 Solution recirculation in absorber

within a second absorber that is exclusively cooled by the solution. This concept is utilized in the residential ammonia/water chiller that is described in Section 9.3.

9.2.3   Comparison Ammonia/Water and Water/Lithium Bromide

When the performance of a single-stage ammonia/water system is compared to that of a water/lithium bromide system, it is found that for the same application the water/lithium bromide system is more efficient. The reasons are hidden in the working fluid properties. Ammonia water systems usually require a rectifier that is not needed in water/salt systems. Further, the specific heat of the ammonia/water solution is about double that for water/salt systems. Thus any inefficiency of the solution heat exchanger causes a larger penalty in ammonia/water systems. There exists also a disadvantage in the desorber. The amount of heat required for just increasing the temperature of the remaining liquid is larger for ammonia/water systems. Last, the latent heat of ammonia is about half as large as that of water. For the same cooling capacity, the ammonia/water system requires higher solution flow rates than water/salt systems.

It can be shown that the term $c_p T/r$ is a good measure for the relative efficiency of the absorption working fluids (Alefeld and Radermacher, 1994). c_p is the specific heat of the liquid phase and r, the latent heat. The smaller the ratio the better the working fluid. The reason is the following. One of the most significant losses within the absorption system itself is the expansion process of the refrigerant before entering the evaporator. The lower the specific heat of the liquid, the less vapor has to evaporate to cool this liquid from the condenser temperature to the evaporator temperature and the more refrigerant is available to provide cooling capacity. In terms of this factor, $c_p T/r$, water/lithium bromide is the better working fluid pair compared to ammonia/water.

9.3   Examples of Ammonia/Water Absorption Systems in Operation

In the following a short overview is presented of ammonia/water absorption systems, that are actually built and in service today.

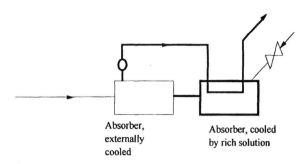

Figure 9.9 Solution-cooled absorber

In the range of small capacities such as residential air-conditioning, 3- and 5-ton (10 to 17 kW) units are manufactured and sold in the U.S. Figure 9.10 shows a schematic of this absorption system. In order to describe it let us start in the lower right-hand corner. This device is the solution pump. The membrane in the middle of the pump acts as a piston that moves the ammonia/water solution from the receiver through the inlet valve to the high pressure side. The membrane itself is moved by applying pressure via oil (lower half of pump) from an oil pump (not shown) that is cycled from low to high pressure levels about once a second. From there the rich solution flows through the rectifier (horizontal device above the solution pump) and then into the solution cooled absorber. After being preheated in the solution cooled absorber, the rich solution enters the desorber. This system does not employ a solution heat exchanger. As discussed later, the change of mass fraction is so large that the temperature glide in the absorber is capable of preheating the solution close to the desorber inlet temperature.

The desorber is direct fired. The burner is arranged vertically on the side of the desorber and the flue gases flow across the finned surface of the desorber tank. The poor solution is picked up by a tube at the bottom of the desorber and recirculated through the desorber to the solution cooled absorber. Instead of a solution heat exchanger, just solution recirculation in the desorber is employed. In the solution-cooled absorber, the poor solution drips over the tube coil of the rich solution while absorbing vapor and is collected at the bottom of the tank. It continues to flow together with the vapor that was not yet absorbed through two parallel passes into a coil, the air-cooled absorber. Here the absorption process is completed and the heat rejected to the surroundings. The outlet of the air-cooled absorber is connected to the solution pump receiver.

The ammonia vapor leaves the desorber and is forced across the rectifier where the heat of rectification is rejected to the rich solution and proceeds to the condenser. The condenser, housed within the same set of fins utilized by the air-cooled absorber, rejects the heat of condensation to the surroundings. The condensate emerges at the same location where the rich solution leaves the air-cooled absorber and circulates through the refrigerant precooler into the evaporator. The refrigerant evaporates inside the coil while the water to be chilled flows across the outside of the coil. The refrigerant vapor returns via the condensate precooler to the solution-cooled absorber. This design appears to be quite complex, but it has proven successful since the 1960s.

Figure 9.11 shows a large capacity ammonia/water system. It is a two-stage system that accommodates two evaporators providing cooling capacities of 300 and 600 refrigeration tons, respectively, at -45 and -34°C for food storage. The plant is located in Verona, California, and driven by waste heat as part of a cogeneration plant. It is housed in a three-story structure that is designed to meet all local seismic requirements. Liquids close to saturation pressure are gravity fed down to the solution and recirculation pumps at the ground level to prevent cavitation. Figure 9.12 shows a schematic of the process. The evaporators, E_I and E_{II}, supply vapor to the respective absorbers which are in series with regard to the solution flow. The two absorbers operate at different pressures and thus two solution pumps are required. Other than the fact that two low pressure levels are employed to provide the two different evaporator temperatures, the system is designed as a single-stage cycle with one desorber, rectifier condenser and solution heat exchanger. Two condensate precoolers are employed, one for each evaporator. The system has been in operation since the mid-eighties.

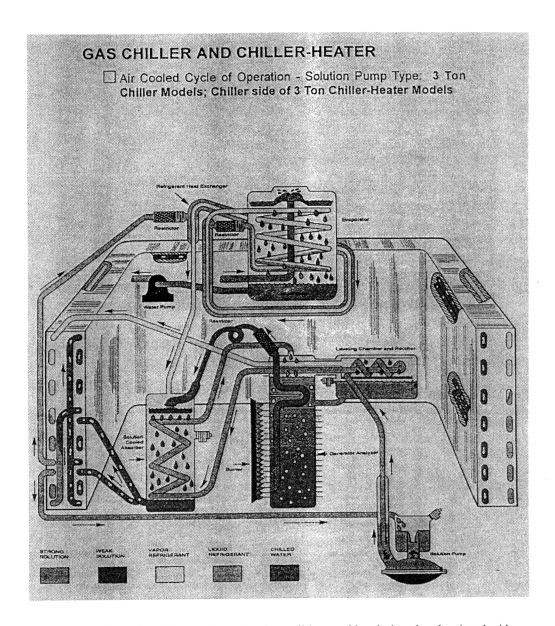

Figure 9.10 Residential ammonia/water air conditioner with solution absorber (used with permission from Robur)

Figure 9.11 Commercial ammonia/water absorption system for cogeneration facility in Verona, California (used with permission from Babcock-Borsig)

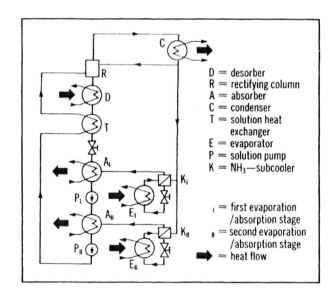

Figure 9.12 Schematic of process used in system of Figure 9.11 (used with permission from Babcock-Borsig)

HOMEWORK PROBLEMS

9.1   For the evaporator data given in Example 9.1 calculate and plot the temperature glide in 11 steps for the temperature ranging from 0 to 88°C. Discuss the shape of the curve. Based on the enthalpy values, estimate what fraction of the total heat of evaporation is available in the first 10 K of the temperature glide.

9.2   Analyze the absorption system according to Example 9.2. However, account for the following non-idealities at first individually and then all of them combined to obtain a more realistic performance.
a) The effectiveness of the solution heat exchanger is 0.7.
b) The effectiveness of the condensate precooler is 0.7.
c) The pump efficiency is 0.5.
d) To provide a driving potential for the absorption process assume that the temperature of the solution in the absorber is actually 2 K higher than assumed in the problem statement.
e) Combine all of the above assumptions into one cycle and find its COP, all amounts of heat exchanged and the pump work.

9.3   Based on Example 9.2, evaluate and plot the change of the COP with the change of the effectiveness of the solution heat exchanger. Discuss the limits, i.e., effectiveness 0.0 and 1.0. In the latter case, what is the difference in the heat capacities of the poor and rich solution?

9.4   Based on Example 9.2, evaluate and plot the change of the COP with the change of the effectiveness of the condensate subcooler. Then vary the purity of the vapor leaving the rectifier and discuss the changes in the influence the condensate subcooler has on the overall system.

9.5   Expand Example 9.6 to include solution recirculation in both, the desorber and the absorber using heat exchanger approach temperatures as the parameter that describes the heat exchange process. Then vary the approach temperature for both independently and discuss the effect on COP and solution heat exchanger capacity.

9.6   Again, using Example 9.6 as the basis analyze the change in COP when the difference in mass fraction between the poor and rich solution is varied from 0.01 to 0.35. Discuss and explain the special situations that occur at the ends of this range.

9.7   Open-ended Project: Write a program for the best possible single-stage ammonia/water system that includes all means to improve efficiency and plot the changes in COP, desorber heat requirement, rectification heat and cooling capacity as the evaporator temperature is changed from -10 to 10°C.

Chapter 10

TWO-STAGE AMMONIA/WATER SYSTEMS

As with water/lithium bromide systems, two-stage ammonia/water absorption systems have the potential for increased efficiency or increased temperature lift. There are 26 different two-stage configurations (Alefeld and Radermacher, 1994). However, the performance potential of only a relatively small number was ever investigated. Nevertheless, since each class of two-stage absorption systems has many different options for implementing the actual machine, the number of configurations that was discussed or actually employed over time is very large even though it covered only few of the classes. For the purpose of this text, we focus on the most common examples. For the sake of consistency the definitions for double effect and two stage are stated once more. The term "double effect" refers to a configuration in which a certain quantity of heat is used twice to generate refrigerant vapor. This is a special case of a two-stage system. In general, two-stage systems can be thought of as being composed of two single-stage absorption systems. The staging allows for increased efficiency (double effect) or for increased temperature lift (double lift). Examples of both variations are discussed in this chapter.

10.1   Double-Effect Ammonia/Water Systems

Figure 10.1 shows an example for a two-stage double-effect ammonia/water system. The left portion of the figure represents the conventional system discussed in the previous chapter. The absorber, termed here absorber 1, the desorber (desorber 1), the solution heat exchanger 1 (SHX1), and the solution pump (pump 1) can be readily identified. The state points follow the same numbering system as previously used. There are further the condenser and evaporator as well as the condensate precooler as discussed before for single-stage systems. The special feature is the heat source for desorber 1. It receives its heat from absorber 2. The vapor to absorber 2 is supplied from the evaporator and the poor solution from desorber 2. In fact absorber 2 and desorber 2 form another solution circuit. The heat requirement of desorber 2 is satisfied by an outside heat source such as steam or flue gas. Desorber 2 produces refrigerant vapor that is condensed in the condenser and the condensate evaporated in the evaporator. Thus the condenser and evaporator do essentially double duty. The evaporator serves now two absorbers and the condenser serves two desorbers.

It is important to realize that one unit of heat supplied to desorber 2 produces refrigerant vapor that in turn produces a certain amount of cooling in the evaporator. When the vapor from desorber 2 is absorbed in absorber 2, the resulting heat of absorption is utilized to generate a second quantity of refrigerant vapor, this time in desorber 1. The vapor stream from desorber 1 is also condensed and evaporated, producing additional refrigeration capacity. In this way, the amount of heat supplied to desorber 2 is used twice to generate refrigerant vapor.

Compared to the double-effect water/lithium bromide system, Figure 7.1, the cycle according to Figure 10.1 has a significantly different appearance. While the water/lithium bromide system has three pressure levels, the ammonia/water system shown here has only two. Of course, the ammonia water system could be arranged in the same manner as the water/lithium bromide system; however, the highest pressure level would be very high. For a

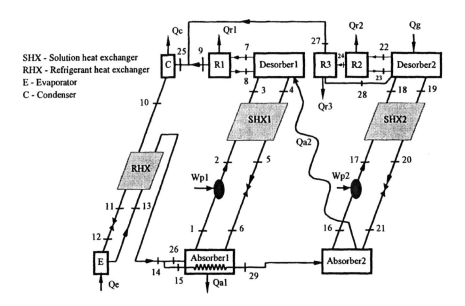

Figure 10.1 Two-stage double-effect ammonia/water absorption system; R2 and R3 are the high and low temperature sections of the rectifier.

typical air-conditioning application it would reach about 7 MPa. This value represents a considerable design challenge. It is much simpler to tack on the second stage to the side, maintaining two pressure levels. In contrast to water/lithium bromide, the solution field of ammonia/water provides the space to do so. Further, as can be seen from the examples in previous chapters, the heat of absorption or desorption is always higher than that of condensation and evaporation (because of the effects of the heat of mixing). In the ammonia/water system of Figure 10.1, the absorber heat is used to fire the desorber. This produces more refrigerant vapor as compared to the water/lithium bromide system in which condenser heat is used. Therefore the performance increase for the ammonia/water system due to staging is potentially larger. This statement is made with some caution since other effects such as additional rectification requirements can offset the gain.

It should be kept in mind that the ammonia vapor streams emerging from the desorbers have different water content. The system will only operate in a stable fashion, when both rectifiers produce vapor of the same water content. This is discussed in more detail in the following example which considers the details of the performance of the cycle according to Figure 10.1

Example 10.1 Two-Stage Double-Effect Ammonia/Water System with Separate Rectifiers

---

Evaluate the two-stage absorption heat pump cycle shown in Figure 10.1. The heat rejected by absorber 2 is completely used to drive the first stage, desorber 1. The working fluid pair is ammonia-water. The rectifier 2 of the high temperature stage is subdivided into two rectifiers that operate on different temperature levels (R1 and R2, Figure 10.1). Rectifier R2 produces vapor at the temperature of the solution leaving the second-stage pump. This heat is contributed to the low temperature desorber 1. The vapor leaving R2 is further rectified in R3 to a mass fraction of 0.99, the same as that of the vapor leaving the first-stage rectifier (R1). This heat is available at a temperature that is too low for additional internal heat exchange and is rejected as waste heat. The isentropic efficiency of the solution pumps is assumed to be 100% while the effectiveness of both the solution heat exchangers and the condensate precooler is 80%. The other input parameters and restrictions are as follows:

1) Evaporator outlet temperature T_{13} = -10°C.
2) Condenser and first stage absorber temperature T_{10} and T_1= 40°C.
3) Refrigerant entering the second-stage absorber is heated using the first-stage absorber heat. T_{29}, the temperature of the refrigerant leaving the first stage absorber is set equal to T_6, the temperature of the poor solution entering absorber 1.
4) The lowest temperature in the second stage absorber T_{16} is 0.1°C higher than the highest temperature T_4 in the first stage generator.
5) The mass fraction difference in the first-stage solution circuit is 0.04, while that in the second-stage solution circuit is 0.06.
6) The first-stage pump mass flow rate is 1.0 kg/sec.
7) Liquid leaving the absorbers and the generators is saturated.
8) Vapor leaving the generators and the rectifiers is saturated.
9) Vapor quality of the solution leaving the evaporator is 0.88.

Find the key properties for all state points. Further, calculate the rate of heat exchanged for all heat exchangers and the rate of work the pumps require. Check the energy balance for consistency and calculate the COP. File: ex 10_1.ees.

Solution:
Using an appropriate software package the following data for the state points are obtained. All state point numbers refer to the respective points in Figure 10.1. Based on these data we find for the amounts of heat exchanged the following values: The amount of heat released in the low temperature absorber (absorber 1) amounts to Q_{abs1} = 153.4 kW, in the evaporator to Q_{eva} = 90.6 kW, in the condenser to Q_{con} = 101.0 kW and in the high temperature desorber (desorber 2) to Q_{des2} = 181.7 kW. These values are all in the range one would expect from experience with single-stage systems. However, in the high temperature absorber only Q_{abs2} = 56.4 kW of heat is released. One would expect that this value is comparable to that of the high temperature desorber. However, most of the heat requirement of desorber 2 results in the evaporation of water that has to be removed through the rectifier. This is not surprising when one considers that the ammonia mass fraction of the solution is very low, x_{18} = 0.0604. The rectifier load, the sum of Q_{rec2} (118.1 kW) and Q_{rec3} (7.6 kW) amounts to 125.7 kW. Nevertheless, desorber 2 produces a refrigerant mass flow rate of 0.020 kg/sec (point 27). Desorber 1 produces a mass flow rate of 0.062 kg/sec which is significantly larger for a much smaller heat input than desorber 2. As one should expect the rectification load on R1 is small (12.2 kW) due to the high ammonia mass fraction of state 7.

As a consequence of the staging, the COP increases to 0.499 which is only slightly higher than that of a single-stage system. The benefit of the second stage is limited severely by the rectification requirement of the high temperature stage. Any measure that would reduce that requirement has great potential to increase the overall efficiency.

The remaining quantities of heat exchanged that were not mentioned so far are the condensate precooler Q_{rhx} = 15.6 kW and the solution heat exchangers Q_{shx1} = 242.3 kW and Q_{shx2} = 94.7 kW, respectively. The pump work input amounts to W_{p1} = 1.48 kW and W_{p2} = 0.46 kW, respectively.

Table 10.1 State points for the ammonia/water system according to Figure 10.1

	h (J/g)	m (kg/s)	P (kPa)	Quality	T (°C)	x (kg/kg)
1	-47.5	1.00	270.0	0.0	40.0	0.3865
2	-46.0	1.00	1540	n/a	40.1	0.3865
3	196.3	1.00	1540	n/a	94.3	0.3865
4	288.9	0.938	1540	0.0	112.2	0.3465
5	30.6	0.938	1540	n/a	54.9	0.3465
6	30.6	0.938	270.0	0.015	49.4	0.3465
7	1525.9	0.066	1540	1.00	103.5	0.9542
8	239.0	0.004	1540	0.0	103.5	0.3865
9	1410.3	0.062	1540	1.00	76.5	0.9900
10	184.1	0.082	1540	0.0	40.0	0.9900
11	-4.9	0.082	1540	n/a	0.4	0.9900
12	-4.9	0.082	270.0	0.044	-11.5	0.9900
13	1094.8	0.082	270.0	0.880	-10.0	0.9900
14	1283.8	0.082	270.0	0.979	13.0	0.9900
15	1283.8	0.020	270.0	0.979	13.0	0.9900
16	431.3	0.334	270.0	0.0	112.3	0.0604
17	432.7	0.334	1540	n/a	112.4	0.0604
18	716.4	0.334	1540	n/a	177.3	0.0604
19	850.2	0.314	1540	0.0	199.5	0.0004
20	548.1	0.314	1540	n/a	130.2	0.0004
21	548.1	0.314	270.0	0.001	129.9	0.0004
22	2424.3	0.082	1540	1.00	183.6	0.3076
23	744.8	0.059	1540	0.0	183.5	0.0604
24	1571.7	0.023	1540	1.00	112.4	0.9310
25	1410.3	0.082	1540	1.00	76.5	0.9900
26	1283.8	0.062	270.0	0.979	13.0	0.9900
27	1410.3	0.020	1540	1.00	76.5	0.9900
28	109.9	0.003	1540	0.0	76.5	0.5300
29	1407.8	0.020	270.0	n/a	49.4	0.9900

COP = 0.499 Q_{des1} = 174.5 kW Q_{abs1} = 153.4 kW Q_{rec3} = 7.6 kW
Q_{eva} = 90.6 kW Q_{des2} = 181.7 kW Q_{abs2} = 56.4 kW Q_{shx1} = 242.3 kW
Q_{con} = 101.0 kW Q_{rec1} = 12.2 kW Q_{rec2} = 118.1 kW Q_{shx2} = 94.7 kW
W_{p1} = 1.48 kW W_{p2} = 0.46 kW Q_{rhx} = 15.6 kW

The performance of the double-effect system of Example 10.1 is not encouraging since the COP is only slightly higher than that of a single-effect system with considerable increase in complexity. This cycle was selected because it can be seen as the starting point for the introduction of the so called "GAX" cycles that are discussed in Chapter 11.

As mentioned in Example 10.1 any reduction of the rectification load in the high temperature stage should increase the performance. One concept that may be of interest is the suggestion of using a bleed line to manage the water balance. This is shown in Figure 10.2.

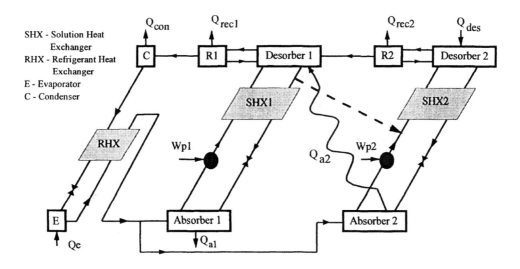

SHX - Solution Heat
    Exchanger
RHX - Refrigerant Heat
    Exchanger
E - Evaporator
C - Condenser

Figure 10.2 Two-stage absorption cycle with bleed line (heavy dashed line)

The vapor from the high temperature desorber passes through the lower temperature desorber where it is partially rectified. Since the vapor leaving the low temperature desorber contains less water than the vapor entering this device, there is a net carryover of water from the high temperature desorber into the low temperature desorber. To compensate for this, a so-called bleed line is introduced as pointed out by the dashed, diagonal line that connects the right leg of the lower temperature solution circuit to the left leg of the high temperature solution circuit. With only a relatively small flow rate through this connection, the water carried over by the vapor stream can be returned. The effectiveness of this measure was demonstrated experimentally in a combined absorption-vapor compression system (Rane et al., 1993).

10.2   Double-Lift Ammonia/Water Systems

Especially for low temperature applications it can happen that the solution field is too narrow to accommodate a single-stage ammonia/water system for the required temperatures. Figure 10.3 shows an example. The single stage-system with the solid lines can be accommodated by the solution field. All major components such as desorber, absorber, condefser, and evaporator fit into the space limited by the vapor pressure lines of pure ammonia and water. This approach works down to evaporator temperatures of about -50°C for minimum absorber temperatures of about 40°C. However, when the evaporator is required to operate at a lower temperature (for example, -70°C) while the absorber temperature is fixed at 40°C, then the lower pressure level decreases and the absorber operating conditions are outside the available solution field, Figure 10.3. Thus another solution must be found. One option is to use again a two-stage configuration (for example, that of Figure 10.2) but now in a different arrangement as compared to Figure 10.1.

The new cycle is shown in Figure 10.4. The evaporator and the condenser are the same as in a conventional single-stage system. However, the condenser heat is now rejected to desorber 1. This is necessary because for the very low evaporator temperatures the condenser pressure is not sufficiently high to allow for condensation at relatively low pressures. However, desorber 1 acts now as the evaporator of a single-stage system that accepts heat at a low

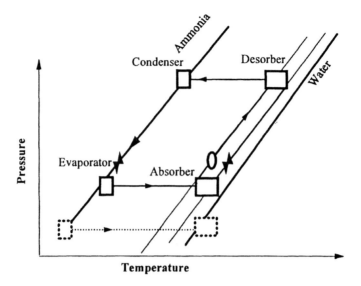

Figure 10.3 Necessity for double-lift systems

temperature level and pumps it to a higher temperature level where it can be rejected to the environment. In this case the heat is rejected by two absorbers, absorbers 1 and 2 in Figure 10.4. The function of desorber 2 remains unchanged. In fact there are only two changes between Figures 10.1 and 10.4. First, the direction of flow in the intermediate solution circuit, desorber 1 and absorber 1, are reversed, which also reverses the function of these two components and those of the pump and expansion valve. Second, the location of the internal heat exchange is modified. In Figure 10.4 the internal heat exchange occurs now between the condenser and one desorber rather than having an absorber heat a desorber as it is the case in Figure 10.1.

When comparing the two absorption system configurations of Figure 10.1 and 10.4, one observes that on the first glance both are identical. Only upon a closer examination are there significant differences. The configuration of Figure 10.1 is a double effect system in which the concept of staging is used to increase the COP. For the configuration of Figure 10.4 the concept of staging is applied in order to obtain a high temperature lift while the COP is essentially that of a single-stage system. Thus staging of absorption cycles can be used for two, mutually exclusive purposes. Either it leads to an increased efficiency or to an increased temperature lift. Although not proven here, this statement is true in general and applies to all absorption system configurations. It is discussed in more detail in Alefeld and Radermacher (1994).

10.3   Two-Stage Triple-Effect Ammonia/Water System

The solution field of ammonia and water is relatively wide and it accommodates a cycle that has potentially very promising features. It is a two-stage triple-effect system as shown in Figure 10.5. In this figure essentially two single stage absorption cycles can be recognized, one inside a second, larger one. The smaller of the two is essentially the conventional single-stage cycle. However, its driving thermal energy is supplied by the second, larger single-stage cycle. This second cycle utilizes an evaporator that is operating at the same temperature level as that of the first. However, the absorber is moved to operate at such high temperatures that the

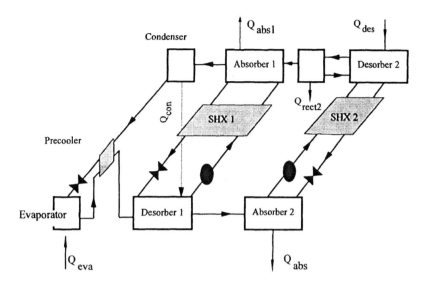

Figure 10.4 Double-lift configuration

absorber heat can be used entirely to drive the desorber of the first cycle. Furthermore, the desorber of the second cycle operates at such a high temperature and pressure level that the heat released by its condenser can be used entirely to once again drive the desorber of the first cycle. Thus the first cycle is fired by two heat sources, the absorber and the condenser of the second. Consequently, one unit of thermal energy supplied to the high temperature (components numbered 2 in Figure 10.5) stage produces a certain quantity of refrigerant that produces cooling in the evaporator. The heat of condensation and absorption of this refrigerant is also used to provide additional refrigerant, and capacity, by firing the smaller stage. Thus one unit of input energy is used three times to produce cooling capacity, which is the criterion for a triple-effect system. This concept was described first by Alefeld (Alefeld, 1983) and reduced to practice using a water/LiBr system as the low temperature single-stage system and a water/zeolite system as the high temperature cycle. A similar system was later patented by DeVault (1990).

Recent work (Ivester, 1994) demonstrated that the configuration of Figure 10.5 can be operated using ammonia/water as the working fluid in both stages for air-conditioning applications. However, the solution field is very tight. All approach temperatures in the internal heat exchangers have to be very small (less than 5 K) and the system should be water cooled to reduce the absorber and condenser temperatures (and with that the desorber temperature) of the low temperature cycle as much as possible. Likewise the evaporator temperature should be as high as possible.

Further, in order to achieve good overall performance, it is important that all major options of internal heat exchange are exploited. This is shown in the next example in detail.

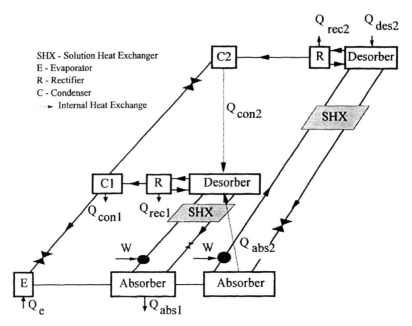

Figure 10.5 Two-stage triple-effect ammonia/water system

Example 10.2 Two-Stage Triple-Effect Ammonia/Water System

Evaluate the performance of a triple-effect two-stage absorption air conditioner according to Figure 10.6. The heats rejected by both the high temperature absorber (absorber 2) and condenser (condenser 2) are used entirely to drive the low temperature desorber (desorber 1), thus generating additional vapor. For this purpose, the lowest temperatures of the high temperature condenser and absorber (state points 12 and 20, respectively) are set 1 K higher than the lowest temperature of the low temperature desorber. The highest evaporator temperature is 14°C and the lowest temperatures in the low temperature condenser and absorber (state points 2 and 10) are 35°C. The mass fraction of the vapor leaving both the rectifiers is 0.99. The mass fraction difference in the low and high temperature solution circuits is 0.06. The mass flow rate of refrigerant leaving the low temperature desorber is 1.0 kg/sec. The quality of vapor leaving the evaporator is 0.982. The isentropic efficiency of both the solution pumps is 100%, the approach temperature in all solution heat exchangers is 3 K. It is assumed that liquid leaving the absorbers, condensers, and the desorbers is saturated. Similarly, vapor leaving the desorbers and rectifiers is saturated. File: ex10_2.ees.

Find the pressure, temperature, mass fraction, vapor quality, and enthalpy of all relevant state points. Calculate the amount of energy exchanged in all heat exchangers and the rectifiers and the work input to the pumps. Check the overall energy balance for the cycle and calculate the COP.

Solution: With the information given above and the usual assumptions that the outlet conditions for all heat exchangers with phase change are saturated, the following properties for the state points are found.

Table 10.2 State points for the two-stage triple-effect ammonia/water system (Figure 10.6)

	h (J/g)	m (kg/s)	P (kPa)	Quality	T (°C)	x (kg/kg)
1	1273.9	1.000	4951	0.982	14.0	0.9950
2	-81.4	8.755	4951	0.0	35.0	0.5297
3	-80.4	8.755	13440	n/a	35.1	0.5297
4	93.2	8.755	13440	0.006	71.3	0.5297
5	127.4	7.755	13440	0.0	81.1	0.4697
6	-63.6	7.755	13440	n/a	39.1	0.4697
7	-63.6	7.755	4951	n/a	39.2	0.4697
8	1407.3	1.008	13440	1.0	70.9	0.9914
9	1379.0	1.000	13440	1.0	63.7	0.9950
10	162.8	1.476	13440	0.0	35.0	0.9950
11	365.4	0.476	13440	0.18	35.1	0.9950
12	127.6	5.914	4951	0.0	74.3	0.3101
13	131.3	5.914	36540	n/a	74.7	0.3101
14	582.8	5.914	36540	0.007	168.6	0.3101
15	655.5	5.437	36540	0.0	181.9	0.2501
16	233.2	5.437	36540	n/a	92.0	0.2501
17	233.2	5.437	4951	0.01	88.8	0.2501
18	1762.4	0.626	36540	1.0	167.7	0.8316
19	1350.1	0.476	36540	1.0	91.3	0.9950
20	365.4	0.476	36540	0.0	74.3	0.9950
21	83.6	0.008	13440	0.0	70.9	0.5297
22	-76.0	8.755	13440	n/a	36.1	0.5297
23	568.9	0.149	36540	0.0	167.7	0.3101
24	194.6	5.914	36540	n/a	89.0	0.3101
25	162.8	1.476	4951	0.119	4.1	0.9950
26	15.0	0.476	4951	0.982	4.0	0.9950

COP $= 1.446$	$Q_{con2} = 469. \text{kW}$	$Q_{abs1} = 1494. \text{kW}$	$Q_{rec2} = 374. \text{kW}$
$Q_{eva} = 1640. \text{kW}$	$Q_{des1} = 1590. \text{kW}$	$Q_{abs2} = 1120. \text{kW}$	$Q_{shx1} = 1481. \text{kW}$
$Q_{con1} = 1313. \text{kW}$	$Q_{des2} = 1135. \text{kW}$	$Q_{rec1} = 38. \text{kW}$	$Q_{shx2} = 2296. \text{kW}$
$W_{p1} = 9.2 \text{kW}$	$W_{p2} = 21.9 \text{kW}$		

For the given conditions, an evaporator capacity of $Q_{eva} = 1640$ kW is found. The high temperature desorber has a heat requirement of $Q_{des2} = 1135$ kW. This heat requirement is high but a consequence of the high water content in the vapor (the ammonia mass fraction is 0.851) leaving desorber 2. Accordingly, the rectifier load is relatively large as well, $Q_{rec2} = 374$ kW. The condenser heat and the absorber heat available from the high temperature stage are $Q_{con2} = 469$ kW and $Q_{abs2} = 1120$ kW, respectively. This amounts to $Q_{des1} = 1590$ kW heat input for the low temperature desorber. It is interesting to note that the high temperature desorber contributes about 32 % of the total refrigerant flow rate while the low temperature desorber contributes 68%, although the heat input to the latter is only 70% of that of desorber 2. The reason is the water content of the vapor. For desorber 1 the ammonia mass fraction is 0.991 and consequently the rectifier heat of desorber 1 is only $Q_{rec1} = 38$ kW. The coefficient of performance amounts to COP = 1.41.

The flow rate of rich solution in the low temperature solution circuit is 8.76 kg/sec for 1 kg/sec of vapor generated in the desorber. In the high temperature solution circuit the flow rate is 12.41 kg/sec. The heat duties in the respective solution heat exchangers are 1481 kW for low

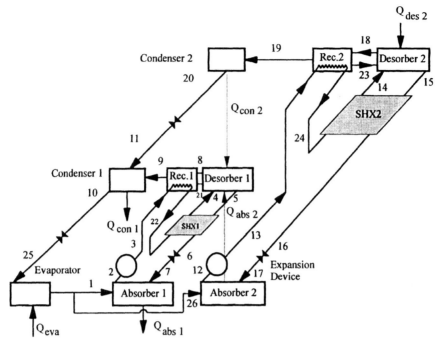

Figure 10.6 Triple-effect two-stage absorption air conditioner

temperature circuit and 4819 kW for the high temperature circuit. The latter has to be larger because of the considerably larger temperature difference between absorber and desorber.

Last, the condenser heat load of the low temperature stage is 1313 kW and the pump power amounts to 9.2 and 21.9 kW for the low and high temperature solution circuit, respectively.

The system has a promising performance, but the overall cycle design is quite complex. As this example shows, the internal heat exchange has to be used quite effectively to reach a COP of 1.41. There are two additional options for internal heat exchange that are not employed in this example and left to the reader to explore. The liquid leaving the high temperature condenser, can be subcooled by preheating the vapor entering the high temperature absorber. Since this measure increases the absorber capacity, it contributes directly to a higher heat input for the low temperature desorber. The COP will increase. The other option is to employ the condensate precooler for the condensate leaving the low temperature condenser. Further, if the approach temperatures are reduced, the performance increases considerably and in the limit of an approach temperature of 0 K the COP reaches about 1.6. This fact emphasizes once again that good internal heat exchange is very important.

Using the same working fluid for both stages has the advantage that in terms of the number of construction materials that are used the system is simpler. Further, one evaporator can serve both stages. However, there are certain performance advantages that can be expected when a cascade is used. For example, when the lower stage is employing water/lithium bromide, then the efficiency of this cycle is higher than that of ammonia/water. For the latter a COP of 0.5 is expected for the lower stage, while that of water/lithium bromide typically reaches 0.7 as discussed previously. The higher temperature stage would still employ ammonia/water. Nevertheless, based on Example 10.2, the lower temperature stage while contributing 62% of the cooling load could, when water/LiBr is used, contribute 40% more capacity increasing the overall COP by 25%, approaching an overall COP of 1.7. It is left to the reader to simulate this cascade system and to confirm the estimate made here.

HOMEWORK PROBLEMS

10.1   Using the pressure-temperature diagram of Figure 3.19, lay out the operating temperatures and pressures for double-effect cycle as shown in Figure 7.1, using ammonia/water as the working fluid. What are the pressure levels you expect for the high temperature desorber? Estimate by what factor the pump power is increased as compared to water/LiBr.

10.2   Develop a model of the cycle of Problem 10.1 and compare the results with the results in Table 7.4

10.3   Based on example 10.1 , vary the mass fraction difference for both solution circuits with the goal to find the optimum COP.

10.4   Consider the one-half effect cycle of Figure 8.1. This cycle can also be operated with an ammonia/water mixture. It is often used when low temperature waste heat is available and refrigeration is needed. Assuming that waste heat will allow for a highest temperature in the desorber of 105°C and that a lowest heat rejection temperature of 35°C in the absorber and condenser is available then:
a) What is the lowest temperature that can be achieved in the evaporator?
b) Write a cycle program for this cycle. Assume typical values for rectifier vapor outlet mass fractions and other heat exchanger outlet conditions.
d) What is the COP for this operating condition?
e) Optimize the mass fractions of the two solution circuits for best COP.

10.5   Using the computer model for Example 10.2, investigate the performance of the triple-effect system in dependence on the concentration difference in the solution circuits.

Chapter 11

GENERATOR/ABSORBER HEAT EXCHANGE (GAX) CYCLES

11.1   Concepts, Configurations and Design Considerations

The GAX cycle is a very elegant way of achieving higher effect performance with a cycle configuration that essentially appears to be a single-stage configuration. Because of its elegance and potential to be implemented in machinery, it is discussed here in detail. The term GAX stands for generator/absorber heat exchange. In the absorption community, the term generator is frequently used for the desorber and therefore the abbreviation GAX is found in the literature quite often. Although desorber/absorber heat exchange is used in a variety of cycles, the term GAX refers to a special case which is the subject of this chapter.

To derive the GAX cycle, we begin with Figure 11.1. It shows one possible configuration of a two-stage ammonia/water system. One of the design variables is the flow rate established by the solution pumps. Let us assume now a designer will reduce that flow rate for both solution pumps while all other parameters are adjusted such that the system capacity remains unchanged. Thus the desorbers have to still produce the same amount of ammonia vapor, although there is less solution supplied to the desorbers. Likewise, the absorbers have to absorb the same amount of vapor as before but with less solution available. The mass balances for both the absorbers and the desorbers can only be satisfied when the change in ammonia mass fraction increases. This requires at the same time that the temperature glide in these heat exchangers increases as well. This is shown schematically in Figure 11.1. The solid lines symbolize the original two-stage cycle, while the dashed lines symbolize the cycle after the solution pumps produce lower flow rates. When we further reduce the pump flow rates the two dashed lines will eventually coincide. This means the mass fractions in both lines are identical.

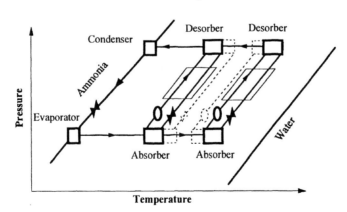

Figure 11.1 GAX concept: This figure shows the schematic of a two-stage cycle; the GAX cycle is obtained when the flow rates of both solution pumps are reduced; the solution circuits approach each other as shown by the dashed lines; when they overlap, they cancel each other.

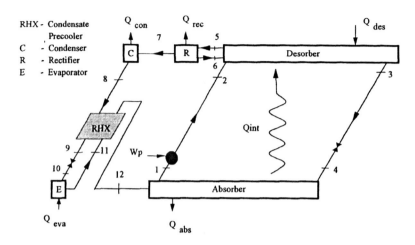

Figure 11.2 GAX cycle

Further, it is observed that for the overlapping lines, the directions of mass flow have opposite orientations. If the flow rate is the same in both pipes, then they cancel each other and they can be omitted altogether. This configuration is shown in Figure 11.2. All major components of the two original solution circuits are still present with some important changes. The original two desorbers are merged into one, as are the absorbers. One pump and one expansion valve canceled each other and are omitted. The solution heat exchangers which are required for each solution circuit are not required anymore. Although there is still a rich and poor solution stream circulating between the desorber and the absorber, the temperature differences between the two remaining solution streams are very large and the respective heat capacities are accommodated better by other means as discussed later. The configuration of Figure 11.2 is termed the generator/absorber heat exchange (GAX) cycle. It was first described by Altenkirch and patented in 1914 (Altenkirch and Tenckhoff, 1914).

The GAX cycle has very interesting characteristics. For example, it can be considered a single-stage cycle. After all, there is only one solution pump and all major components that occur once in a single-stage cycle, such as absorber, desorber, evaporator and condenser, are present only once. Just the solution heat exchanger is now replaced by the GAX heat exchanger. An important implication of this similarity is that the first cost of the GAX cycle should be comparable to that of a single-stage system.

On the other hand, the GAX cycle can be considered a two-stage or double-effect system because there is definitely an absorber heating a desorber. A certain quantity of driving heat is reused to generate additional cooling capacity with potentially significant increases in efficiency. This thought can be taken a step further. The GAX cycle could be imagined as originating from a three-stage system. This system would be based on the configuration of Figure 11.1; however, theoretically, a third solution circuit could have been added to the right (to higher temperatures) of the original system as shown in Figure 11.3. In this three-stage system, the existing high temperature desorber would then be heated by the new, even higher temperature absorber. Correspondingly, the new desorber would receive heat from the outside source. After undergoing a merging process similar to that described in Figure 11.1, the resulting cycle would be the GAX cycle of Figure 11.2. The point is that the GAX cycle is a hybrid cycle which can exhibit two or three stage features. Since the classification of the GAX cycle is not clear and will probably be debated forever, it may be appropriate to suggest that

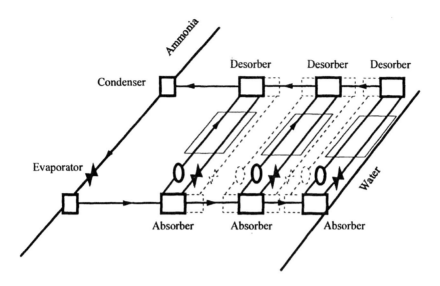

Figure 11.3 GAX origins

the GAX cycle is a higher effect configuration (because of its potential for high COPs) with single-stage features (because of the apparent simplicity). Inspired by this line of thought, there are several development projects underway with the goal to develop a commercial, mass-produced version of the GAX system. One custom built system is in operation and is briefly described at the end of this chapter.

It was mentioned above that the heat capacity of the remaining solution streams can be accommodated in other ways than using a solution heat exchanger. First, the poor solution stream, state point 3, Figure 11.2, is considered. The outlet for the poor solution from the desorber is at the highest temperature that occurs in the entire cycle. From a thermodynamic point of view, the energy contained in the poor solution is best used at as high a temperature level as possible. Thus this solution should reject its heat in a solution recirculation configuration as shown in Figure 9.7 to the high temperature portion of the desorber. As an alternative, the energy can be carried into the high temperature end of the absorber and then be transferred into the desorber at the lower desorber temperature range. A detailed analysis of the GAX system reveals that both methods are equivalent from an energy point of view. This means the desorber heat requirement does not change whether or not the solution recirculation is employed. What does change, however, is the temperature level at state point 3. The temperature drops when solution recirculation is used, which can be an important design benefit. Solution recirculation can also be employed in the absorber as shown in Figure 11.4. Here, the cold, rich solution at state point 17 is recirculated through the absorber and preheated essentially to the desired desorber inlet temperature. Thus the amount of heat rejected by the absorber to the outside is reduced as is the desorber heat requirement.

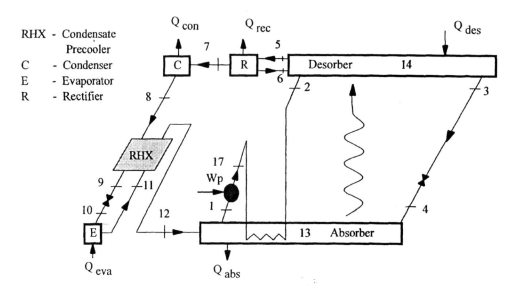

Figure 11.4 GAX cycle with absorber solution recirculation

Example 11.1 GAX Cycle

Evaluate the conventional GAX cycle as illustrated in Figure 11.4. The working fluid pair is ammonia-water. The highest evaporator temperature is 5°C while the lowest absorber and condenser temperature is 40°C. The vapor quality of the solution leaving the evaporator is 0.94. Saturated liquid is assumed to leave the absorber (state point 1), the generator (state point 3), and the condenser (state point 8). Saturated vapor leaves the generator (state point 5) and the rectifier (state point 7). The rectifier is analyzed as a reversible device that produces a vapor mass fraction of 0.995. The mass fraction difference between the rich and the poor solutions in the solution circuit is 0.35. The mass flow rate through the solution pump (m_1) is 1.0 kg/sec. The pump efficiency is 50% and the condensate precooler has an effectiveness of 0.8. Before entering the desorber, the rich solution is preheated (absorber solution recirculation) in the low temperature part of the absorber (i.e., that part of the absorber which does not heat the desorber) with an approach temperature of 0 K. Furthermore, the approach temperature at either end of the GAX heat exchanger is set to 0 K as well. As a result, the rich solution enters the desorber saturated. File: ex11_1.ees.

1) Find the pressure, temperature, enthalpy, vapor quality and the concentration for each relevant state point.

2) What are the rates of heat exchange in the evaporator, the condenser, the absorber, the generator, the rectifier, and between the absorber and the generator?

3) Check the overall system energy balance and calculate the COP of the system.

Solution:

Using mass and energy balances on the GAX cycle of Figure 11.4 and the given input values, the following results are obtained for the individual state points. State point 14 in Figure 11.4 represents that state in the desorber at which the temperature is equal to the highest absorber temperature. Thus between states 2 and 14, the desorber can accept heat from the absorber.

State point 13 represents that point in the absorber at which its temperature is equal to the low temperature end of the desorber. Thus between points 13 and 4 the absorber can supply heat to the desorber. When all approach temperatures are zero, the temperatures of points 13 and 2 and 14 and 4 are the same.

Table 11.1 State points for the ammonia/water system according to Figure 11.4

	h (J/g)	m (kg/sec)	P (kPa)	Quality	T (°C)	x (kg/kg)
1	-60.6	1.0	478.4	0.0	40.0	0.489
2	139.8	1.0	1548.0	n/a	83.7	0.489
3	610.1	0.591	1548.0	0.0	163.3	0.139
4	610.1	0.591	478.4	0.095	123.7	0.139
5	1440.4	0.418	1548.0	1.0	83.7	0.984
6	140.0	0.010	1548.0	0.0	83.7	0.489
7	1365.4	0.409	1548.0	1.0	67.1	0.995
8	187.6	0.409	1548.0	0.0	40.0	0.995
9	53.6	0.409	1548.0	n/a	12.2	0.995
10	53.6	0.409	478.4	0.009	3.1	0.995
11	1198.0	0.409	478.4	0.940	5.0	0.995
12	1332.1	0.409	478.4	0.999	30.1	0.995
13			478.4		83.7	
14			1548.0		123.7	
17	-58.0		1548.0	n/a	40.4	0.489

COP = 1.110	Q_{des} = 422. kW	Q_{abs} = 369. kW	Q_{rec} = 42. kW
Q_{eva} = 468. kW	Q_{con} = 482. kW	W_p = 2.6 kW	Q_{rhx} = 55. kW
Q_{gax} = 400. kW	$Q_{sol-rec}$ = 198. kW		

It should be noted that the GAX cycle has only 16 state points. This is an indication of the apparent simplicity of the concept. State points 15 and 16 in the above table are omitted but will be used for other cycle configurations in Example 11.2. The total amount of heat released from the absorber amounts to Q_{abstot} = 768 kW with Q_{abs} = 369 kW being rejected as waste heat. The remainder, $Q_{available}$ = 400 kW is being transferred to the desorber. The total desorber heat requirement is Q_{destot} = 821 kW with 400 kW supplied by the absorber. When the amount of heat that the desorber could accept in the temperature range that is covered by the absorber is calculated, then this value is found to be 468 kW. It is about 68 kW higher than what the absorber actually supplies. The remaining 421 kW (including the 68 kW missing from the absorber) for the desorber have to be supplied from an outside heat source. This makes very clear that 49% of the desorber heat requirement is contributed by the absorber. The evaporator capacity is Q_{eva} = 468 kW. This amounts to a COP of 1.11. This value is higher than what is typically expected in actual machinery due to a number of idealizations in the calculations that may be not very well matched in an actual machine (this is addressed in the next example). Nevertheless, these idealizations are important to reduce this example's complexity initially. It is noted that the COP of a GAX cycle is a strong function of the temperature lift. The rectifier heat amounts to Q_{rec} = 42 kW and the condenser heat to Q_{con} = 482 kW. The amount of heat exchanged in the condensate precooler is Q_{rhx} = 55 kW. The pump work is 2.6 kW.

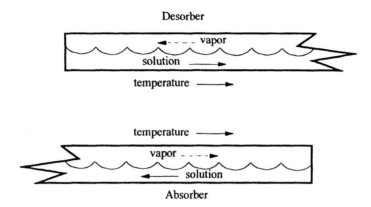

Figure 11.5 Heat and mass transfer in the desorber/absorber heat exchanger

The GAX cycle evaluation of Example 11.1 is based on a number of assumptions for the sake of simplicity. These are addressed now. First, it is assumed that the approach temperatures for the solution recirculation and the desorber/absorber heat exchange are negligible. This is not realistic. Furthermore, the heat and mass exchange processes in the desorber/absorber heat exchange are quite complex. This is exemplified in Figure 11.5 which shows, enlarged from Figure 11.2, the desorber and absorber portions that are meant to exchange heat. The wavy lines symbolize the liquid/vapor interface in each of the heat exchangers. The temperature increases for each from the left to the right as indicated by the arrows. The solution streams are in counter-current flow to each other and so are the vapor streams. The vapor streams are also in counter-current flow with respect to the solution in the same heat exchanger. This is the optimum arrangement. From a heat transfer point of view, the solution streams that release or require the heat that is being transferred should be counter-current so that we have a counter current flow heat exchanger. At the same time, each of the vapor streams should be counter-current to the respective solution stream in order to minimize entropy production associated with the coupled heat and mass transfer between vapor and liquid streams. These issues are discussed in (Bassols, 1988; Kang and Christensen, 1994).

To fulfill these requirements with an actual heat exchanger represents a considerable challenge. For absorption and desorption processes, a falling film heat exchanger design would be ideal. In this case it may even be possible to maintain the vapor/liquid counterflow relationship. However, for the GAX heat exchanger, this would mean a parallel flow heat exchanger from a temperature point of view, which is not ideal. One possible solution is to employ a hydronic heat transfer loop (i.e., a water circulation loop) between the absorber and desorber as is indicated in Figure 11.6. The disadvantage of such a loop is that two temperature differences are introduced for heat transfer and the pump requires additional power. Example11.2 below demonstrates that the additional temperature difference is a considerable penalty.

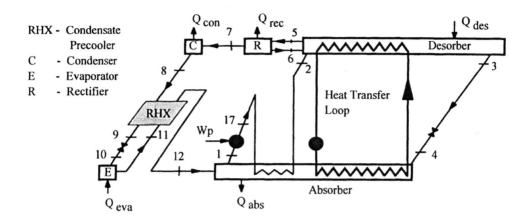

Figure 11.6 GAX cycle with hydronic heat transfer loop

Example 11.2 Effect of Approach Temperature

Based on the results of Example 11.1, vary all approach temperatures from 0 to 14 K, simultaneously for all three heat exchangers (absorber solution recirculation and both ends of the GAX heat exchanger). File: ex11_2.ees.

Solution:

Using the computer program that describes the GAX cycle of Example 11.1, the approach temperatures are varied and the program is executed for a number of different cases. The results are summarized in Figures 11.7 and 11.8. Figure 11.7 shows the COP of the GAX cycle versus the approach temperature difference. As this difference increases, the COP decreases quite dramatically. Considering that possibly a hydronic heat transfer loop may be used to transfer heat from the absorber to the desorber, an approach temperature of 14 K may not be unrealistic. However, for this value, the COP has dropped from 1.10 to about 0.75. Of course, not all approach temperatures would be the same in an actual machine but this example gives an indication of how sensitive the GAX cycle performance is with regard to the effectiveness of the internal heat transfer.

Figure 11.8 shows various heat transfer rates in the GAX cycle and how they change as the approach temperatures increase. First, observe that the evaporator capacity is constant as controlled by the inputs selected for this model. The amount of heat required by the desorber in the temperature range that overlaps with that of the absorber decreases with increasing approach temperatures. This is a consequence of the decreasing temperature interval in which heat can be exchanged. Similarly, the amount of heat available from the absorber decreases as well. Although both quantities decrease, the mismatch between the two increases as shown by Q_{diff}. Since the amount of heat available in the absorber is already small, it decreases faster when limited by temperature constraints than the amount of heat required by the desorber.

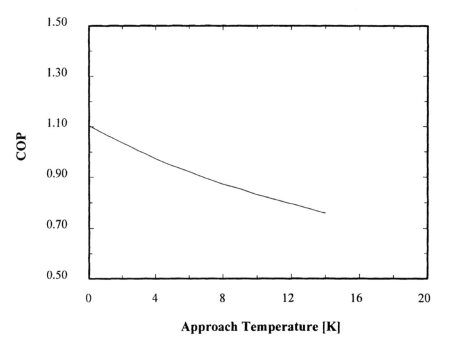

Figure 11.7 COP versus approach temperature

Since the amount of heat exchanged internally between absorber and desorber is decreasing at constant evaporator capacity, the net heat input to the desorber must increase. Likewise, the absorber heat rejection to an external sink must increase as well. These considerations explain the decrease in COP.

For the GAX cycle to work as designed, it is necessary that the temperature of the hot end of the absorber be warmer than the lowest desorber temperature. For a typical air-conditioning application this condition is fulfilled, but when the same cycle is meant to operate as a heat pump that heats a building, then the evaporator temperature varies as dictated by the outdoor temperature.

Inspection of Figure 11.9 shows that below a certain outdoor temperature the overlap of temperature ranges in the desorber and absorber ceases to exist. At this point there is no benefit from the GAX heat exchange and the cycle can be thought to conceptually degenerate to single-effect performance. For such a high-lift operating condition the GAX heat exchanger must be replaced by a solution heat exchanger to achieve normal single-effect performance. Conceptually the GAX cycle has the feature that it converts into a single-effect cycle when the temperature lift (temperature difference between evaporator and condenser temperature) exceeds a certain maximum.

It should be noted that in Figure 11.9 the lowest absorber temperature was allowed to drop below that of the condenser. In an actual application that would not be acceptable. The lowest absorber temperature should stay constant in order to deliver heat at a useful temperature level. Therefore, a more realistic representation is shown in Figure 11.10. It can be seen that as the evaporator temperature drops the lowest desorber temperature actually increases. This is forced by the decreasing mass fraction of the rich solution.

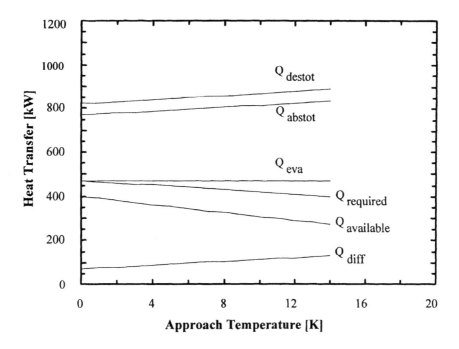

Figure 11.8 Results for Example 11.2

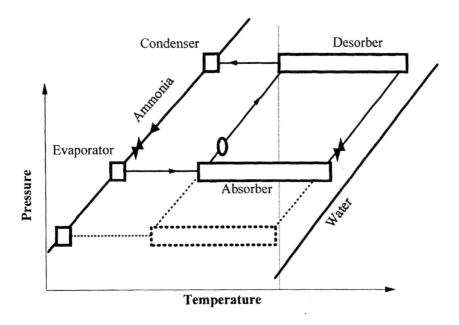

Figure 11.9 Transition from GAX to single-stage cycle

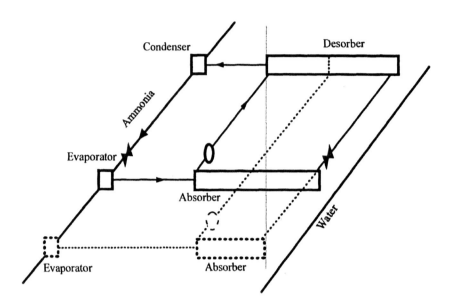

Figure 11.10 Transition from GAX to single-stage cycle at constant lowest absorber temperature

11.2 Branched GAX Cycle

In Section 11.1, it was mentioned that the heat requirement of the low temperature end of the desorber is not fully satisfied by the absorber. Under most operating conditions, the amount of heat provided by the high temperature end of the absorber is less than the heat requirement of the low temperature end of the desorber. Details of this situation are described in Scharfe et al. (1986). The degree to which this happens varies considerably with the actual operating conditions. This mismatch between heat requirement and heat supply is caused by the following.

Considering the pressure-temperature diagram in Figure 3.19, it can be observed that the lines of constant mass fraction lie closely together at high mass fractions. They are much more widely spaced at low mass fractions. Further the mass fraction variable is highly non-linear in the following sense: when adding 1 kg of ammonia to 9 kg of pure water, the mass fraction of the resulting solution is 0.10. It changed from 0.00 to 0.10. However, when 1 kg of pure ammonia is added to 9 kg of ammonia/water solution with a mass fraction of 0.90, the mass fraction changes only by 0.01 from 0.90 to 0.91. In the first case the saturation temperature (at constant pressure) changes quite considerably while in the second, hardly at all. Consequently, the resulting temperature glide is considerable in the first case and close to zero in the second. However, in both cases, assuming that the added ammonia is absorbed from the vapor phase into the solution, approximately the same amount of latent heat is released.

Coming back to the desorber/absorber heat exchange, the reader will notice from Figures 11.1 and 11.2 that the absorber portion that supplies heat to the desorber operates with an

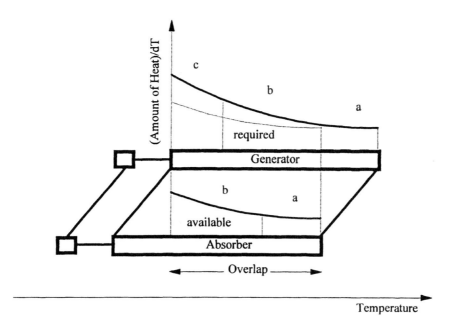

Figure 11.11 Mismatch: amount of heat available and amount of heat required

average mass fraction that is lower than the average mass fraction present in that desorber portion that receives the heat from the absorber. Thus, for the same amount of heat, the temperature glide in the absorber tends to be larger than the temperature glide in the desorber. The amount of heat the absorber can provide to the desorber is limited by the temperature interval in which heat exchange is possible. In the same temperature interval the desorber has a larger heat capacity. This situation is summarized in Figure 11.11. In this schematic of the GAX cycle the amount of heat per temperature interval dT is superimposed above the absorber (sections a and b) and above the desorber (sections a, b and c). The two curves are the same for both heat exchangers, but the one for the desorber is shifted to higher temperatures as required by the solution field. The area above the absorber and below the curved line marked a and b represents the amount of heat the absorber is able to deliver to the desorber. The absorber line is also plotted above the desorber as a dotted line. However, the equivalent sections to a and b in the absorber are moved to higher temperatures for the desorber. Because of the curvature, the amount of heat the absorber is able to supply (dashed line) is always lower than the amount of heat the desorber is able to accept (full line marked b and c).

The mismatch between absorber and desorber heat capacity for GAX heat exchange is explained here in terms of average mass fraction differences because this is the dominant contribution. Other variables play a role too, such as variations in the heat of mixing and variations in the specific heat of the solution.

Once the mechanism of the mismatch is understood, a solution can be proposed. The amount of heat provided by the absorber to the desorber can be increased by increasing the mass flow rate in this portion of the absorber only. This is accomplished with the so-called branched GAX cycle as shown in Figure 11.12. A second solution pump is used. This pump

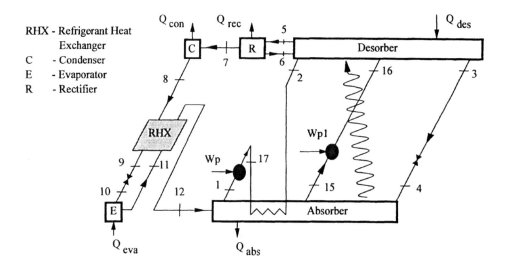

Figure 11.12 Branched GAX cycle

can be seen as leftover from the transition in Figure 11.1 that led to the GAX cycle. Thus the two solution streams that were assumed to cancel each other in Figure 11.1 do not do so entirely. It turns out that for most cases the flow rate in the liquid line with the pump should be larger than in the expansion valve leg. In this case the solution flow rate in the high temperature ends of both desorber and absorber are increased, and the high temperature end of the absorber can supply more heat to the low temperature end of the desorber. At the same time, the heat requirement of the high temperature desorber is increased, but less external heat needs to be supplied to the low temperature end of the desorber. This configuration with a second solution pump is termed the branched GAX cycle (Herold et al., 1991; Rane and Erickson, 1994). In general, there can be more than one additional solution pump used for further improvements; however, there is a diminishing return.

Example 11.3 Branched GAX Cycle

Evaluate the branched GAX cycle shown in Figure 11.12. This cycle differs from the GAX cycle in that it has an additional pump which transfers working fluid from the absorber to the desorber. This serves the function of balancing the energy mismatch between the absorber and the generator. The assumptions made in analyzing this cycle are the same as those used for the GAX cycle, Example 11.1. The additional pump draws liquid from the absorber, point 15, at a temperature equal to the saturation temperature of the rich solution entering the desorber, point 2. The flow rate is determined by the requirement that the heat needed by the desorber and the heat available from the absorber must match. The input parameters are as follows:

1) Pump isentropic efficiency is 50% and condensate precooler effectiveness is 80%.
2) $T_{11} = 5°C$; $T_8 = T_1 = 40°C$.

3) The mass fraction of vapor leaving the rectifier (x_7) is 0.995.
4) The difference between the rich and the poor solution concentrations in the solution circuit is 0.35.
5) The mass flow rate through the main pump (m_1) is 1.0 kg/sec.
6) The vapor quality of the two-phase solution leaving the evaporator (Q_{11}) is 0.94.
File: ex11_3.ees.

Solution:
The example is evaluated first for a given approach temperature of 0 K for all heat exchangers in connection with absorber and desorber. This value is then varied similarly to Example 11.2 With the information given above and the usual assumptions that the outlet conditions for all heat exchangers with phase change are saturated, the following properties for the state points are found.

Table 11.2 State points for the branched GAX cycle according to Figure 11.12 (approach temperatures 0 K)

	h (J/g)	m (kg/sec)	P (kPa)	Quality	T (°C)	x (kg/kg)
1	-60.6	1.000	478.4	0.0	40.0	0.489
2	139.8	1.000	1548.0	n/a	83.6	0.489
3	610.1	0.697	1548.0	0.0	163.3	0.139
4	610.1	0.697	478.4	0.095	123.7	0.139
5	1440.4	0.436	1548.0	1.0	83.7	0.984
6	140.0	0.010	1548.0	n/a	83.7	0.489
7	1365.4	0.427	1548.0	1.00	67.1	0.995
8	187.6	0.427	1548.0	0.0	40.0	0.995
9	51.1	0.427	1548.0	n/a	12.2	0.995
10	51.1	0.427	478.4	0.009	3.1	0.995
11	1198.0	0.427	478.4	0.94	5.0	0.995
12	1334.5	0.427	478.4	0.999	30.1	0.995
13			478.4		83.7	
14			478.4		123.7	
15	189.4	0.124	478.4	0.0	83.7	0.261
16	191.8	0.124	1548.0	n/a	84.1	0.261
17	-58.0	1.000	1548.0	n/a	40.4	0.489

COP = 1.174 Q_{des} = 417. kW Q_{abs} = 363. kW Q_{rec} = 44. kW
Q_{eva} = 489. kW Q_{con} = 503. kW W_p = 2.6 kW Q_{rhx} = 58. kW
Q_{gax} = 472. kW $Q_{sol-rec}$ = 198. kW W_{p1} = 0.3 kW

Based on the above state points and then varying the approach temperature simultaneously, the following results are obtained. First, the results are discussed for an approach temperature of 0 K. The total amount of heat released from the absorber amounts to Q_{abstot} = 835 kW with Q_{abs} = 363 kW being rejected as waste heat. The total absorber heat increased and the portion

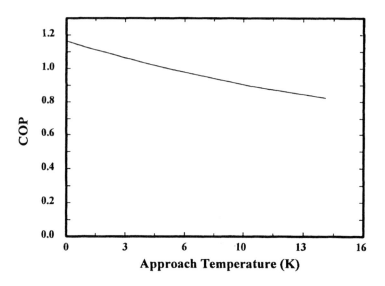

Figure 11.13 COP versus approach temperature for branched GAX cycle of Example 11.3

rejected as waste heat decreased slightly as compared to the GAX cycle. Both factors increase the amount of heat used internally ($Q_{available}$ = 472 kW is being transferred to the desorber) as compared to Example 11.1 This is about 20% more than for the GAX cycle. The total desorber heat requirement is Q_{destot} = 889 kW with 472 kW supplied by the absorber. The desorber demand increased by 68 kW, which is less than the 72 kW in additional heat supply from the absorber. Now the absorber supplies 100% of the heat required by the low temperature portion of the desorber. The desorber receives 417 kW from an outside heat source which is 5 kW less than before. Now 53% of the desorber heat requirement is supplied by the absorber, 4% more than for the GAX system. The evaporator capacity, Q_{eva} = 489 kW, increased by 21 kW. This amounts to a COP of 1.17, a 5% improvement over the GAX cycle. The rectifier heat amounts to Q_{rec} = 44 kW and the condenser heat to Q_{con} = 503 kW. The amount of heat exchanged in the condensate precooler is Q_{rhx} = 58 kW. The main solution pump work is 2.6 kW and the branch pump requires 0.35 kW which reflects the fact that the flow rate in the branch is about 25% of that in the main solution pump. Rectifier heat and precooler heat duty are increased. This is a consequence of the higher refrigerant mass flow rate which increased by about 5%.

Figure 11.13 shows the COP as it varies with the approach temperature difference. At 0 K it is 1.17 and drops to 0.85 for a 14 K approach temperature. This compares quite favorably with the values of 1.10 and 0.76 for the conventional GAX cycle. The performance improvement due to the second solution pump is in the range of 5%. Figure 11.14 shows all relevant quantities of heat as a function of the approach temperature. As for the GAX cycle, the evaporator capacity is constant. However, now the difference between the heat required and delivered is zero. As for the GAX cycle, the absorber heat and the desorber heat increase with increasing approach temperatures. Nevertheless, there is still a net improvement in COP.

The degree to which the branched GAX performance exceeds that of the GAX depends on the mismatch of the heat supplied by the absorber and that required by the desorber in the same

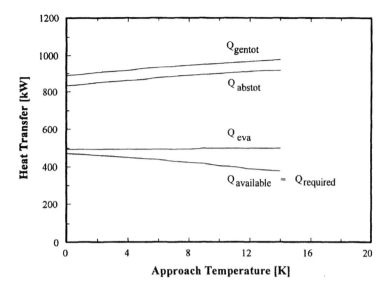

Figure 11.14 Results for Example 11.3

temperature interval. This mismatch changes with operating conditions. In Examples 11.1 and 11.3, the difference in mass fraction between the poor and rich solution was set to 0.35. When this number is varied, the heat deficit in the desorber/absorber heat exchanger varies as well. This is shown in Figure 11.15. The difference between the two curves is the deficit. It should be noted that depending on the operating and design conditions this difference will become zero and may even reverse the sign (right end of curves in Figure 11.15).

The performance of the GAX cycle and the branched GAX cycle was investigated in detail by Herold et al. (1991) and is summarized in two figures. Figure 11.16 shows the cooling COP versus the temperature lift for both cycles and for the single-stage system. As the temperature lift decreases, the COP for the single-effect cycle is essentially constant but it increases for the GAX cycle and even more so for the branched GAX cycle. It is more revealing to study the same graph for heating, Figure 11.17. As the temperature lift increases (from the right to the left) the performance of the branched GAX cycle decreases faster than that of the GAX cycle, while again the single-effect COP is essentially constant. But it does become apparent that the three curves converge on that of the single-effect system. Thus the internal heat exchange between absorber and desorber converts into a solution heat exchanger and both GAX cycles into a single-effect system.

11.3 GAX Cycle Hardware

A 250-kW, gas-driven GAX cycle is installed in a government building in Maastricht, The Netherlands (Bassols et al. 1994). The working fluid is NH_3 /H_2O. With the exception of the directly fired generator, all the heat exchangers are of the plate-fin type. The building, which has a total installed capacity of 250 kW, is located on the bank of the river Maas; the heat pump is being used simultaneously to heat the building and to cool the computer room. The cooling demand is not sufficient to cover the low temperature heat demand so additional low temperature heat is drawn directly from the river water.

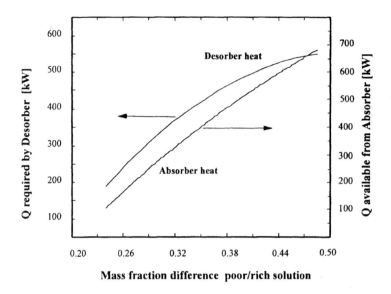

Figure 11.15 Energy mismatch between absorber and desorber

In installing the heat pump, particular attention was paid to the control of the entire heating system with the objective of keeping the return temperature as low as possible and ensuring that the heat pump can be operated continuously at maximum coefficient of performance (COP). Over the period from September 1993 to June 1994, the absorption heat pump has been in operation for 4600 hours with a seasonal average COP of 1.53. It has brought about savings equivalent to 52,000 m^3 of natural gas.

The process being used is shown schematically in Figure 11.18. The high temperature desorber (G2) is gas fired and the vapor leaves the desorber passing through the rectifier (RC) and a partial condenser (PC) to the refrigerant condenser (C). From here the condensate passes the condensate precooler (CC) which is an integral part of the evaporator and the enters the evaporator (E). The vapor passes the condensate precooler and enters the high temperature absorber together with the poor solution (A1). The high temperature absorber is integrated with the low temperature desorber (G1). The absorption process is completed in the low temperature absorber (A2). From here, the rich solution is pumped in solution recirculation through A2 and then enters the solution heat exchanger (SHE).

Optimum heat transfer from the absorber to the desorber can only be achieved by running the process in such a way that the difference between the condensation and evaporation temperatures is kept low. This makes it necessary to develop heat exchangers with a large surface area and a high heat transfer coefficient.

Plate-fin heat exchangers are used in all components except the directly fired desorber. The plate fin design allows the use of different components for each individual process (evaporation, condensation and absorption), thus enabling optimum adaptation of the process to the surface geometry. This allows optimum adjustment of an absorption heat pump over a large operating range. It makes it possible to run processes with evaporation temperatures from -25 to +20°C and condensation temperatures from 30 to 55°C. For each process, the con-

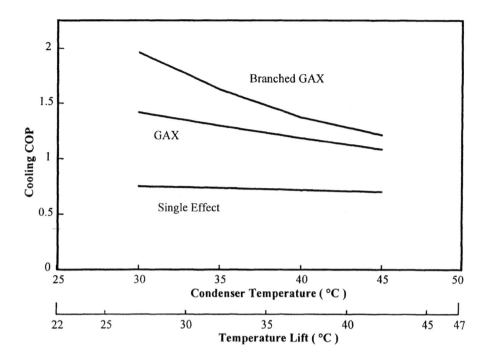

Figure 11.16 Cooling COP versus temperature lift

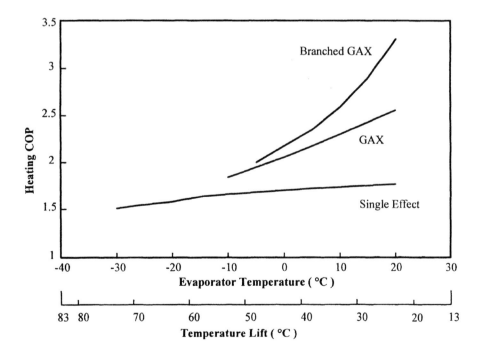

Figure 11.17 Heating COP versus temperature lift

AWP 250 ABSORPTION HEAT PUMP

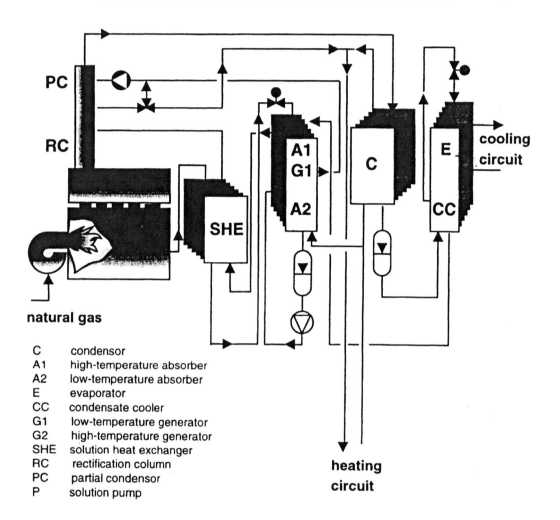

natural gas

C	condensor
A1	high-temperature absorber
A2	low-temperature absorber
E	evaporator
CC	condensate cooler
G1	low-temperature generator
G2	high-temperature generator
SHE	solution heat exchanger
RC	rectification column
PC	partial condensor
P	solution pump

Figure 11.18 Schematic of GAX cycle

Figure 11.19 Photograph of installed GAX system

centrations of the rich and weak solutions must be adjusted to achieve maximum heat exchange between absorber and generator. The system includes a microprocessor controller, which continuously monitors and optimizes the process.

Since a small temperature lift is required during most periods when the system is in operation, it is possible to achieve a high percentage of high-efficiency operating hours. At low heat source temperatures (i.e., large temperature lift), it is possible to maintain operation with the efficiency of a single-stage process. Figure 11.19 shows a photograph of the installed system.

HOMEWORK PROBLEMS

11.1   Using the pressure-temperature-diagram of Figure 3.19, describe the operating limitations of the GAX cycle: When the absorber and condenser have to have a temperature of at least 40°C, to what temperature can the evaporator temperature drop until the GAX effect ceases to exist? What is the highest desorber temperature under these conditions?

11.2   Consider only the desorber of the GAX cycle of Figure 11.4. For the operating conditions given in Example 11.1 calculate and plot the enthalpy of the solution in the desorber versus the temperature. Discuss the results.

11.3   In Example 11.1 vary the approach temperature for the solution recirculation in the absorber so that the recirculation eventually ceases to exist. Plot the COP as a function of the approach temperature and discuss the result.

Chapter 12

DIFFUSION ABSORPTION CYCLE

12.1   Introduction

The diffusion-absorption refrigeration cycle was pioneered around 1920 by two Swedes named von Platen and Munters. The cycle is unique in that it runs without any mechanical work input. This is achieved by pumping the fluids using a bubble pump driven by heat. Another unique feature of the cycle is that it is essentially noise-free. These two characteristics provide solid niche markets for the cycle in the recreational vehicle and the hotel room refrigerator markets. The cycle has come a long way since its origin and is currently manufactured in numerous locations throughout the world including Sweden, U.S., Mexico, Argentina, China and India.

The diffusion-absorption cycle utilizes ammonia-water-hydrogen as working fluid. The roles of ammonia and water are familiar from absorption cycle experience. The hydrogen is used as a capping gas to equalize the pressure throughout the cycle to allow the low-head bubble pump to operate as the liquid circulator. In the diffusion-absorption cycle, the partial pressure of the ammonia gas varies from point to point instead of the overall system pressure. In reality, there are small variations in system pressure in the machine that are quite important for operation. The cycle utilizes a regenerative gas heat exchanger between the evaporator and absorber which is driven by gravity-induced pressure differences.

The diffusion-absorption cycle has inherent irreversibilities that are larger than those found in typical absorption cycles. In particular, there is increased mass transfer resistance on the vapor side of the processes due to the presence of the hydrogen. There is also an additional heat exchanger called the auxiliary gas heat exchanger. These factors explain why the cycle performance is relatively low and why the cycle has never emerged from the shadows to capture more than the niche markets.

A diffusion absorption refrigerator (DAR) uses a three-component working fluid consisting of the refrigerant (ammonia), the absorbent (water) and the auxiliary gas (hydrogen). The refrigerant serves as a transporting medium to carry energy from a low temperature source to a high temperature sink. The water absorbs the refrigerant at low temperature and low partial pressure and releases it at high temperature against high partial pressure. The auxiliary gas provides pressure equalization for the working fluid between the condenser and evaporator. The number of possible working fluid combinations is infinite, but in practice, the combination in the widest commercial use is ammonia-water-hydrogen. Helium can also be used as the auxiliary gas with a performance penalty.

The introduction of an auxiliary gas into an ammonia-water absorption cycle was first proposed by Geppert (1900). He suggested using air as a pressure equalizer. However, the use of air decreases the coefficient of performance (COP) of a DAR to such an extent that made it impossible for commercial use. Von Platen and Munters utilized hydrogen as the auxiliary gas but still found a low COP in early versions. In order to increase COP, von Platen and Munters (1928) suggested that there should be a gas circuit between the evaporator and absorber. This circulation reduces the diffusion resistance due to the auxiliary gas considerably. With these changes, the diffusion-absorption refrigerator was widely used for domestic refrigerator

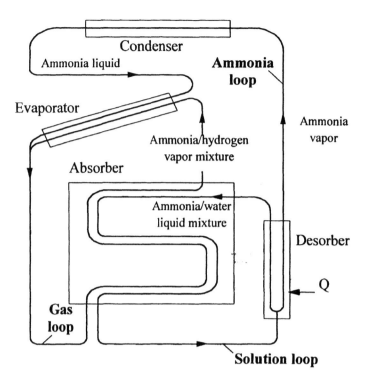

Figure 12.1 Three loops in diffusion absorption refrigerator

applications during the 1930s and 1940s, but was eventually replaced by more efficient vapor compression refrigerators.

The diffusion-absorption refrigerator cycle is related to an ammonia/water absorption heat pump cycle since it uses ammonia as the refrigerant and water as the absorbent. Watts and Gulland (1958) reviewed the patents related to this area and concluded that the development of the system was conducted by shrewd design rather than by a scientific approach. Stierlin (1971) contributed significantly to the gas circuit and made numerous contributions to the overall development. Eber (1975) discussed a compact auxiliary gas heat exchanger for the diffusion absorption refrigerator.

12.2   Cycle Physics

The DAR with ammonia, water and hydrogen as the working fluid has been used for the last 60 years. Millions of units have been put into service as refrigerators and freezers. The elements of one such system are shown in Figure 12.2 and include a desorber, a condenser, an evaporator, an absorber, an auxiliary gas heat exchanger and a rectifier. The three fluid loops discussed with reference to Figure 12.1 can also be identified on Figure 12.2. The ammonia loop includes all the components since ammonia circulates through all the components. Ammonia-water solution circulates through the solution loop (circuit), which includes the desorber, bubble pump, absorber and solution heat exchanger. The auxiliary gas circulates through the gas loop, which includes the evaporator, absorber and auxiliary gas heat exchanger. The operation of the cycle is described next starting with the absorber.

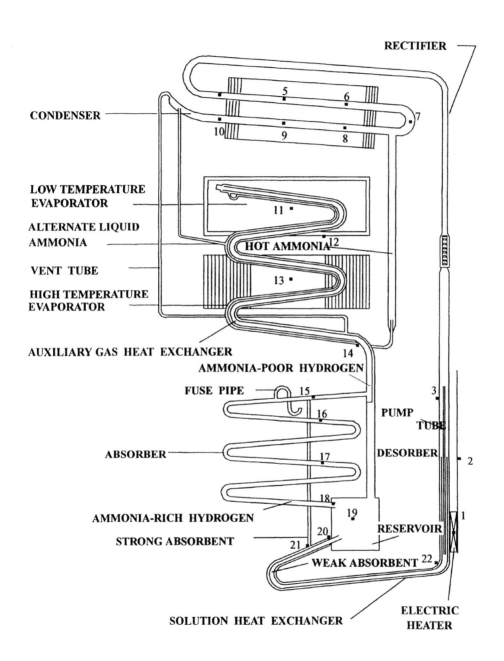

Figure 12.2 Diffusion-absorption refrigerator hardware schematic

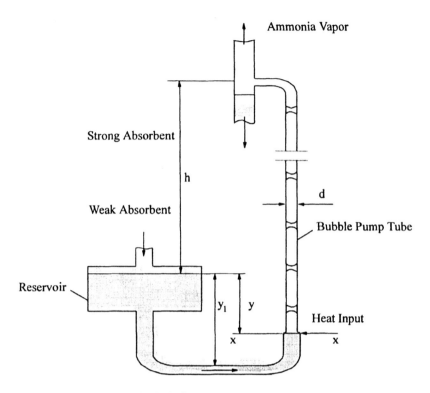

Figure 12.3 Schematic diagram of a bubble pump

The liquid reservoir at the base of the absorber contains approximately 0.35 mass fraction ammonia. This ammonia-rich liquid (weak absorbent) comes from the absorber where liquid flows down by gravity in counter-flow to the hydrogen-ammonia vapor mixture. From the absorber, the weak absorbent flows through the solution heat exchanger and arrives at the bubble pump. Due to the heat input, some ammonia vapor is driven out of the solution and forms bubbles which push liquid up in the bubble pump. The vapor passes through the rectifier where water is removed and returned to the generator. The purified ammonia vapor flows to the condenser. After the bubble pump the intermediate absorbent flows through the outer tube of the bubble pump and passes through the boiler, where more ammonia is vaporized. Finally, the strong absorbent, which typically contains 0.1 to 0.2 mass fraction ammonia as it leaves the boiler, flows through the inner tube of the solution heat exchanger.

The vapor from the desorber is at a temperature of around 200°C. The vapor contains a small quantity of water vapor which can be removed by passing the mixed vapor through the rectifier where the purification takes place accompanied by heat rejection. The small amount of water in the ammonia drains to the generator along with the reflux ammonia. The water content of the vapor leaving the rectifier is close to zero. This vapor enters the condenser, which is a finned tube with natural convection air cooling. The ammonia vapor condenses by rejecting heat to the air and the condensate flows to the evaporator.

Ammonia-poor hydrogen enters the lower section of evaporator and flows upward, in counter-flow to the downward flowing liquid ammonia. The auxiliary gas atmosphere accommodates the partial pressure of the ammonia vapor in accordance with Dalton's law. As the ammonia evaporates into the hydrogen the partial pressure of the ammonia gas rises and the evaporation temperature also rises.

The ammonia-rich gas mixture leaves the bottom of the evaporator and passes down through the gas heat exchanger to the absorber. In the absorber, the ammonia is absorbed from the gas by the liquid solution. The auxiliary gas, which is almost insoluble in the liquid, passes up from the top of the absorber, through the outer tube of auxiliary gas heat exchanger, and into the evaporator together with some residual ammonia vapor. The hydrogen and ammonia gas circulation loop is driven by natural convection caused primarily by the large density differences associated with the ammonia fraction in the vapor. The hot and cold legs of the loop are brought together in the auxiliary gas heat exchanger.

The liquid circulation in the cycle is driven by the heat-powered bubble pump shown schematically in Figure 12.3. Thus, it can be considered the heart of the system. The height difference between the solution level in the reservoir and the level at which the bubbles are formed in the pump is a governing parameter. The determination of the required height difference was investigated by Narayankhedkar and Maiya (1985). Stierlin (1967) presented a new bubble pump/generator which is claimed to significantly improve performance. By introducing the rectification heat into the liquid from the absorber before it goes into the generator, the COP of the cycle increases.

12.3   Choice of the Auxiliary Gas

The DAR technology has traditionally used hydrogen as auxiliary gas. The total amount of hydrogen in the DAR system is relatively small (~10 g) and DAR systems have an excellent safety record. However, the hydrogen does burn with a high temperature flame when the DAR is exposed to excessively high temperatures such as those encountered in a building fire. Leakage of the hydrogen and ammonia into conditioned spaces is another concern although the amounts are small enough as to not cause problems. Stierlin (1988) reported such leaks happen in the first five years at a total rate of approximately 2% of units. In order to exclude safety hazards associated with hydrogen, the hydrogen can be replaced by helium. The comparisons of hydrogen and helium as auxiliary gas have been investigated by (Stierlin and Ferguson, 1988; Kouremenos and Stegou-Sagia; 1988, Narayankhedkar and Maiya, 1985; Herold and Chen, 1994).

The results of these investigations indicate that hydrogen performs better than helium. The difference is apparently due to localized natural convection cells which form in the absorber as the ammonia is absorbed. Since the auxiliary gas remains in the vapor phase, the gas becomes less dense as absorption of ammonia occurs. This causes localized mixing cells to form which bring additional ammonia close to the interface and augment the overall absorption process. The density difference between hydrogen and helium causes a significant difference in the strength of these convective cells. These differences were found to partially explain the measured performance differences for hydrogen and helium (Herold and Chen, 1994).

Another effect which contributes is the circulation rate in the gas loop. The gas loop circulation is driven by natural convection caused by density differences due to ammonia content differences. Thermal effects, although also present, are secondary effects. As the gas passes through the evaporator and picks up ammonia, its density increases significantly. The density difference between the ammonia-poor gas rising through the hot leg and the ammonia-rich gas falling through the cold leg provides the driving force.

The gas loop provides an interesting fluid flow problem primarily from two aspects: 1) the up-flowing leg is driven by a pressure difference while the down-flowing leg is driven by gravitational forces and 2) the direction of flow of such a natural convection loop depends on small forces associated with asymmetrical design details.

12.4   Total Pressure of the System

Due to the presence of the auxiliary gas, the system pressure in a DAR is uniform throughout the machine except for small hydrostatic and viscous effects. This is in sharp contrast to a conventional absorption heat pump, where the pressure difference between the high pressure side (desorber, rectifier and condenser) and low pressure side (absorber and evaporator) is substantial. In the DAR, differences in vapor pressure of the working fluid, governed by temperature and liquid mass fraction, manifest as differences in the partial pressure in the vapor. The auxiliary gas partial pressure varies throughout the cycle in a complementary relationship to the working fluid vapor pressure so that the pressure differences in the cycle are small.

Small hydrostatic pressure differences are important in a DAR both on the vapor and liquid sides of the machine. These pressure differences drive the liquid and vapor flows which transfer energy from one component to another. On the liquid side, the bubble pump uses buoyancy-induced pressure differences to circulate liquid from the desorber to the absorber. On the vapor side, the auxiliary gas heat exchanger carries a recirculating flow of auxiliary gas which picks up ammonia in the evaporator and gives up ammonia in the absorber. The density difference between the ammonia-rich and -poor streams is the driving force for this recirculating flow.

The amount of auxiliary gas charged into the system has a direct influence on the operating conditions in the cycle. In particular, the total pressure of the system governs the temperature at which the condenser operates. When the machine starts from room temperature, the internal flow rates are zero and the working fluid exists as a two-phase mixture consisting of the three components, NH_3, H_2O and H_2. Initially, the vapor phase is largely auxiliary gas with a small partial pressure of NH_3 and almost no H_2O. As heat is applied to the bubble pump (desorber), some of the liquid vaporizes and the bubbles rise through a small diameter tube and "pump" liquid up to the top of the absorber.

Until the vapor pressure of the liquid in the desorber exceeds the total pressure in the machine, the vapor leaving the desorber must diffuse through the auxiliary gas blanket. In this regime, the condensation process is controlled by diffusion through the auxiliary gas. When the vapor pressure exceeds the total pressure, a pressure driven flow is started which sweeps the auxiliary gas out of the condenser. In this regime, the vapor is largely NH_3, with a small amount of H_2O, and the condensation coefficients are much higher than with auxiliary gas present. The actual steady-state operating regime in the condenser is some hybrid of these two regimes. The majority of the condensation process will take place in the regime where the auxiliary gas concentration is low. The auxiliary gas concentration will adjust according to the load on the system.

Therefore, the temperature of the first drop of condensate is determined by the auxiliary gas charge. If the system is charged to 25 bar, and the assumption is made that the rectifier section strips out all the water, then the first condensate temperature will correspond to the saturation temperature of pure NH_3 at 25 bar (58°C). This means that no matter what the temperature of the sink is, the first condensate will form at a temperature corresponding to the auxiliary gas charge. This operating characteristic of a DAR influences performance. If the auxiliary gas charge is large, leading to a high first condensate temperature, then the performance is degraded. However, if the auxiliary gas charge is too small, the system performance drops to zero in high ambient temperature situations. This can occur if the NH_3 saturation pressure, corresponding to the ambient temperature, is greater than the system pressure. In such an operating condition, the bubble pump fails and the system shuts down. Thus, the total pressure of the system, which is a function of auxiliary gas charge pressure, should be just sufficient to provide the required temperature range across the condenser.

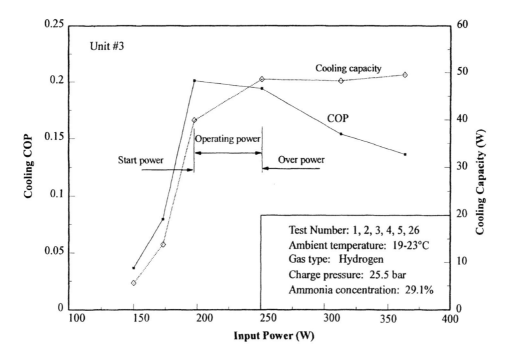

Figure 12.4 Measured DAR performance versus input power

12.5   Cycle Performance

The steady-state performance characteristics of a domestic refrigerator-sized DAR unit are shown in Figure 12.4. This unit was a production unit typical of units sold in 1994 (Norcold, 1994). The unit has an air-cooled rectifier without any attempt at integrating the rectifier heat. Tests were run at typical ambient temperatures with the load supplied by heat leak into an insulated box in a typical refrigerator configuration. No attempt was made to control the temperature of the cooled space. Instead, the unit was run continuously and the air temperature inside the unit was measured.

The COP and capacity values reported were determined based on a heat transfer model of the insulated box. The model was calibrated against a series of heat input tests where a known amount of energy was dissipated inside the insulated box and the resulting temperatures were recorded. This allowed the heat transfer characteristics of the insulated box to be determined. Of course, the direction of energy flow is different between the calibration tests and actual refrigeration operation. However, because the thermal resistance of the system is dominated by the resistance of the insulation in the wall, and because the heat transfer in the wall is largely independent of the direction of energy flow, the model was found to work dependably.

Both COP and capacity are plotted in Figure 12.4 versus the power input to the heater. The system was heated by an electric heater for these tests. Below approximately 175 W, the bubble pump did not start and the system did not run. This was due to heat loss from the bubble pump and the rectifier to the ambient. Once the start-up threshold power was exceeded, both COP and capacity rise rapidly to their maximum values.

The COP decreases as input power is increased beyond 200 W. This is due to a series of factors but it is primarily due to liquid ammonia leaving the evaporator. When input power is high, the ammonia flow rate is high. However, the refrigerator cannot utilize all of the ammonia due to the dynamics of the insulated box. Increased capacity tends to decrease the air temperature inside the box but this reduces the temperature difference between the evaporator and the air. Thus, at high power input, only a fraction of the ammonia that enters the evaporator is evaporated. The remainder flows through the gas heat exchanger back to the reservoir where it mixes back into solution to start another circuit. This liquid carryover represents a loss to both COP and capacity since it does not provide refrigeration.

The unit tested has a nominal power input of 300 W. Based on Figure 12.4, it is seen that this causes the machine to operate at full capacity but sub-optimum COP. However, the high ambient operating characteristics are such that the machine does not attain full capacity, which is needed most at high ambient, until 300 W.

Appendix A

PROPERTIES OF LiBr/H$_2$O

Table A.1 Solubility of pure LiBr in water (Boryta, 1970)

Temp. °C	S.D[a]	Solubility x %	S.D.[a]	Temp. °C	S.D.[a]	Solubility x%	S.D.[a]
-53.60	0.12	45.20	0.11	24.29	0.07	60.63	0.11
-49.32	0.09	48.03	0.07	33.14	0.05	62.50	0.08
-42.12	0.19	49.63	0.56	38.26	0.19	63.96	0.01
-36.32	0.58	50.09	0.53	44.27	0.10	65.17	0.09
-32.96	0.27	50.50	0.12	50.35	0.03	65.82	0.07
-29.17	0.20	51.20	0.47	57.58	0.02	66.16	0.04
-25.24	0.11	51.70	0.11	63.42	0.16	66.55	0.16
-16.11	0.16	51.95	0.17	70.90	0.21	67.37	0.15
-13.47	0.05	53.70	0.07	71.69	0.05	67.39	0.09
-8.94	0.04	54.75	0.18	82.68	0.11	68.32	0.11
-4.54	0.01	55.92	0.09	83.11	0.08	68.27	0.16
+1.11	0.13	56.81	0.55	91.36	0.08	68.99	0.06
5.10	0.15	57.22	0.12	91.82	0.03	69.05	0.08
9.93	0.03	58.08	0.13	102.02	0.05	70.08	0.19
18.99	0.06	58.67	0.02	101.05	0.04	70.04	0.19

[a] Standard deviation.

Table A.2 Differential diffusion coefficient of aqueous lithium bromide solutions at 26°C (Stokes, 1950)

Mass Fraction LiBr	Diffusion Coefficient (cm^2 sec.$^{-1}$ × 10^{-5})
(a)	1.379
0.4	1.300
0.9	1.279
1.7	1.285
2.6	1.296
4.3	1.328
5.9	1.360
8.3	1.404
12.0	1.473
15.4	1.542
18.7	1.597
21.8	1.650
24.9	1.693

(a) Nurnst limiting value at infinite dilution

Table A.3 Heat of formation of aqueous lithium bromide solutions (Rossini et al., 1952)

Description	x %	State	H$_f^0$ (25°C) kcal/mole
std state	100.00	aq	-95.45
in 3 H$_2$O	82.40	aq	-91.34
4 H$_2$O	81.02	aq	-92.75
5 H$_2$O	79.68	aq	-93.484
8 H$_2$O	75.92	aq	-94.296
10 H$_2$O	73.61	aq	-94.506
12 H$_2$O	71.43	aq	-94.639
15 H$_2$O	68.39	aq	-94.762
20 H$_2$O	63.86	aq	-94.887
25 H$_2$O	59.90	aq	-94.963
50 H$_2$O	45.71	aq	-95.121
100 H$_2$O	31.02	aq	-95.216
200 H$_2$O	18.88	aq	-95.276
400 H$_2$O	10.59	aq	-95.327
800 H$_2$O	5.64	aq	-95.360
1600 H$_2$O	2.91	aq	-95.384
3200 H$_2$O	1.48	aq	-95.402
6400 H$_2$O	0.74	aq	-95.414
∞ H$_2$O	0.00	aq	-95.45

Table A.4 Heats of solution and dilution for lithium bromide in water at 25°C (Lange and Schwartz, 1928; Lower, 1961; Uemura and Hasaba, 1964)

Concentration moles LiBr per 100 moles H$_2$O	x %	Q$_t$ kcal mole LiBr	Q$_d$ cal mole H$_2$O
0.07	0.33	11.652	-
0.16	0.76	11.613	0.1
0.25	1.19	11.590	0.2
0.50	2.35	11.546	0.4
1.0	4.59	11.484	1.2
2.0	8.79	11.386	3.3
3.0	12.63	11.318	6.3
4.0	16.16	11.236	10.7
5.0	19.42	11.166	17.3
7.0	25.23	11.022	35.1
10.0	32.52	10.798	79.3
12.0	36.64	10.632	121.8
15.0	41.96	10.361	213.0
17.0	45.04	10.160	309.0
20.0	49.08	9.795	508.0
25.0	54.65	9.074	1010.0
30.0	59.12	8.236	1508.0
32.0	60.67	7.904	1672.0

Table A.5 Differential heat of dilution, Q$_d$, of aqueous lithium bromide solution (Lower, 1961)

Temp °C	x % LiBr	Q$_d$ = kcal/kg H$_2$O		
		40	50	60
0		7.8	28.0	-
20		9.0	32.5	86.5
40		11.0	36.0	91.0
60		12.0	39.0	95.0
80		13.0	42.0	98.0
100		15.0	44.5	102.0
120		17.0	48.0	106.0

Table A.6 Integral heat of solution, Q$_t$, of aqueous lithium bromide solutions (Lower, 1961)

Temp. °C	x % LiBr	Q$_t$ = kcal/kg solution		
		40	50	60
0		46	53.5	-
20		48	56.0	55.5
40		49	57.0	56.5
60		51	57.8	56.7
80		52	58.5	56.8
100		53	59.5	57.0
120		54	61.5	57.7

Table A.7 Average contact angle (°) between pure lithium bromide brine and select metals (Foote)

Metal	buffed surface[a]	treated surface[b]
316 stainless steel	10.1	13.6
409 steel	17.4	11.3
steel	10.7	11.6
copper	13.2	13.0
90/10, Cu/Ni	14.3	11.1
95/5, Cu/Ni	13.7	12.2

(a) Wet polished with Scotch-brite® buffing wheel.
(b) Buffed surfaces exposed to boiling lithium bromide brine for 24 hours

Table A.8 Surface tension of lithium bromide aqueous solutions (dynes/cm) (Bogatykh and Evnovick, 1963; Uemura and Hasaba, 1972)

°C \ x%	0	5	10	15	20	25	30	35	40	45	50	55	60	65	70	75	80	85	90
5	78.19	77.82	76.38	75.42	74.37	73.62	72.83	72.05	71.22	70.45	69.60	68.90	68.15	67.38	66.58	65.75	64.92	64.30	63.60
10	79.73	78.70	77.70	76.65	75.53	74.72	73.81	73.03	72.21	71.30	70.33	69.78	68.98	68.23	67.43	66.62	65.80	65.19	64.51
15	81.13	80.10	78.92	77.82	76.73	75.80	74.83	74.08	73.22	72.23	71.18	70.56	69.90	69.12	68.30	67.48	66.68	65.48	65.33
20	82.44	81.25	80.08	78.92	77.83	76.85	75.96	75.20	74.38	73.32	72.71	71.52	70.88	70.14	69.33	68.50	67.61	66.94	66.26
25	83.79	82.61	81.35	80.22	79.00	78.12	77.15	76.32	75.50	74.52	73.50	72.72	71.90	71.13	70.27	69.40	68.54	67.84	67.16
30	85.12	83.92	82.75	81.50	80.22	79.36	78.46	77.58	76.69	75.70	74.89	74.00	73.04	72.22	71.33	70.50	69.60	68.82	68.14
35	86.92	85.72	84.50	83.18	81.83	80.93	80.04	79.15	78.23	77.42	76.45	75.48	74.51	73.60	72.72	71.75	70.92	70.85	69.40
40	88.97	87.68	86.37	85.12	83.79	82.68	81.76	80.81	79.92	78.95	78.06	77.04	76.07	75.22	74.31	73.34	72.44	71.59	70.81
45	91.32	89.84	88.53	87.20	85.86	84.78	83.83	82.79	81.92	80.85	79.88	78.82	77.87	77.02	76.03	75.10	74.08	73.22	72.34
50	93.80	92.22	90.84	89.34	88.03	86.98	85.94	84.88	83.87	82.78	81.81	80.73	79.71	78.72	77.79	76.73	75.77	74.82	73.94
55	96.25	94.72	93.21	91.75	90.23	89.20	88.02	87.02	85.93	84.78	83.78	82.64	81.58	80.60	79.58	78.50	77.42	76.44	75.57
60	98.79	97.21	95.58	94.00	92.47	91.22	90.18	89.03	88.02	86.88	85.80	84.58	83.35	82.35	81.31	80.28	79.10	78.22	77.25

Enthalpy curve fit to McNeely (1979) enthalpy data from 0 < x < 0.4 mass fraction lithium bromide

A_0	-33.1054264	B_0	1.0090734
A_1	0.13000636	B_1	-0.01377507
A_2	0.00097096	B_2	8.5131E-5

$$h = A_0 + A_1X + A_2X^2 + T(B_0 + B_1X + B_2X^2)$$

where X is in % lithium bromide, T is in °F, h is in BTU/lb

Vapor pressure curve fit to McNeely (1979) PTX data from 0 < x < 0.45 mass fraction lithium bromide

A_0	0.99996643	B_0	-0.0639021
A_1	0.00074446	B_1	0.12222883
A_2	-1.316E-5	B_2	-0.0050312
A_3	9.209E-7	B_3	0.00020138

$$A = A_0 + A_1X + A_2X^2 + A_3X^3$$
$$B = B_0 + B_1X + B_2X^2 + B_3X^3$$

$$T_{dew} = (T-B)/A$$

where X is in % lithium bromide, T is in °C. From T_{dew}, determine the vapor pressure of the solution by evaluating the vapor pressure of pure water at the temperature T_{dew}.

Appendix B

PROPERTIES OF NH₃/H₂O

VISCOSITY, cP
VISCOSITE, mPa.s

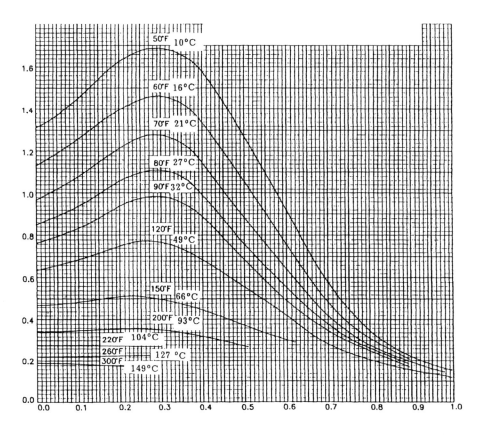

AMMONIA WEIGHT FRACTION
POURCENTAGE D' AMMONAC EN MASSE

Figure B.1 Viscosity of ammonia/water liquid. Source: Thermodynamic and Physical Properties of NH₃ - H₂O (reproduced by permission from International Institute of Refrigeration, see reference section under (IIR, 1994) for complete citation).

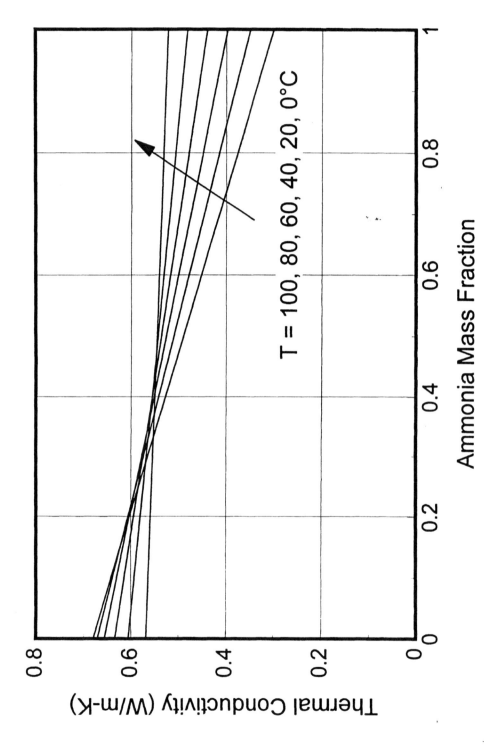

Figure B.2 Predicted thermal conductivity of ammonia/water liquid (Wang, 1992).

SURFACE TENSION, dyn / cm
TENSION SUPERFICIELLE, mN / m

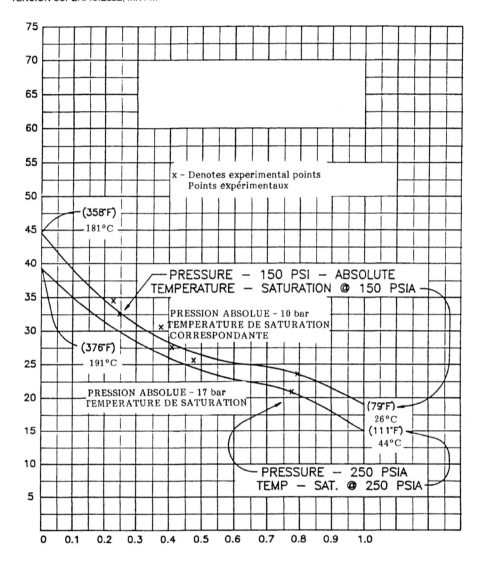

AMMONIA WEIGHT FRACTION
POURCENTAGE D' AMMONAC EN MASSE

Figure B.3 Surface tension of ammonia/water liquid. Source: Thermodynamic and Physical Properties of NH$_3$ - H$_2$O (reproduced by permission from International Institute of Refrigeration, see reference section under (IIR, 1994) for complete citation).

Appendix C

ABSORPTION CYCLE MODELING

C.1 Introduction

The discussion of modeling in this appendix is based on a generic, water/lithium bromide, single-effect cycle without advanced features. This allows focus on the aspects unique to modeling. It is left to the reader to put these modeling concepts together with the advanced component discussions in the main text, to model advanced cycles. Several examples of complete cycle models are included on the companion disk. Included in those examples is the cycle discussed in this appendix.

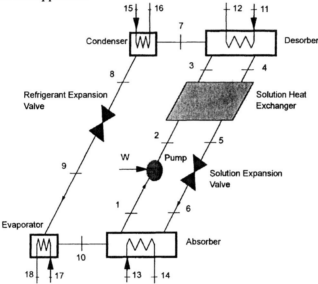

Figure C.1 Schematic of single-effect cycle with external heat transfer processes

The models of interest are steady-state system models. The objective is to create a model that is a useful representation of a real absorption cycle. Thus, the inputs and outputs are chosen to obtain as much connection to real practice as possible. A schematic of the generic single-effect cycle is included as Figure C.1. The inputs and outputs associated with the model are indicated in Table C.1. In the following sections, this cycle is examined from various perspectives with the goal of assembling the overall system model. For simplicity in the presentation, it is useful to concentrate on the physics of the cycle itself and to postpone consideration of the external heat transfer processes until after a full understanding of the cycle is obtained. This approach requires a cycle schematic as shown in Figure C.2. Both of the schematics in Figures C.1 and C.2 represent the same cycle but in the case of Figure C.2, the heat transfer loops are replaced by heat transfer rates (Q's). This simplified view avoids some clutter in the analysis. The entire material on cycle modeling is based on Figure C.2 except in Section C.5 which emphasizes the external heat transfer.

Table C.1 Model inputs and outputs

Inputs	Outputs
Solution pump flow rate	Refrigerant and solution loop flow rates
External heat transfer fluid flow rates	Thermodynamic state points
External heat transfer fluid inlet temperatures	Heat transfer rates
Heat exchanger sizes	System performance

C.2 Mass Balance Considerations

Mass balances can be routinely written on each of the components in the absorption cycle in Figure C.2, as shown in Table C.2, Equations C.2 to C.9. However, a complete understanding of the unique mass balance issues associated with absorption cycles is critical if one is to be successful in modeling new cycles. Because the working fluids are mixtures, it is necessary to obtain a mass balance for each of the mixture species. Once the species mass balances are satisfied, the overall mass balance is satisfied by definition. Thus, for each component in the cycle, one expects to write a number of mass balances equal to the number of species in the working fluid. For binary mixtures such as water/lithium bromide and ammonia/water, this means two mass balances for each component. The analyst can write either

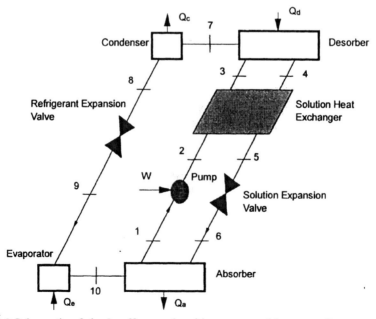

Figure C.2 Schematic of single-effect cycle without external heat transfer processes

mass balances for each of the species or the overall mass balance plus one species mass balance. It is important to avoid including a redundant third mass balance since such an equation does not contribute any additional information and, if included, tends to lead to a discontinuity during attempts at solution of the set of equations.

Redundancy in mass balances can also be easily introduced when one is dealing with cycles because of the closed loop nature of the flow systems. This is seen by examining some specific cases. Assuming a binary working fluid, there are two variables for each stream. These are the mass flow rate and the mass fraction (i.e., the composition). From a mass balance perspective, the specification of these two variables completely defines the streams. For the purpose of the mass balance discussion in this section the mass fractions of the streams are viewed as being defined by other constraints, namely, the other equations in the model such as the energy balances and heat transfer equations. From this perspective, then, the mass fractions of all the streams are known. Furthermore, it is assumed that the solution pump mass flow rate is known. This corresponds to turning on a pump with a known flow rate characteristic. If it is assumed that the flow rate in stream 1 is known, then there are nine unknown flow rates to be determined in Equations C.2 to C.9.

For the components that involve only a single inlet and single outlet, the mass balance considerations are largely trivial. These include the pump, both expansion valves, the condenser and evaporator. Furthermore, the solution heat exchanger also falls into this category because the solution streams on the hot and cold sides do not mix. For all of these components, the overall composition of the working fluid does not change and the mass flow rate in equals the mass flow rate out.

The redundancy in Equations C.2 to C.9 can be demonstrated by combining Equations C.1 to C.3 and C.6 to C.8 with Equation C.4 in such a way as to eliminate the mass flow rates m_1, m_2, m_5, m_6, m_8, m_9, and m_{10}. The resulting equations are identical to Equations C.5. A similar redundancy occurs in all cycles because of recirculating flows. The redundancy becomes more clear if all the trivial mass balances are removed from the system of equations. This can be done by ignoring the components with a single inlet and single outlet. With this in mind, the single-effect cycle mass flows can be represented as in Figure C.3. There are only two components in the single-effect cycle that have non-trivial mass balances. It is left to the reader to show that the mass balances on the two components in Figure C.3 are identical. Although this may seem obvious for the simple systems shown in Figures C.2 and C.3, more complex systems can introduce significant confusion as to which equations should be included in a well-posed model. By breaking the problem down and analyzing the mass balance equations separately, some confusion can be avoided. The bottom line conclusion is that there are only two independent mass balance equations in a single-effect, $H_2O/LiBr$ absorption machine.

A slightly more complicated flow scheme is illustrated in Figure C.4. This flow scheme is similar to that found in a parallel-flow double-effect absorption cycle. In this case there are seven flow streams. It is assumed that the mass flow rates of streams 1 and 3 are fixed by their respective pumps. Thus, there are five unknown flow rates. It can be easily shown that there are also five independent mass balance equations. The resulting set can be written as

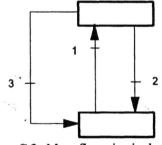

Figure C.3 Mass flows in single-effect cycle

$$m_2 + m_3 + m_6 = m_1 + m_4$$

$$m_2 x_2 + m_3 x_3 + m_6 x_6 = m_1 x_1 + m_4 x_4$$

$$m_3 = m_4 + m_5 \qquad\qquad \textbf{C.1}$$

$$m_3 x_3 = m_4 x_4 + m_5 x_5$$

$$m_5 + m_6 = m_7$$

Any other mass balances that can be written on this system can be shown to be dependent. In this case, there are five independent mass balance equations and five unknown flow rates (as before, it is assumed here that the mass fractions of all streams are known).

The number of independent mass balances can be determined for a new cycle by the following procedure. Consider only components with non-trivial mass balances. Start with cycle components that carry only refrigerant (i.e., a pure fluid) and count one mass balance for each. Then, for each cycle component carrying solution, count two mass balances. If all the streams connected to a given component have appeared in a previous mass balance, count zero for that component. The mass balances that are counted are the ones that should be included in the model, as well. This procedure yields 2 and 5, respectively, as the number of independent mass balances for the single and double effect cycles pictured in Figures C.3 and C.4.

It should be noted that although the discussion in this section is based on the assumption that the mass fractions of all streams are known, the overall result (i.e., the number of independent mass balances) is independent of that assumption. This assumption was convenient to introduce the discussion of mass balance redundancy because it allowed the mass balance considerations to be uncoupled from the remaining equations in the model. In the following sections this restriction is lifted to allow the full coupling expected in an actual absorption machine where the mass balances may influence the mass fractions and the energy balance may influence the mass flow rates.

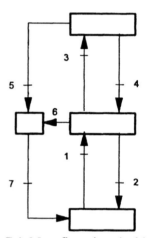

Figure C.4 Mass flows in a double-effect cycle

Example C.1 Resorption Cycle

Determine the number of independent mass balances that can be written for the resorption cycle in Figure 8.5.

Solution: All components carry solution (there are no pure refrigerant components). Also, the mass flow rate of the vapor flowing from the desorber to the absorber at point 7 must equal the mass flow rate at point 14. Thus, these two flows are counted as one. Starting with absorber 1, count two mass balances. Count zero for desorber 1 since all the flows connected to it already appeared in the absorber 1 mass balances (assuming $m_7 = m_{14}$). Then, count two mass balances for absorber 2 and zero for desorber 2. The final result is four independent mass balances for the resorption cycle.

The counting procedure introduced here for water/lithium bromide cycles needs to be modified slightly to deal with more general binary mixture working fluids such as ammonia/water. Basically, the only difference is that two mass balances are included for all components. Instead of counting only one equation for the components involving pure refrigerant, two equations are counted for all non-trivial components. This is illustrated in the following example.

Example C.2 Mass Balances in Ammonia/Water Cycles

Determine the number of independent mass balances in the ammonia/water single effect cycle pictured in Figure 9.3.

Solution: Start with the rectifier and count two. Then count two for the desorber. Since all mass flows in the system are connected to these components, count zero for the absorber. The final result is that there are four independent mass balance equations for this cycle.

Observation: The additional mass balance equations for the ammonia/water cycle, as compared to the water/lithium bromide cycle, result from the existence of the additional component (the rectifier). However, as is found for all such cycles, the solution loop introduces redundant mass balances if the modeler blindly writes mass balance equations for each component.

A simple check of the mass balance equations is to develop a preliminary model of the cycle where it is assumed that the mass fractions of all streams are known. Under this assumption, it should be possible to solve the set of mass balances. If the set is not complete, or if redundant mass balances are included, difficulties will be encountered in the solution. Once the number of independent mass balances is determined, using the methods described in this section, the next step is to write the energy balances for the model as described in the following section.

C.3 Energy Balances

For each component of the system, an energy balance can be written. In general, the energy balance accounts for the energy flowing into (or out from) the system with mass flows and any heat transfer or work interactions. The component energy balances for the cycle shown in Figure C.2 are given in Table C.2 as Equations C.10 to C.17. The energy flows are assumed positive in the direction of the arrows shown on the figure. The enthalpy values needed to evaluate these equations must be obtained from the property data sources for the working fluid of interest. The enthalpy values depend on the thermodynamic state and the state depends on the operating conditions of the adjoining state points. Thus, the system is coupled and must, in general, be solved simultaneously.

Table C.2 Mass and energy balances

Mass Balances	Energy Balances

Mass Balances

Pump

$$m_1 = m_2$$
$$x_1 = x_2 \tag{C.2}$$

Solution Heat Exchanger

$$m_2 = m_3 \qquad x_2 = x_3$$
$$m_4 = m_5 \qquad x_4 = x_5 \tag{C.3}$$

Solution Expansion Valve

$$m_5 = m_6$$
$$x_5 = x_6 \tag{C.4}$$

Absorber

$$m_{10} + m_6 = m_1$$
$$m_{10}x_{10} + m_6x_6 = m_1x_1 \tag{C.5}$$

Desorber

$$m_3 = m_4 + m_7$$
$$m_3x_3 = m_4x_4 + m_7x_7 \tag{C.6}$$

Condenser

$$m_7 = m_8$$
$$x_7 = x_8 \tag{C.7}$$

Refrigerant Expansion Valve

$$m_8 = m_9$$
$$x_8 = x_9 \tag{C.8}$$

Evaporator

$$m_9 = m_{10}$$
$$x_9 = x_{10} \tag{C.9}$$

Energy Balances

Pump

$$m_1h_1 + W = m_2h_2 \tag{C.10}$$

Solution Heat Exchanger

$$m_2h_2 + m_4h_4 = m_3h_3 + m_5h_5 \tag{C.11}$$

Solution Expansion Valve

$$h_5 = h_6 \tag{C.12}$$

Absorber

$$m_{10}h_{10} + m_6h_6 = m_1h_1 + Q_a \tag{C.13}$$

Desorber

$$m_3h_3 + Q_g = m_4h_4 + m_7h_7 \tag{C.14}$$

Condenser

$$m_7h_7 = m_8h_8 + Q_c \tag{C.15}$$

Refrigerant Expansion Valve

$$h_8 = h_9 \tag{C.16}$$

Evaporator

$$m_9h_9 + Q_e = m_{10}h_{10} \tag{C.17}$$

C.4 Heat Transfer Processes

Heat transfer processes are of fundamental importance in modeling the performance of an absorption cycle. The primary function of the cycle is to transfer heat from a low temperature to a high temperature and this function is achieved by proper choice of the size and type of heat exchanger components. The cycle modeling approach described in Section 4.10 gives the modeler a tool to examine the effect of heat exchanger size on cycle performance.

There are five heat exchangers in the cycle described in Figure C.1. These are the desorber, absorber, condenser, evaporator and solution heat exchanger. The first four exchange heat between the cycle and its surroundings while the solution heat exchanger transfers heat internally within the cycle. These components have been described in some detail in earlier chapters. With the exception of the solution heat exchanger, these components are combined heat and mass exchangers. The mass transfer and heat transfer processes are closely coupled. As discussed in Section 4.10, a model of the overall transfer process must somehow account for both the heat transfer and mass transfer.

The simplified approach described in Section 4.10 utilizes an overall transfer characteristic of the heat and mass transfer device to describe the component. The transfer characteristic is defined in terms of the temperature-driving potential but this choice is arbitrary and the model could be equally well written in terms of the mass transfer-driving potential. The overall transfer characteristic of a particular device will certainly depend on both heat transfer and mass transfer resistances but because the two processes are coupled, the model can be written in terms of either driving potential.

In general, the heat transfer models allow the heat transfer characteristic of the device to be modeled by a single equation. For example, a UA formulation for any of the heat exchangers will involve equations of the form

$$Q = UA\ \Delta T_{lm} \hspace{4cm} \text{C.18}$$

with the log mean temperature difference defined in terms of the temperature differences at the two ends of the device (as discussed in Section 4.10).

C.5 Equation and Variable Counting

The equations representing an absorption cycle include those already described in Sections C.1 to C.4 and also include assumptions about the thermodynamic state at various points within the cycle. Because of the number of equations involved, particularly for cycles with more complexity than a single-effect cycle, it is necessary for the modeler to count the equations to ensure that a well-posed problem is being solved. Equation solvers do not always give good diagnostics regarding the equation set. If a problem is under- or over-specified, the solution obtained may not be correct.

Equation counting is an art in the sense that there is not one answer to the problem. Given a particular cycle, two modelers might be expected to come up with a different count as to the number of equations needed to model the cycle. Both can be right. The choice of variables to include in the count explains the different answers. The rule which ties all such analyses together (assuming they are correct) is that the number of equations for a well-posed problem must equal the number of variables.

Consider the cycle in Figure C.1. It is perhaps best to first count the variables in the system. Based on the mass balance discussion in Section C.2, there are three mass flow rates internal to the cycle. This count ignores all devices with a single inlet and single outlet and assumes that the flow rate in equals the flow rate out of such devices.

There are ten internal state points indicated by number on the schematic. Assuming a binary fluid at each state point, it is known from the Gibbs phase rule that three properties are needed to specify the thermodynamic state at each point. However, a number of assumptions are commonly made which reduce the number of variables significantly. It is common to assume that there are only two pressures in the system. Thus, instead of ten pressures, there are only two. If the fluid pair under consideration is $H_2O/LiBr$, then the refrigerant is assumed to be pure H_2O. Thus, for the four state points in the refrigerant circuit, the mass fraction is known to be zero. Furthermore, the single inlet/single outlet components in the solution loop do not cause any mass fraction changes. Thus, there are only two independent solution mass fractions in the entire cycle. So, instead of ten mass fractions, there are only two. The pump work is also determined by the model and so it must be counted as a variable.

Associated with each of the external water loops there are three variables (inlet and outlet temperature and flow rate). The variables are summarized in Table C.3. The total number of variables for the single-effect cycle is 30. In the discussion that follows, these variables will be specified either as inputs or by the equations that represent the model. To understand the number of required inputs, the next step is to count the equations.

Some of the equations have already been discussed in Sections C.2 to C.4. From that discussion we know that there are two independent mass balances, there are eight energy balances and there are five heat transfer equations. In addition to those, an equation is usually included to compute the solution pump work based on the pressure difference and the flow rate. This is a total of 16 equations. By subtracting the number of equations from the number of variables, the number of inputs required to close the system is found to be 14.

The temperature of the vapor stream exiting the desorber (point 7) has not been specified by any of the equations in the model. Thus, the user must specify a value for this temperature to close the system. From heat transfer considerations, the vapor temperature should fall between the highest and lowest temperatures present in the desorber. As the vapor evolves from the liquid surface, it will be in local equilibrium. But as the liquid moves through the desorber, the temperature changes. Thus, the temperature of the mixed vapor stream leaving the desorber can take a range of values depending on the hardware design. From an equation counting viewpoint, the specification of the temperature at point 7 can be thought of as either an input (if a value is specified by the analyst) or as an equation if the temperature is selected based on the some other temperature in the system. Whether counted as an input or an equation, the temperature at point 7 must be specified.

It is common in a system level analysis to assume that the working fluid leaving the desorber, absorber and condenser is saturated liquid (points 4, 1 and 8) and that the fluid leaving the evaporator is saturated vapor (point 10). This amounts to specifying four vapor qualities (note that the specification of the saturated liquid states is done by specifying a vapor quality of zero). With the four vapor qualities specified and the temperature at point 7 specified, nine additional inputs are required.

A typical set of inputs corresponding closely to the physical inputs to an actual cycle, are 1) the solution pump flow rate and 2) the flow rate and inlet temperature of each external water loop. With these inputs, assumptions and the indicated set of equations, a well-posed problem is formed.

Variations on the set of variables and inputs are possible. For example, it is possible to specify the capacity of one of the heat exchangers instead of the mass flow rate of the external loop for that heat exchanger. If one wishes to include the component heat transfer rates as variables in the variable count, then the energy balance equations on those components must be reformulated in terms of the component heat transfer rates. When this is done, there are five additional energy balance equations which come along with the five heat transfer rates. The

Table C.3 Variables, equations and inputs in the single-effect model

Variables	
Internal	
Pressure	2
Mass fraction	2
Mass flow rates	3
Temperatures	10
Pump work	1
External	
Inlet temperatures	4
Outlet temperatures	4
Mass flow rates	4
Total Variables	**30**

Equations	
Mass balance	2
Energy balance	8
Heat transfer	5
Pump work	1
Total Equations	**16**

Inputs	
Mass flow rates	5
Inlet water temperatures	4
Vapor quality	4
Temperature	1
Total Inputs	**14**

conclusion is then that there are 35 variables and 21 equations which still require 14 inputs. Thus, it is seen that the number of equations and variables are not unique. This explains why the equation and variable counts coming from different absorption cycle software packages are not the same even when the cycle solved is identical. It is an important exercise for the user of these packages to count the equations independently if a full understanding of the solution is to be obtained. For those who wish to generate models for new cycles, the equation counting step represents an important discipline since it is not always obvious, during model development, why a new model is not running properly. If the modeler can convince him/herself that the correct equation set is being used, then the problem can be more easily isolated.

Equation counting can be quite confusing unless a rigid methodology is established to assist in deciding which variables to include and whether to consider certain constraints as equations or as inputs. For example, no mention has been made in this description of the property relations for the working fluid. This choice is made here because it simplifies the view of the problem and allows more insight. It is not necessary to consider the enthalpy as a variable since it is determined once the thermodynamic state is determined (as per the Gibbs phase rule). However, this formalism is just one of many possible choices in how to count equations. As mentioned already, the art of equation counting in such a system provides much room for individual expression.

Example C.3 Resorption Cycle Equation Counting

Determine the number of equations, variables, and the number of inputs required to close the system for the resorption cycle pictured in Figure 8.5.

Solution: The formalism introduced in this section results in the following tables.

Variables, equations and inputs in the resorption cycle model

Variables	
Internal	
Pressure	2
Mass fraction	4
Mass flow rates	5
Temperatures	14
Pump work	2
External	
Inlet temperatures	4
Outlet temperatures	4
Mass flow rates	4
Total Variables	**39**

Equations	
Mass balance	4
Energy balance	10
Heat transfer	6
Pump work	2
Total Equations	**22**
Inputs	
Mass flow rates	6
Inlet water temperatures	4
Vapor quality	4
Temperature	2
Mass fraction difference	1
Total Inputs	**17**

Observations: The comments here refer to each entry in the tables, in order starting from the top arranged according to the column headings.

Variables-Internal: The cycle is assumed to be a two-pressure machine. The mass fraction in each of the four solution legs is counted as a variable. The mass fractions in the vapor lines are assumed to be known as zero and are not counted as variables. There are five internal mass flow rates (assuming $m_7 = m_{14}$ is counted as only one). There are 14 internal state points and the temperature at each is counted.

Variables-External: Each of the four heat transfer loops has a flow rate and an inlet and outlet temperature associated with it. Thus, there are 12 external variables.

Equations: There are four independent mass balance equations as discussed already in Example C.1. There are ten components, each of which generates an energy balance. There are six heat exchangers, each of which generates a heat transfer equation. There are two solution pumps, each of which generates a pump equation.

Inputs: The total number of inputs required is determined by subtracting the number of equations from the number of variables. In this case, 17 inputs are required. Six mass flow rates are input including the four heat transfer loop flow rates and two solution flow rates. The inlet temperature to the four heat transfer loops are inputs. The vapor quality at the outlet of each of the absorbers and desorbers are input. The temperature of the vapor leaving the desorbers must be specified in some manner. Here, these are counted as two inputs.

The final input listed was chosen to close the system. Although such an input is not needed for a traditional single-effect cycle, it is needed for the resorption cycle as shown by this formal count. The exact nature of the additional input was chosen after the count revealed that it was needed. It is a useful exercise to try replacing this input by one of the redundant mass balance equations. The result is a system with the correct number of equations but the system is not well-posed and will not give a meaningful solution. This type of dilemma is met often when writing the model for a new cycle. The equation counting formalism is one more tool to help the modeler keep organized.

C.6   Convergence Issues and Importance of Selecting an Initial Guess

The systems of equations that result from the modeling process described in this appendix are difficult to solve because of their basic nature. They are large sets consisting of multiple equations even for the simplest cycle configuration. They are non-linear by nature due to the fact that the fluid properties are non-linear functions of the P-T-X (pressure-temperature-mass fraction) variables but also because the mass flow rates, which are variables, are multiplied by state point properties in the mass and energy balances. Furthermore, the heat transfer equations are usually non-linear. Large sets of non-linear equations always present a challenge to solve. The iterative solution methods that are available to handle such sets of equations are sensitive to several characteristics. The two main problems that crop up are 1) multiple solutions and 2) sensitivity to initial guess values.

Multiple solutions are possible in non-linear systems such as those resulting from absorption cycle models. It is sometimes possible to anticipate these and to include additional constraints on the variables that avoid convergence to an unwanted solution. One source of unwanted solutions is the set that comes about when the heat transfer potentials in a heat exchanger are inverted with respect to the actual solution. For example, in a condenser the temperatures internal to the cycle are always greater than the temperatures external to the cycle. However, if the solver does not know this, it may be possible to find a solution where the condenser becomes a heat input device. Although this does not make sense physically, it is possible unless it is strictly forbidden by imposing constraints on the variables. It is possible to avoid such unwanted solutions if the iteration is started close to the actual solution. This implies that a "good" initial guess is available.

Iterative solvers require initial guesses for each of the unknown variables. A good initial guess is simply one that allows the iteration to converge on the correct solution. It is not always easy to obtain a good initial guess. If a converged solution is available for a given model, one very useful technique is to step away from that solution in small steps using the output from one run as the initial guess for the next run. If the steps taken are sufficiently small, the model will converge at each step and by taking many small steps, a new solution can be obtained far away from the original solution. This process can be very tedious if the desired operating conditions are far away from the operating conditions of the available converged solution. In general, some variables are more sensitive than others. The character of the particular model must be determined by experimentation.

If a converged solution is not available, as is the case when a new cycle is being modeled, a manual solution of the cycle using pencil and paper is recommended. Such a solution may take a few hours to prepare but it may also be the quickest path to the result. It is not necessary to hand calculate the exact conditions of interest (although the closer the better). Once any converged solution is obtained, it is usually a simple matter to step away from it to the conditions of interest.

The suggested approach for finding a full solution (to use as an initial guess), including heat exchanger models, is to draw a reasonable solution on a Duhring plot and then to calculate the internal state points assuming variables as needed. For example, for such a calculation, it is usually necessary to assume values for the mass fractions of the solution streams and the pressures. Once a consistent set of variables is determined for the state points internal to the cycle, than the heat transfer rates can also be calculated. These can then be used to back out a reasonable set of variables for the heat transfer loops. The end result is a fully consistent solution to the set of equations. This set is then used as the initial guess to get the process started.

C.7 Equation Solvers

The set of equations resulting from the system level modeling approach described in this appendix are non-linear, coupled, algebraic equations. They are non-linear due to the fact that the thermodynamic properties are, in general, non-linear functions of the independent variables in the problem. In particular, the enthalpy is a non-linear function of temperature, pressure and mass fraction. The equations are coupled as are the components in the cycle. The operating conditions in one component influence the downstream component (or components) through changes in the flow rate and state of the working fluid. The equations reflect this coupling and it implies that the equations must be solved simultaneously. The equations are algebraic as opposed to differential. Algebraic equations result from dealing with a steady-state model with lumped component models. Once the equations have been identified in this manner, the type of solver needed is clear. Many routines have been developed for this type of system with the most familiar being the Newton-Raphson method.

The Newton-Raphson method can be formulated in multiple dimensions so that each unknown variable in the cycle represents one dimension. The method must be set up so that the number of equations equals the number of unknowns. An initial guess value is required for each of the unknowns. At each iteration, the method calculates a correction to each of the unknowns designed to bring the entire system closer to a solution. The problem can be easily formulated in vector notation where the unknowns are a vector. Then, the correction at each iteration is a correction vector. The correction vector is computed by linearizing the set of equations around the current guess value of the solution. The linearization is done numerically and involves evaluating the derivative of each equation with respect to each of the independent variables. The method is summarized below.

The system of equations to be solved can be represented in vector form as

$$\bar{F}\,(\bar{x})\ =\ 0 \qquad\qquad \text{C.19}$$

so that each individual equation can be written as

$$F_i(\bar{x}) = 0 \qquad i = 1, 2 \cdots n \qquad\qquad \text{C.20}$$

where n is the number of equations. For a well-posed system, n is the number of independent variables as well. The iterative solution process begins by specifying an initial guess for the unknowns. The iteration step values of the independent variables are represented by a superscript so that the initial guess is denoted $\bar{x}^o$. In general, the equations will not be exactly satisfied at the initial guess value which is denoted as

$$\bar{F}(\bar{x}^o) = \bar{F}^o = \neq 0 \qquad\qquad \text{C.21}$$

If the system were linear, the following equation would allow the analyst to solve for a correction vector that would locate the solution in one step. However, for the more general case of non-linear equations, the method amounts to linearizing the system of equations about the current guess and solving for a correction vector. Because of the non-linearity, the procedure requires multiple steps to converge. For each of the n equations, we write

$$\bar{\nabla}F_i \cdot \bar{x}_c^1 = -F_i^o \qquad i = 1, 2, \cdots n \qquad\qquad \text{C.22}$$

This results in n equations in the n unknown components of the correction vector. This system is linear and can be solved explicitly for the correction vector. The updated guess for the solution is then

$$\bar{x}^1 = \bar{x}^o + \bar{x}_c^1 \qquad\qquad \text{C.23}$$

This iterative procedure is repeated until the solution is sufficiently close to satisfying the original equations.

Several difficulties can be encountered in the solution of such sets of non-linear algebraic equations. The primary problems are 1) multiple solutions and 2) lack of convergence. Non-linear equations can, in general, have multiple solutions. Thus, when an iterative solver converges on a solution, it may not be the solution of interest to the analyst. Care must be taken to examine the solution and to establish that it does represent a valid solution to the problem of interest. Another problem is divergence of the iterative scheme. Particularly when the guess values for the independent variables are far away from a solution, it is common for non-linear equations to exhibit divergence using the Newton-Raphson scheme. When the scheme works, it usually converges very rapidly. However, when it fails other measures must be taken to find a solution.

Numerous alternative solution techniques have appeared in the literature. These methods include various schemes to accelerate convergence and to avoid divergence. A classical scheme to avoid divergence is the Marquardt-Levenberg algorithm which represents a hybrid between the Newton-Raphson method and a steepest descent method. Another powerful scheme for loosely coupled systems is to block the equations into groups that can be solved independently. The computational effort required to perform an iteration is related to n . Thus, any reduction in n gives significant acceleration of convergence. Blocking essentially splits the problem into minimum size sub-problems that can be solved more efficiently while still leading to the solution of the global set of equations.

Appendix D

CONCENTRATION MEASUREMENTS BY TITRATION

Various techniques are available to measure the concentration of an absorption fluid. The most basic methods are those that go back to the chemistry of the solution. In the present section, titration methods are presented for both water/lithium bromide and ammonia water concentration determinations. The accuracy of the methods can be quite high when good laboratory technique is used. The drawback of these methods is that they tend to be time consuming, especially when high accuracy is sought.

Ammonia/Water

The determination of the relative content of ammonia and water in a mixture can be done in a very straightforward way due to the fact that ammonia forms a base (weak base) when mixed with water. This enables the analyst to titrate with a strong acid and to note the amount of acid needed to shift the pH from base to acid. By careful tracking of the mass of the various constituents, the amount of acid added can be related to the original concentration of ammonia. This technique is described in various basic chemistry references (for example, Skoog and West, 1979).

For ammonia/water, the base nature of the solution results from the following reaction that occurs upon mixing

$$NH_3 \ + \ H_2O \ = \ NH_4^+ \ + \ OH^- \qquad\qquad D.1$$

The reaction in Equation D.1 has an equilibrium constant of 1.8×10^{-5} (Becker and Wentworth, 1972). This implies that the majority of the ammonia is intact with only a small fraction dissociated at equilibrium. For example, if one adds 10 moles of ammonia to 90 moles of water, the equilibrium mole fraction of OH^- is approximately 0.0013 which is why ammonia/water is called a weak base. As the acid is added to the mixture during titration, each free hydrogen ion combines with one hydroxyl ion in a neutralization reaction as follows

$$H^+ \ + \ OH^- \ = \ H_2O \qquad\qquad D.2$$

Thus, as the acid is added, the mixture pH shifts from high to low values. This shift is best illustrated by an example.

Example D.1   Determination of the Concentration of Ammonia/Water by Titration

---

The starting point is a vessel containing 1000 g of pure water. Mix the ammonia water sample with the water to get a dilute ammonia/water sample with a low vapor pressure (so that no ammonia will be lost by evaporation). The mass of the ammonia/water sample, of unknown concentration, is 10 g. By titration, the amount of 1 N HCl required to make the indicator methyl orange change from yellow to red is 36.5 ml. Determine the concentration of the ammonia/water sample.

Solution: The concentration of the hydrogen ions in the HCl is given, typically, in terms of normality defined as the number of gram equivalents per liter. In this case, the number of gram equivalents is the number of moles of hydrogen ions. Thus, the number of moles of hydrogen ions added to achieve neutralization is 0.0365 moles. This implies that there were 0.0365 moles of ammonia present in the sample. This would be a mass of 0.6216 g of ammonia. Since the original sample had a mass of 10 g, the mass fraction of ammonia can be calculated to be 0.06216 (6.216 % ammonia). Note that the 1000 g of water used as an absorbent does not enter the calculation.

It is useful to examine the pH as a function of the amount of HCl added. Based on the conditions given in this example, the pH varies as shown in Figure D.1. In the region where pH is greater than 7 (i.e., the base region), the pH is calculated as (14 - pOH). The concentration of OH^- is determined from the base dissociation reaction indicated in Equation D.1. From Figure D.1 it can be seen that the pH changes abruptly near the neutralization point. This results primarily from the logarithmic definition of pH. However, it is a key to the accuracy of the present method since the indicator used does not change color at the neutralization point but over the range of pH from 4.4 to 3.1 (Becker and Wentworth, 1972, Table 13.1). Since the pH change is so sensitive to the amount of acid added in that range, the method gives a very accurate indication of the ammonia concentration.

This calculation can be reduced to a simple recipe as

$$x_{NH_3} = \frac{V\, N_{HCl}\, M_{NH_3}}{m_s}$$

where x_{NH3} -   mass fraction of NH_3
       V -   volume of titrated HCl (liter)
       N_{HCl} -   normality of HCl solution (moles/liter)
       M_{NH3} -   molecular weight of NH_3 (17.031 g/mole)
       m_s -   sample mass (g)

Using the indicated units, no additional factors are needed in the calculation.

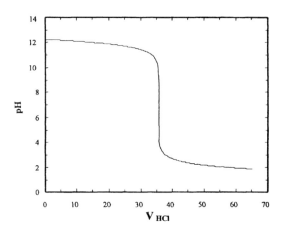

Figure D.1 Volume of HCl versus pH

Water/Lithium Bromide

The determination of the amount of LiBr in an aqueous solution can be accomplished by titration of the solution with silver nitrate. The silver binds to the bromide ion and precipitates as a solid suspension. By including an indicator in the solution, the point at which all the bromide has been removed from solution can be accurately found. The indicator recommended is dichlorofluorescein ($C_{20}H_{10}C_{12}O_5$) 0.1% in alcohol solvent, which is greenish-yellow in solution. As the end point of the titration is passed, i.e., as the last bromine ion is bound by the silver, the surface color of the precipitate changes to pink due to adsorption of the indicator on the solid surface. Prior to the end point, the indicator does not adsorb due to the electric charge associated with adsorbed bromine ions. The titration procedure is a little more involved than that for ammonia/water due to several factors. These are summarized next.

The sample needs to be buffered by sodium bicarbonate ($NaHCO_3$) to prepare it for the indicator. Dextrin ($C_6H_{10}O_5$, a starch) is used to help make the end point clear. As the end point of the titration is reached, pink flashes may be seen which disappear upon shaking. The end point is reached when the pink color change remains upon shaking. It needs to be noted that the precipitate is photosensitive and will turn purple or black if exposed to light. This requires that the procedure be done rapidly to avoid confusion caused by light effects.

Example D.2 Determination of the Concentration of Water/Lithium Bromide by Titration

---

Assume that a 7 g sample of water/lithium bromide solution has been titrated with 1.0 N silver nitrate and the volume required to obtain the color change was 40.3 ml. Determine the salt concentration in the sample.

Solution: The starting point for the titration procedure is a vessel containing 100 ml of distilled water. To this, add approximately 7 g of sample, 0.1 g of dextrin powder, 0.5 g of sodium bicarbonate and 0.1 ml of indicator (dichlorofluorescein). This mixture should be well mixed. Titrate with 1.0 N silver nitrate solution until the color changes to pink.

The number of silver nitrate molecules used during the titration equals the number of LiBr molecules in the original sample. The number of silver nitrate molecules is the volume times the molar concentration (normality) and equals 0.0403 moles. Based on the molecular weight of LiBr of 86.848, this number of moles would have a mass of 3.5 g. Then, since the mass of the sample is known to be 7 g, the mass fraction is determined to be 0.5 (50% LiBr by mass).

This calculation can be reduced to a simple recipe as

$$x_{LiBr} = \frac{V N_{AgNO_3} M_{LiBr}}{m_s}$$

where x_{LiBr} -   mass fraction of LiBr
        V -   volume of titrated $AgNO_3$ (liter)
        N_{AgNO3} -   normality of $AgNO_3$ solution (moles/liter)
        M_{LiBr} -   molecular weight of LiBr (86.848 g/mole)
        m_s -   sample mass (g)

Using the indicated units, no additional factors are needed in the calculation.

Appendix E

INTRODUCTION TO ENGINEERING EQUATION SOLVER

E.1 Overview

EES (pronounced "ease") is an acronym for Engineering Equation Solver. The basic function provided by EES is the numerical solution of non-linear algebraic and differential equations. In addition, EES provides built-in thermodynamic and transport property functions for many fluids, including water, dry and moist air, most CFC and HCFC refrigerants, and others. Included in the property data base are thermodynamic properties for lithium bromide/water and ammonia/water mixtures. The combination of a robust non-linear equation solver and absorption fluid properties makes EES a very powerful tool for analysis and design of absorption systems.

A demonstration version of EES for PC-compatible computers using the Microsoft Windows operating system is included on a disk provided with this book. Almost all of the example problems in the book have been solved using EES and the solutions are included on the disk. This appendix provides instructions for installing the EES program on your computer and using it to solve absorption cycle problems.

The capabilities of the version of EES provided on the disk have been necessarily reduced to allow its distribution with the book. The file Save, Copy and Paste commands have been deactivated and the number of simultaneous equations that can be solved has been limited from 1000 in the original version to 250. This version of EES is useful only for solving and modifying the example problems. A complete version of the program for either Macintosh or PC-compatible computers can be purchased from the software publisher at the following address.

F-Chart Software
4406 Fox Bluff Rd.
Middleton, WI 53562
Phone: (608) 836-8531
FAX: (608) 836-8536
http://www.fchart.com

E.2 Installing EES on Your Computer

This version of EES is designed to operate with the Microsoft Windows operating system, version 3.1 or newer. The program and example files are provided in a compressed form on a standard 3.5 inch high-density disk. An installation program is provided which must be run from the Windows operating system. To start the installation program, choose the **Run...** command from the **File** menu in the Windows Program Manager. In the box under the words "Command Line:", enter A:\ Setup, as shown.

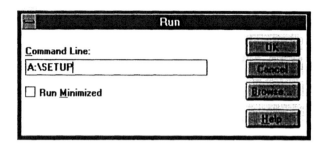

where A: is your floppy drive designation. The installation program will provide a series of prompts which will lead you through the complete installation of the EES program and the files containing the solutions to the examples in this book. The default installation program will create a Windows Group called EES Demo containing a single icon representing the EES program. Double-clicking on this icon will start the program. Pressing the F1 key at any point in the program will bring up a Help window.

E.3 EES Menus

EES commands are distributed among nine pull-down menus. A brief summary of their functions follows.

The **File** menu provides commands for loading, merging and saving work files and libraries, and printing. Saving files is disabled in this demonstration version.

The **Edit** menu provides the editing commands to cut, copy, and paste information within EES or between EES and other application. The Copy and Paste commands are disabled in the demonstration version.

The **Search** menu provides the commands to find and/or replace text in the Equations window.

The **Options** menu provides two types of commands. The first group provides commands for setting the guess values and bounds of variables and information on built-in and user-supplied functions. A second group allows the unit system and default information and formatting options to be changed.

The **Calculate** menu contains the commands to check, format or solve the equation set.

The **Table** menu contains commands to set up or alter the contents of the Parametric and Lookup Tables. The Parametric Table, similar to a spreadsheet, allows the equation set to be solved repeatedly while varying the values of one or more variables.

The **Plot** menu provides commands to modify an existing plot or prepare a new plot of data in the Parametric, Lookup, or Array tables. Curve-fitting capability is also provided.

The **Windows** menu provides a convenient method of bringing any of the EES windows to the front and of organizing the windows.

The **Help** menu provides commands for accessing the online help documentation. The F1 key also displays help information.

E.4 A Simple Problem

Start EES and enter the following two equations on separate lines in the **Equations** window.

Note that EES makes no distinction between upper- and lowercase letters and the ^ sign (or ******) is used to signify raising to a power. If you wish, you may view the equations in mathematical notation by selecting the **Formatted Equations** command from the **Windows** menu.

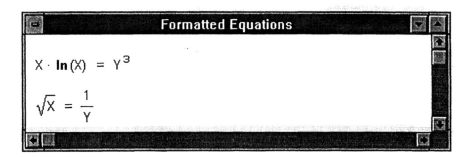

Select the **Solve** command from the **Calculate** menu. A dialog window will appear indicating the progress of the solution. When the calculations are completed, the button will change from Abort to Continue. Click the Continue button. The solution to this equation set will then be displayed.

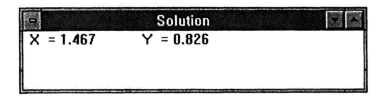

EES has many other features and capabilities as you will see. However, its major capability is its ability to easily solve a system of equations. Note that the order in which the equations are entered does not matter. Unlike FORTRAN and other programming languages, EES allows the equations to be entered with unknown variables on the left or right of the equal sign. The remainder of this appendix will demonstrate other capabilities of the EES program to solve example problems from the book.

E.5 Example Problem 2.2

The zero-order model represented by Equations 2.17 to 2.22 is a simple representation of an absorption cycle operating at finite capacity. However, considerable computational effort is needed to solve this problem since the equation set is non-linear and repeated calculations are needed to identify the optimum heat exchange area distribution. Start EES and enter the following equations.

```
┌──────────────────────────────────────────────────────────────┐
│ ▭            Equations - d:\ex2_2.ees              ▼ ▲         │
│ "Example 2.2"                                            ▲     │
│                                                                │
│ Q_h=[T_h-T_hi]/R_h                            "Eqn 2.17"       │
│ Q_c=[T_ci-T_c]/R_c                            "Eqn 2.18"       │
│ Q_e=[T_e-T_ei]/R_e                            "Eqn 2.19"       │
│ Q_h+Q_e=Q_c                                   "Eqn 2.20"       │
│ COP=Q_e/Q_h; COP=T_ei/T_hi*[(T_hi-T_ci)/(T_ci-T_ei)] "Eqn 2.21"│
│ T_hi-T_ci=T_ci-T_ei                           "Eqn 2.22"       │
│ "Problem information"                                          │
│ U=0.5              "kW/m2-K  - same value for all heat exchangers"│
│ A_h+A_c+A_e=10    "Total area is 10 m2"                        │
│ R_h=1/(U*A_h)                                                  │
│ R_c=1/(U*A_c)                                                  │
│ R_e=1/(U*A_e)                                                  │
│ T_h=200+273.15; T_c=50+273.15; T_e=-20+273.15    "All T's in K"│
│ ◄▭                                                        ►    │
└──────────────────────────────────────────────────────────────┘
```

Note that any information between quotation marks (or braces {}) is an optional comment. Variable names must start with a letter and consist of 30 or fewer letters and numbers. The underscore character ( ) is an optional special character which instructs EES to display the characters to the right of the underscore as a subscript in the **Formatted Equations** and **Solution** windows. Multiple equations can be placed on one line if they are separated by a semi-colon.

If you try to solve the equations at this point, EES will display an error message because there are 17 variables but only 15 equations. The missing information is the areas of the three heat exchangers. We know that the total area of all three exchangers is 10 m^2, but the purpose of the problem is to find the areas of each of the heat exchangers which maximizes the cooling capacity. As a start, set all of the areas equal by entering the following two equations into the Equations window.

A_h=10/3; A_c=10/3

Note that it is not necessary to enter all three heat exchanger areas since we know that the sum is 10 m^2. In fact, EES would consider the problem to be over-defined if you entered, in addition, a value for A_e and it would present an error message.

You could try to solve the problem now, but if you do EES will highlight the line containing Equation 2.21 and display "Evaluation of this equation resulted in division by zero". The problem here is that, by default, the guess values of all variables are 1.0. These guess values can be changed using the **Variable Info** or **Default Info** commands in the **Options** menu. Since T_ci and T_ei both have a guess value of 1.0, the denominator on the right-hand side of Equation 2.21 is zero which produces the error message. Select the **Variable Info** command in the **Options** menu. A dialog window will appear in which the guess values, lower and upper bounds, display format, and units for each defined variable can be set. Use the scroll bar to bring the temperature variables T_ci, T_ei, and T_hi into view. Now, enter reasonable guess values for these temperatures. T_ci must be greater than T_c in order for heat flow to be in the proper direction. Set T_ci to 350 K. Similarly, T_hi and T_ei must be less then T_h and T_e, respectively. Set these temperatures to 460 and 240 K, respectively. You may wish to change the display format and enter the units for variables here as well. Note that the values you enter for T_ci, T_hi and T_ei are not critical, as these are just guess values that EES will use to initiate the iterative solution. EES will calculate the correct values. Shown below is the **Variable Info** dialog window with the guess values for these three temperatures changed. Click the OK button to store the initial guesses and dismiss the dialog window.

Variable Information						
Variable	Guess	Lower	Upper	Display		Units
Q_e	1.000	-inf	inf	F	3	
Q_h	1.000	-inf	inf	F	3	
R_c	0.600	-inf	inf	F	3	
R_e	0.600	-inf	inf	F	3	
R_h	0.600	-inf	inf	F	3	
T_c	323.2	-inf	inf	F	1	K
T_ci	350.0	-inf	inf	F	1	K
T_e	253.2	-inf	inf	F	1	K
T_ei	240.0	-inf	inf	F	1	K
T_h	473.2	-inf	inf	F	1	K
T_hi	460.0	-inf	inf	F	1	K
U	0.500	-inf	inf	F	3	

OK   **Update**   **Cancel**

Select the **Solve** command from the **Calculate** menu (or press F2) and the solution for equal heat exchange areas will be displayed.

Solution		
A_c = 3.333	A_e = 3.333	A_h = 3.333
COP = 0.535	Q_c = 44.444	Q_e = 15.491
Q_h = 28.953	R_c = 0.600	R_e = 0.600
R_h = 0.600	T_c = 323.2 [K]	T_{ci} = 349.8 [K]
T_e = 253.2 [K]	T_{ei} = 243.9 [K]	T_h = 473.2 [K]
T_{hi} = 455.8 [K]	U = 0.500	

By default, EES displays all variables in fixed decimal point notation with three numbers to the right of the decimal point. You can change the variable display in the **Variable Information** dialog window, as was done for the temperatures. However, you can also make these changes by simply entering the information in the small dialog window which appears when you click on a variable name in the Solution window.

The cooling capacity (Q_e) when the areas of the three heat exchangers are equal is 15.49 kW. However, this is likely to not be the optimum distribution of heat exchanger area. You could simply change the values of A_h and A_c and repeat the calculations to see if you can find a set of values which produces a larger cooling capacity. However, EES can automate this optimization process with its internal optimization algorithms.

To have EES find the optimum distribution of heat exchanger area, first delete {or comment out} the A_h=10/3 and A_c=10/3 equations. Next, select the **Min/Max** command from the **Calculate** menu. Click the Maximize button and select Q_e from the list on the left. There are two degrees of freedom in this problem. Select any two of the three heat exchanger areas as the independent variables from the list on the right. The dialog window should now appear as below.

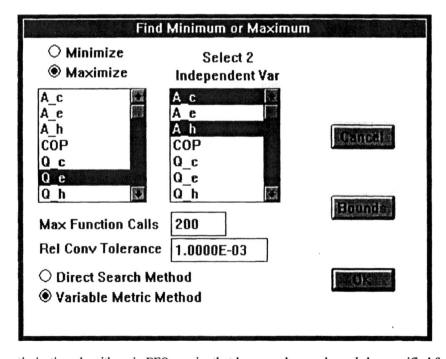

The optimization algorithms in EES require that lower and upper bounds be specified for the independent variables. Click the Bounds button to bring up a subset of the **Variable Info** dialog window in which you can enter the bounds. Enter reasonable values for the bounds. Since the total area of all three heat exchangers is 10 m², it is unlikely that any one of the areas will be less than 1 m² or greater than 9 m².

Variable Information					
Variable	Guess	Lower	Upper	Display	Units
A_c	3.333	1.0000E+00	9.0000E+00	F 3	
A_h	1.000	1.0000E+00	9.0000E+00	F 3	

[OK]   [Update]   [Cancel]

Click the OK button to dismiss the **Variable Information** dialog window. Then click the OK button in the **Min/Max** dialog window to initiate the optimization calculations. When the calculations are completed, the Solution window will show that the maximum cooling capacity is 17.486 kW and the areas of the high, low and intermediate heat exchangers are 2.706, 2.286, and 5.008 m², respectively.

Solution		
Maximization of Q_e[A_c,A_h]		
A_c = 5.008	A_e = 2.286	A_h = 2.706
COP = 0.530	Q_c = 50.477	Q_e = 17.486
Q_h = 32.991	R_c = 0.399	R_e = 0.875
R_h = 0.739	T_c = 323.2 [K]	T_{ci} = 343.3 [K]
T_e = 253.2 [K]	T_{ei} = 237.9 [K]	T_h = 473.2 [K]
T_{hi} = 448.8 [K]	U = 0.500	

E.6 Example Problem Showing Property Functions

One of the most significant features of EES is that it includes functions for the thermodynamic and transport properties of fluids. Property data for many pure fluids, (e.g., steam, ammonia, R-12, R134a, carbon dioxide, and many others) are built into the program. This simple example demonstrates the use of these functions.

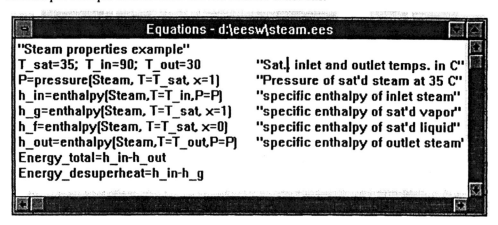

Shown above are all of the equations needed to solve the problem for steam. The first line simply sets the temperatures. The condenser pressure is found using the built-in **pressure** function. Thermodynamic property functions for pure real fluids require three parameters. The first parameter is the name of the fluid, Steam in this case. You could also use Water, which does exactly the same thing as Steam, or Steam_NBS which implements a more accurate but computationally more intensive equation of state. (A complete list of the built-in property functions and fluid names can be viewed using the **Function Info** command in the **Options** menu.) The following two parameters can be any two independent thermodynamic variables. The variables are identified with a single letter to the left of the equal sign; the value of the variable is provided to the right of the equal sign. The T=T_sat parameter in the **pressure** function indicates that a temperature is being supplied and its value is equal to the value of variable T_sat. The x=1 parameter indicates that the vapor quality is unity. The **pressure** function will return the saturation pressure corresponding to T_sat. The value returned by the **pressure** function will depend on the selected unit system. The version of EES provided with this book was set for SI units with temperatures in °C and pressures in kPa. These unit system choices can be viewed or changed with the **Unit System** command in the **Options** menu. The **enthalpy** function operates in a similar manner. Selecting the **Solve** command from the **Calculate** menu produces the following solution.

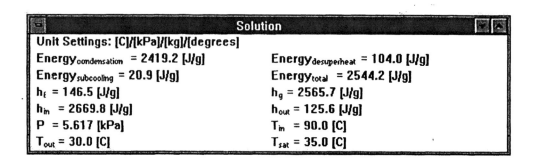

Improved accuracy can be obtained by replacing the word "Steam" with "Steam_NBS" in all of the thermodynamic function references. The **Replace** command in the **Search** menu provides a convenient means of making these changes. Use the Replace command to change the property name to "Ammonia" to complete the example problem.

E.7 Example Problem 4.2

EES is particularly useful for absorption system analyses because thermodynamic property data for ammonia-water and lithium bromide-water mixtures are provided with the program. Unlike the built-in property functions used in the previous section, the ammonia/water and lithium bromide/water properties are not an integral part of the EES program but are instead provided by external routines. This example demonstrates the use of the ammonia-water property functions. The problem also demonstrates the use of array variables which automatically produces a table of properties for each state point.

The desorption problem considered in Example 4.1 can be solved simply with a mass balance, an energy balance, and ammonia-water property data for the inlet and outlet states, as indicated in the following set of equations.

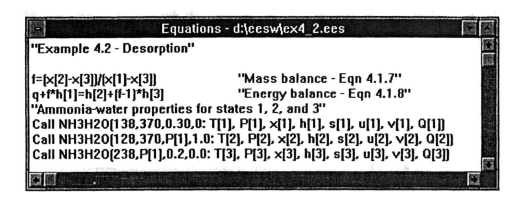

The ammonia/water properties for each state are determined through a call to the external NH3H2O program. Three independent properties are needed to fix each state. The NH3H2O program is an example of an EES procedure. An EES procedure can accept one or more inputs and return one or more calculated values. The calling format for the NH3H2O procedure is:

CALL NH3H2O(Code, In1, In2, In3 : T, P, x, h, s, u, v, Q)

The four parameters to the left of the colon (Code, In1, In2, and In3) are inputs. The parameters to the right of the colon are outputs calculated by the NH3H2O program. The NH3H2O routine operates in SI units with T=[K], P=[bar], x=[ammonia mass fraction], h=[kJ/kg], s=[kJ/kg-K], u=[kJ/kg], v=[m3/kg], and Q=[vapor mass fraction]. For saturated states, $0<=Q<=1$. Subcooled states are indicated with Q=-0.01; superheated states have Q=1.01.

The first input, Code, is a three digit number which indicates which three properties are supplied in In1, In2, and In3, as indicated in the table.

Property	T	P	x	h	s	u	v	Q
Code	1	2	3	4	5	6	7	8

For state 1, the entering liquid state, the Code value is 138 indicating that the temperature (1), mass fraction (3), and quality (8) are provided in the following three inputs. The exiting vapor state 2 uses Code 128 since the temperature, pressure, and quality are specified.

Array variables are used in this example for the properties of each state. Array variables contain a numerical index enclosed in square brackets. T[1] is the temperature of state 1, h[2] is the specific enthalpy of state 2, and so on. In EES, an array variable is just like any other variable except that the array index will display as a subscript in the **Formatted Equations** window and the value is displayed in tabular form in the **Arrays** window. Selecting the **Solve** command will produce the following table showing state point information and the solution to this problem.

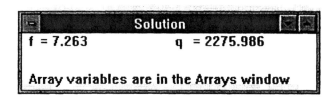

E.8 A Final Example

This final example demonstrates the use of the lithium bromide/water property functions and the plotting capability in EES. The problem is to plot the enthalpy of a lithium bromide/water solution as a function of composition at temperatures of 10 and 100 °C for mass fractions between 0.4 and 0.75.

Enter the following equations into EES.

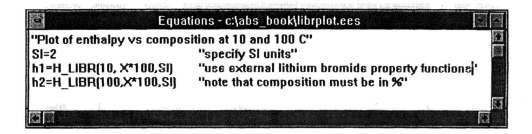

The property data for water/lithium bromide solutions are provided in a set of external functions, the names of which end with _LiBr. The H_LiBr function returns the enthalpy [Btu/lb or kJ/kg] of a lithium bromide-water solution as a function of temperature [°F or °C] and concentration [lithium bromide mass percentage]. The third parameter determines the unit system. If this parameter is 1, the function accepts and returns values in English units, and if it is 2, as in the example above, the function operates in SI units.

The problem cannot be solved at this point since the mass fraction of lithium bromide (X) is not specified. In this problem, we want to vary the value of X between 0.4 and 0.75 and plot the enthalpies. This is most easily done using a Parametric table, which is similar to a spreadsheet.

Select **New Table** from the **Parametrics** menu. The following dialog window appears.

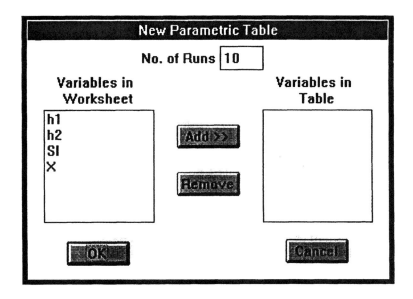

The list on the left shows all of the variables appearing in the equations. The list on the right holds the variables which are to appear in the Parametric table. Click on X and then on the Add button. Repeat for h1 and h2 and then click the OK button. A table with three columns and 10 rows is generated.

You could now simply enter values for X in the first column of the Parametric table starting with 0.40 ranging to 0.75 or you could let the EES do this for you. To have the EES enter the values of X, select **Alter Values** from the **Parametrics** menu. Click on X and enter

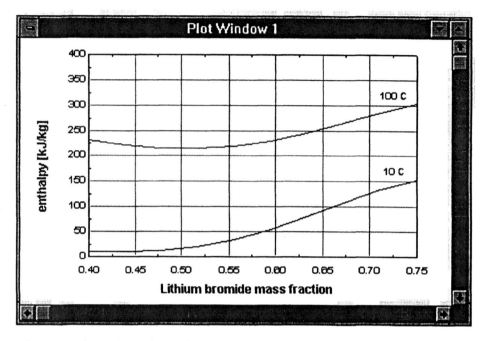

0.4 for the starting value and 0.75 for the last value. EES fills in values of X.

Now, select **Solve Table** from the **Calculate** menu. For each row in the table, EES uses the specified value of X and determines the values of h1 and h2 (for temperatures of 10 and 100 °C, respectively).

Parametric Table			
	X	h1 [kJ/kg]	h2 [kJ/kg]
Run 1	0.400	9.5	230.7
Run 2	0.439	10.5	221.6
Run 3	0.478	13.0	215.6
Run 4	0.517	20.3	214.4
Run 5	0.556	34.1	219.0
Run 6	0.594	54.5	229.5
Run 7	0.633	80.2	245.3
Run 8	0.672	108.3	264.6
Run 9	0.711	134.6	285.2
Run 10	0.750	153.2	303.9

Once the values are determined, they can be plotted using the **New Plot Window** command in the **Plot** menu. Select X to appear on the x-axis and h1 for the y-axis. Set the minimum and maximum values for each scale and other plotting options and then click the OK button. A plot will appear. Next select **Overlay Plot** from the **Plot** menu and repeat the procedure, but this time select h2 for the y-axis.

Once the plot is produced you can customize it in many ways. You can use the **Add Text** command to label the plots. Double-clicking on a text item, an axis-scale or the plot rectangle will bring up a dialog window where additional formatting changes can be made.

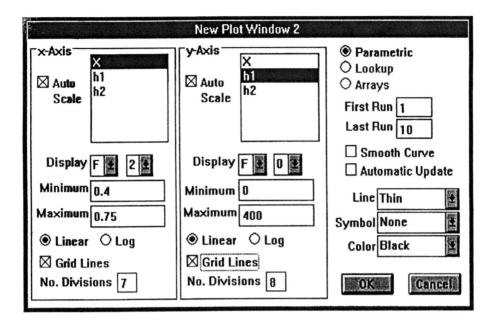

E.9 Conclusion

This appendix has described a number of the more important capabilities of EES with respect to modeling absorption systems. EES is a very versatile program with many other features. Additional information on the program operation can be obtained from the **Help** menu or from the program manual. Like any tool, EES requires some effort to master. The effort expended in learning how to efficiently use EES will prove to be a worthwhile investment in that it will allow you to easily solve problems that would have otherwise been intractable.

Appendix F

OVERVIEW OF THE ABSIM SOFTWARE PACKAGE

F.1   Overview

ABSIM (an acronym for ABsorption SIMulation) is a modular computer code for simulation of absorption systems. It was developed with DOE/ORNL sponsorship by Dr. Gershon Grossman. This modular code is based on unit subroutines containing the governing equations for the system's components and on property subroutines containing thermodynamic properties of the working fluids. The components are linked together by a main program which calls the unit subroutines according to the user's specifications to form the complete cycle. When all the equations for the entire cycle have been established, a powerful mathematical solver routine is employed to solve them simultaneously. Eleven absorption fluids are presently available in the code's property database, and twelve units are available to compose practically every absorption cycle of interest including the ones given in this book. The code in its present form may be used not only for evaluating new cycles and working fluids, but also to investigate a system's behavior in off-design conditions, to analyze experimental data and to perform preliminary design optimization. The code is user-oriented and requires a relatively simple input containing the given operating conditions and the working fluid at each state point. A graphical user interface enables the user to draw the cycle diagram on the computer screen, enter the input data interactively, run the program and view the results either in the form of a table or superimposed on the cycle diagram. Special utilities enable the user to plot the results and to produce a PTX diagram of the cycle. The code has been employed successfully to simulate a variety of single-effect, double-effect and dual loop absorption chillers, heat pumps and heat transformers employing the working fluids LiBr-H_2O, H_2O-NH_3, LiBr/$ZnBr_2$-CH_3OH, NaOH-H_2O and more. The same code has been used to simulate the rather complex Generator-Absorber Heat Exchange (GAX) and advanced GAX cycles employing ammonia-water, in several cycle variations, and a variety of triple-effect chillers employing lithium bromide/water.

To obtain a free copy of the latest developmental version of ABSIM with a user's manual and several examples, please contact:

Abdi Zaltash
Building Equipment Research Program
Oak Ridge National Laboratory
P.O. Box 2008, Building 3147, MS-6070
Oak Ridge, Tennessee 37831-6070

Phone: (423)-574-4571
Fax: (423)-574-9338
E-mail: alz@ornl.gov

F.2   ABSIM Program Structure

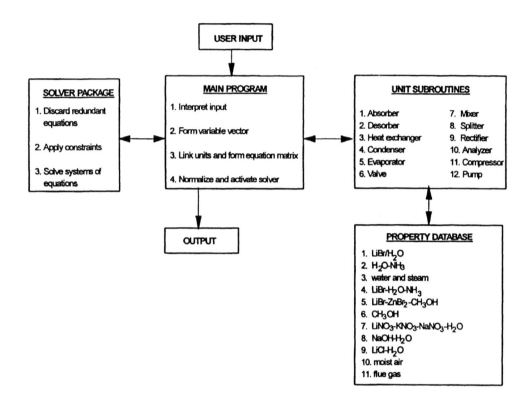

REFERENCES and BIBLIOGRAPHY

The following listing contains the references as referred to in the text together with a bibliography of additional material that may be of interest to the reader. The references are sorted by chapters, first author's last name and year of publication. Subheadings are used for major groups of references in some of the chapters.

Chapter 2

Alefeld, G., 1987, "What Needs to be Known About Fluid Pairs to Determine Heat Ratios of Absorber Heat Pumps and Heat Transformers?": Proc. of the 1987 IEA Heat Pump Conf., Orlando, Florida, April 28-30, 1987, Ch. 26, pp. 375-387.

Alefeld, G., Radermacher, R., 1994, Heat Conversion Systems, CRC Press, Boca Raton, FL.

Altenkirch, E., 1913, "Reversible Absorptionsmaschinen", Zeitschrift fuer die gesamte Kaelteindustrie, Vol. 20, Januar, pp 1-8 and 115-119 and 150-161.

Bejan, A., 1982, Entropy Generation through Heat and Fluid Flow, Wiley-Interscience, New York.

Ellington, R.T., Kunst, G., Peck, R.E., Reed, J.F., 1957, "The Absorption Cooling Process", IGT Res. Bull., Vol. 14.

Herold, K.E., 1989, "Performance Limits for Thermodynamic Cycles", ASME Bound Volume AES-Vol. 7, pp. 15-22.

Herold K.E. , Radermacher R. , 1989, "Absorption Heat Pumps", Mechanical Engineering, Vol. III, No. 8, pp. 68-73, August.

Herold, K.E., Radermacher, R., 1990, "Optimum Allocation of Heat Transfer Surface in an Absorption Heat Pump", Reno, NV: Proceedings of 25th IECEC (International Energy Conversion and Engineering Conference), Vol. 2, pp. 217-221.

Herold, K.E., Haslach, H.W., 1994, "Reversible Performance of Cycles with Variable Temperature Heat Transfer Interactions", International Journal of Mechanical Engineering Education, Vol. 22, No. 2, pp. 91-99.

Klein, S.A., 1992,"Design Considerations for Refrigeration Cycles," Int. J. Refrig., Vol.15, No. 3, pp. 181-185.

Ziegler, F., Alefeld, G., 1987, "Coefficient of Performance of Multistage Absorption Cycles", Int. J. Refrig., Vol. 10, pp. 285-295.

Second Law Analysis

Abrahamsson, K., Jernqvist, A., Aly, G., 1994, "Thermodynamic Analysis of Absorption Heat

Cycles", New Orleans: Proceedings of the International Absorption Heat Pump Conference ASME AES-Vol. 31, January 19-21, pp. 375-384.

Anand, D.K., Kumar, B., 1987, "Absorption Machine Irreversibility Using New Entropy Calculations", Solar Energy, Vol. 39, No. 3, pp. 243-256.

Andrews, D.H., 1956, "A Theoretical Study of the Thermodynamic Relations Underlying the Absorption Refrigeration Cycle", AGA Final Report - Project ZS-10.

Ataer, O., Gogus, Y., March 1991, "Comparative Study of Irreversibilites in an Aqua-Ammonia Absorption Refrigeration System", Int. J. Refrig., Vol. 14, pp. 86-92.

Briggs, S.W., 1971, "Second Law Analysis of Absorption Refrigeration", Chicago, IL: Proceedings AGA & IGT Congress on Natural Gas Research & Technology.

Carmody, S. A., Shelton, S.V., 1994, "Direct Second Law Analysis of Advanced Absorption Cycles Utilizing an Ideal Solution Model", New Orleans: Proceedings of the International Absorption Heat Pump Conference, ASME AES-Vol. 31, January 19-21, pp. 369-374.

Colenbrander, C.B., van der Horst, J.F., Meijnen, A.J., 1988, "Exergy-Economic Analysis of a Heat Transformer", London: Proceedings of Absorption Heat Pumps Workshop, April 12-14, pp. 55-65.

Curran, H.M., 1978, "Thermodynamic Equivalence of Thermally Driven Refrigeration Systems", Solar Energy, Vol. 21, pp. 451-452.

Didion, D., Garvin, D., Snell, J., August 1990, "A Report on the Relevance of the Second Law of Thermodynamics to Energy Conservation", U.S. Dept. of Commerce , Vol.1, pp. 46.

Gupta, C.P., Sharma, C.P., 1976, "Entropy Values of Lithium Bromide-Water Solutions and Their Vapors", ASHRAE Trans., Vol. 82, Pt. 2, pp. 35-46.

Herold, K.E., Moran, M.J., 1987, "Recent Advances in the Thermodynamic Analysis of Absorption Heat Pumps," Proceedings of the 4th International Symposium on Second Law Analysis of Thermal Systems, Rome, Italy, pp. 97-101.

Herold, K.E., 1991, "The Entropic Average Temperature Concept as a Tool for Analyzing Cycles Involving Mixtures," Proceedings of International Conference on the Analysis of Thermal and Energy Systems, Athens, Greece, pp. 23-34.

Koehler, W.J., Ibele, W.E., Soltes, J., Winter, E.R., 1987, "Entropy Calculations for Lithium Bromide Aqueous Solutions and Approximation Equation", ASHRAE Trans., Vol. 93, pp. 2379-2388.

Kouremenos, D.A., 1989, "Second Law Analysis of Non-Equilibrium Fluid Streams", Forsch. Ingenieurwes., Vol. 55, No. 1, pp. 10-15.

Perez-Blanco, H., Pan, L., July 1985, "Comparative First- and Second-Law Analysis of an Absorption Cycle", ORNL/TM - 9595, pp. 1-38.

Tozer, R.M., James, R.W., 1994, "Thermodynamics of Absorption Refrigeration: Ideal Cycles", New Orleans: Proceedings of the International Absorption Heat Pump Conference, ASME AES-Vol. 31, January 19-21, pp. 393-400.

Tripp, W., January 1966, "Second-Law Analysis of Absorption Refrigeration Systems", ASHRAE J., Vol. 18 pp. 49-57.

Chapter 3

Alefeld, G., Radermacher, R., 1994, Heat Conversion Systems, CRC Press, Boca Raton, FL.

Bosnjakovic, F., translated by Blackshear, P.L., Jr., 1965, Technical Thermodynamics, Holt, Rinehart & Winston, New York.

Di Guillio, R.M., Lee, R.J., Jeter, S.M., Teja, A.S., 1990, "Properties of Lithium Bromide-Water Solutions at High Temperatures and Concentration. I. Thermal Conductivity", ASHRAE Trans., Vol. 96, Pt. 1, pp. 702-708.

Fenton, D.L., Noeth, A.F., Gorton, R.L., 1991, "Absorption of Ammonia into Water", ASHRAE Trans., Vol. 97, Pt. 1, pp. 204-213.

Foote Technical Data, Bulletin 145, Foote Mineral Company.

Gibbs, J.W., 1876, "On the Equilibrium of Heterogeneous Substances", Trans. Connecticut Academy of Arts and Sciences, Vol. 3, pp. 108-520.

Harr, L., Gallagher, J.S., 1978, "Thermodynamic Properties of Ammonia", J. of Phys. Chem. Ref. Data, Vol. 7, No. 3, pp. 635-792.

Harr, L., Gallagher, J.S., Kell, G.S., 1984, NBS/NRC Steam Tables, Hemisphere, New York.

Herold, K.E., 1985, "Thermodynamic Properties of Lithium Bromide/Water Solutions with Application to an Absorption Temperature Boosting Heat Pump", Ph.D. Dissertation, Ohio State University.

Ibrahim, O.M., Klein, S.A., 1993, "Thermodynamic Properties of Ammonia-Water Mixtures", ASHRAE Trans., pp. 1495-1502.

Lee, R.J., DiGuillio, R.M., Jeter, S.M., Teja, A.S., 1990, "Properties of Lithium Bromide-Water Solutions at High Temperatures and Concentration. II. Density and Viscosity", ASHRAE Trans., Vol. 96, Pt. 1, pp. 709-728.

Lewis, G.N., Randall, M., 1961, revised by Pitzer, K.S., Brewer, L., Thermodynamics, 2nd edition, McGraw-Hill, New York.

Macriss R.A., Zawacki T.S., 1989, "Absorption Fluid Data Survey: 1989 Update", ORNL Report, ORNL/Sub84-47989/4.

Moran, M.J., Shapiro, H.N., 1992, Fundamentals of Engineering Thermodynamics, John Wiley & Sons, New York.

Moran, M.J., Herold, K.E., 1989, "Thermodynamic Properties of Solutions from Fundamental Equations of State with Application to Absorption Heating/Cooling System Analysis: A Review," Proceedings of International Symposium on Thermodynamic Analysis and Improvement of Energy Systems, Beijing, China, June 5-8, pp. 445-464.

Niebergall, W., 1959, Sorptions-Kaltemaschinen, Handbuch der Kaltetechnik, Vol. 7, Springer-Verlag, Berlin.

Patterson, M.R., Perez-Blanco, H., 1985, "Sensitivity of Absorption Cycle Calculations to Fluid Property Errors Calculated Stochastically", Int. Energy Conversion Eng. Conf., pp. 1739-1747.

Reid, R.C., Prausnitz, J.M., Poling, B.E., 1987, The Properties of Gases and Liquids, 4th edition, McGraw-Hill, New York.

Rowlinson, J.S., Swinton, F.L., Liquids and Liquid Mixtures, Butterworths, 3rd Edition, 1982.

Stokes, R.H., 1950, J. Am. Chem. Soc., Vol. 72, pp. 2243.

Uemura, T., Hasaba, S., 1964, Technical Report Kansai University, Vol. 6, pp. 31-55.

Weast, R.C., 1994, Handbook of Chemistry and Physics, 75th edition, CRC Press, Boca Raton, FL.

Ziegler, B., Trepp, Ch., 1984, "Equation of State for Ammonia-Water Mixtures", Int. J. Refrig., Vol. 7, No. 2, pp. 101-106.

Heat and Mass Transfer

Bergles, A.E., 1976, "Survey and Augmentation of Two-Phase Heat Transfer", ASHRAE Trans., Vol. 82, Pt. 1, pp. 891-905.

Bird, R.B., Stewart, W.E., Lightfoot, E.N., 1960, Transport Phenomena, John Wiley and Sons, New York.

Breber, G., Palen, J.W., Taborek, J., 1980, "Prediction of Horizontal Tubeside Condensation of Pure Components Using Flow Regime Criteria", J. Heat Transfer, Vol. 102, No. 3, pp. 471-476.

Brown, G.G., 1956, "Unit Operations", Vapor-Liquid Transfer Operations, John Wiley & Sons, pp. 322-345.

Chaddock, J., Buzzard, G., 1986, "Film Coefficients for In-Tube Evaporation of Ammonia and R-502 With and Without Small Percentages of Mineral Oil", ASHRAE Trans., Vol. 92, Pt. 1A, pp. 22-40.

Combs, J.G., 1977, "An Experimental Study of Heat Transfer Enhancement for Ammonia Condensing on Vertical Fluted Tubes", M.S. Thesis , University of Tennessee, Knoxville.

Denny, V.E., Jusionis, V.J., 1972, "Effects of Noncondensable Gas and Forced Flow on Laminar Film Condensation", Int. J. Heat Mass Transfer, Vol. 15, pp. 315-326.

Denny, V.E., Jusionis, V.J., 1972, "Effects of Forced Flow and Variable Properties on Binary Film Condensation", Int. J. Heat Mass Transfer, Vol. 15, pp. 2143-2153.

Grossman, G., 1983, "Simultaneous Heat and Mass Transfer in Film Absorption Under Laminar Flow", Int. J. Heat Mass Transfer, Vol. 26, No. 3, pp. 357-371.

Hijikata, K., Nagasaki, T., Lee, S.-K., 1992, " Numerical Analysis of Heat and Mass Transfer in a $LiBr/H_2O$ Film Agitated by Ridged Fins", ASME Proceedings, Winter Annual Meeting November 8-13, Anaheim, CA.

Idema, P.D., Liem, S.H., Van Der Wekken, B., 1985, "Heat and Mass Transfer in Vapor and Liquid Phases of a H_2O/NH_3 Regenerator", Trondheim: Proceedings of the International Institute of Refrigeration.

Incropera, F.P., DeWitt, D.P., 1990, Fundamentals of Heat and Mass Transfer, 3rd edition, John Wiley & Sons, New York.

Li, S.S., Ohadi, M.O., Radermacher, R., Dessiatoun, S., 1994, "Review of the Available Correlations for Prediction of Two-Phase Heat Transfer of Ammonia", IIAR Technical Report (International Institute of Ammonia Refrigeration).

Minkowycz, W.J., Sparrow, E.M., "Condensation Heat Transfer in the Presence of Noncondensables, Interfacial Resistance, Superheating, Variable Properties and Diffusion", Int. J. Heat Mass Transfer, 1966, Vol. 9, pp. 1125-1144.

Nusselt, W., 1916, "The Surface Condensation of Water Vapor", (in German), Z. Ver. Dtsch. Ing., Vol. 60, pp. 541-546, 596-575.

Panchal, C.B., Arman, B., 1991, "Analysis of Condensation and Evaporation of Ammonia/Water Mixtures in Matrix Heat Exchangers", ASME Paper 91-HT-16, presented at National Heat Transfer Conference, Minneapolis.

Rohsenow, W.M., 1952, "A Method for Correlating Heat Transfer Data for Surface Boiling Liquids", Trans. ASME, Vol. 74, pp. 969.

Shaw, M.M., 1976, "A Correlation for Heat Transfer During Boiling Flow Through Pipes", ASHRAE Trans., Vol. 82, Pt. 2, pp. 66-86.

Shaw, M.M., 1975, "Visual Observations in an Ammonia Evaporator", ASHRAE Trans., Vol. 81, Pt. 1, pp. 295.

Sparrow, E.M., Minkowycz, W.J., Saddy, M., 1967, "Forced Convection Condensation in the Presence of Noncondensables and Interfacial Resistance", Int. J. Heat Mass Transfer, Vol. 10, pp. 1829-1845.

Uddholm, H., Setterwall, F., 1988, "Model for Dimensioning a Falling Film Absorber in an Absorption Heat Pump", Int. J. Refrig., Vol. 11, pp. 41-45.

Van der Wekken, B.J.C., Wassenaar, R.H., 1988, "Simultaneous Heat and Mass Transfer Accompanying Absorption in Laminar Flow Over a Cooled Wall", Int. J. Refrig., Vol. 11, pp. 70-77.

Other Working Fluids

Bjurstrom, H., 1991, "Slurries as Absorbent Fluids", Tokyo: Proceedings of Absorption Heat Pump Conference, September 30-October 2, pp. 171-176.

Bothe, A., Nowaczyk, U., Schmidt, E.L., Steimle, F., 1988, "New Working Fluid Systems for Absorption Heat Pumps", London: Proceedings of Absorption Heat Pumps Workshop, April 12-14, pp. 13-22.

Davidson, W.F., Erickson, D.C., 1985, "New High Temperature Absorbent for Absorption Heat Pumps", ORNL/Sub/85-22013/1, Oak Ridge National Laboratory Final Report on contract DE-AC05-84OR21400.

Doi, A., Nakao, K., Ikeuchi, M., Fujita, T., 1991, "An Optimum Cycle Period of a Chemical Heat Pump Using $CaCl_2/CH_3NH_2$", Tokyo: Proceedings of Absorption Heat Pump Conference, September 30-October 2, pp. 339-344.

Furutera, M., Furukawa, T., Mizuuchi, M., 1991, "Development of Compact Laminated Heat Exchanger for R22-DEGDME Absorption Refrigerating Machine", Tokyo: Proceedings of Absorption Heat Pump Conference, September 30-October 2, pp. 195-200.

Gierow, M., Jernqvist, A., 1994, "Selection of Working Pair for Sorption Heat Pumps: A Computer Simulation Study", New Orleans: Proceedings of the International Absorption Heat Pump Conference, ASME AES-Vol. 31, January 19-21, pp. 303-310.

Hanna, W.T., Wilkinson, W.H., 1982, "Absorption Heat Pumps and Working Pair Developments in the United States Since 1974", Berlin: Proceedings of New Working Pairs for Absorption Processes Workshop, April 14-16, pp. 71-84.

Heiks, J.R., Garrett, A.B., 1954, "The Magnesium Bromide-Lithium Bromide-Water System", J. Am. Chem. Soc., Vol. 76, pp. 2587-2590.

Herold, K.E., Howe, L.A., Radermacher, R., Erickson, D.C., November 1990, "Development of an Absorption Heat Pump Water Heater Using an Aqueous Ternary Hydroxide Working Fluid", Int. J. Refrig., Vol 14, pp. 156-167.

Hodgett, D.L., 1982, "Absorption Heat Pumps and Working Pair Developments in Europe Since 1974", Berlin: Proceedings of New Working Pairs for Absorption Processes Workshop, April 14-16, pp. 57-70.

Howe, L.A., Erickson, D.C., 1991, "High-Temperature Industrial Absorption Heat Pumping", Tokyo: Proceedings of Absorption Heat Pump Conference, September 30-October 2, pp. 165-170.

Jank, R., 1982, "Current and Historical Heat Pump Activities in Berlin", Berlin: Proceedings of New Working Pairs for Absorption Processes Workshop, April 14-16, pp. 11-16.

Kamoshida, J., Hirata, Y., Isshiki, N., Katayama, K., 1991, "Absorption Cycle with Many Kinds of Solutions for Storing Power/Upgrading Temperature", Tokyo: Proceedings of Absorption Heat Pump Conference, September 30-October 2, pp. 279-284.

Knoche, K.F., Molitor, K., Seitz, C.W., Zerres, H., Sagifi, M., 1988, "Periodically Operating Heat Pump with Working Fluid Methanol/Water-Lithium Bromide", London: Proceedings of Absorption Heat Pumps Workshop, April 12-14, pp. 145-154.

Knoche, K.F., Seitz, C.W., 1991, "Absorption Heat Pump with Periodically Operating Absorber", Tokyo: Proceedings of Absorption Heat Pump Conference, September 30-October 2, pp. 151.

Knoche, K.F., Lang, R., Straub, R., 1991, "Modular Absorption Heat Pump with a Thermally Driven Solution Pump", Tokyo: Proceedings of Absorption Heat Pump Conference, September 30-October 2, pp. 159-164.

Lai, H., Li, C., Zheng, D., 1991, "Dynamic Performance Simulation of $CaCl_2/CH_3OH$ Chemical Heat Pump", Tokyo: Proceedings of Absorption Heat Pump Conference, September 30-October 2, pp. 345-350.

Le Goff, P., Labidi, J., Ranger, P.M., Jeday, M.R., Matsuda, H., 1991, "New Concepts and New Materials for Heat Pumps Operating at Two Temperatures", Tokyo: Proceedings of Absorption Heat Pump Conference, September 30-October 22, pp. 19-30.

Macriss, R.A., 1982, "Overview and History of Absorption Fluid Development in the USA (1927-1974)", Berlin: Proceedings of New Working Pairs for Absorption Processes Workshop, April 14-16, pp. 37-56.

Murakami, K., Sato, H., Watanabe, K., 1991, "A Study of Bubble-Point Pressures for the Binary HCFC-22/DTRG Mixtures", Tokyo: Proceedings of Absorption Heat Pump Conference, September 30-October 2, pp. 267-272.

Okano, T., Asawa, Y., Fujimoto, M., Nishiyama, N., Sanai, Y., 1994, "Development of an Air-Cooled Absorption Refrigerating Machine Using a New Working Fluid", New Orleans: Proceedings of the International Absorption Heat Pump Conference, ASME AES-Vol. 31, January 19-21, pp. 311.

Ramshaw, C., Winnington, T., 1988, "An Intensified Absorption Heat Pump (ROTEX)", London: Proceedings of Absorption Heat Pumps Workshop, April 12-14, pp. 258-268.

Sawada, N., Tanaka, T., Mashimo, K., 1994, "Development of Organic Working Fluids and Application to Absorption Systems", New Orleans: Proceedings of the International Absorption Heat Pump Conference, ASME AES-Vol. 31, January 19-21, pp. 315-320.

Wang, S., Luo, B., Zheng, D., 1991, "Enthalpy-Concentration Relationship for Binary Aqueous Solutions of Glycols", Tokyo: Proceedings of Absorption Heat Pump Conference, September 30-October 2, pp. 273-278.

Zhuo, C. Z., Machielsen, C.H.M., 1994, "Experimental Measurements of an Absorption Heat Transformer with the Working Pair TFE-PYR", New Orleans: Proceedings of the International Absorption Heat Pump Conference, ASME AES-Vol. 31, January 19-21, pp. 321-326.

Ziegler, F., 1991, "Compression-Absorption Cycles with R22/E181 and NH_3/H_2O" Tokyo: Proceedings of Absorption Heat Pump Conference, September 30-October 2, pp. 91-96.

Chapter 4

Asawa, Y., Yokoyama, T., Fujimoto, M., Abe, K., 1991, "Absorption Enhancement by Adding External Force to a Falling Film Liquid Solution", Tokyo: Proceedings of Absorption Heat Pump Conference, September 30-October 2, pp. 201-206.

Bassols-Rheinfel, J., 1988, "Experimental/Numeric Optimization of the Heat Exchange Surfaces of a Sorption Heat Pump", London: Proceedings of Absorption Heat Pumps Workshop, April 12-14, pp. 419-427.

Benzeguir, B., Setterwall, F., Uddholm, H., 1991, "Use of a Wave Model to Evaluate the Falling Film Absorber Efficiency", Int. J. Refrig., Vol. 14, pp. 292-296.

Conlisk, A.T., 1994, "The Use of Boundary Layer Techniques in the Design of a Falling Film Absorber", New Orleans: Proceedings of the International Absorption Heat Pump Conference, ASME AES-Vol. 31, January 19-21, pp. 163-170.

Cosenza, F., Vliet, G.C., 1990, "Absorption in Falling Water/LiBr Films on Horizontal Tubes", ASHRAE Trans., Vol. 96, Pt. 1, AT-90-30-3.

Fitt, P.W., Hassoon, H., 1988, "A Circulating Bubble Absorber", London: Proceedings of Absorption Heat Pumps Workshop, April 12-14, pp. 383-390.

Furukawa, M., Enomoto, E., Sekoguchi, K., 1994, "Boiling Heat Transfer in High Temperature Generator of Absorption Chiller/Heater", New Orleans: Proceedings of the International Absorption Heat Pump Conference, ASME AES-Vol. 31, January 19-21, pp. 517-524.

Greiter, I., Wagner, A., Weiss, V., Alefeld, G., 1994, "Heat/Mass Transfer in a Horizontal-Tube Falling-Film Absorber with Aqueous Solutions", New Orleans: Proceedings of the International Absorption Heat Pump Conference, ASME AES-Vol. 31, January 19-21, pp. 225-232.

Grossman, G., 1991, "The Combined Heat and Mass Transfer Process in Film Absorption", Tokyo: Proceedings of Absorption Heat Pump Conference, September 30-October 2, p. 43-52.

Grossman, G., 1991, "Film Absorption Heat and Mass Transfer in the Presence of Non-Absorbable Gases", Tokyo: Proceedings of Absorption Heat Pump Conference, September 30-October 2, pp. 389-394.

Haselden, G.G., Malaty, S.A., 1959, "Heat and Mass Transfer Accompanying the Absorption of Ammonia in Water", Trans. Instn Chem. Eng., Vol. 37, pp. 137-146.

Haselden, G.G., Sutherland, J.P., 1960, "A Study of Plate-Efficiency in the Separation of Ammonia-Water Solutions", International Symposium on Distillation, pp. 27-32.

Infante Ferreira, C. A., Keizer, C., Machielsen, C.H.M., 1984, "Heat and Mass Transfer in Vertical Tube Bubble Absorbers for Ammonia/Water Absorption Refrigeration Systems", Int. J. Refrig., Vol. 7, pp 348-357.

Isshiki, N., Ogawa, K., Hosaka, H., Sasaki, N., 1988, "Studies on Mechanism and Enhancement of Absorption Heat and Mass Transfer", London: Proceedings of Absorption Heat Pumps Workshop, April 12-14, pp. 399-408.

Isshiki, K., Sasaki, N., Funato, Y., 1991, "R & D of CCS (Constant Curvature Surface) Tubes for Absorption Heat Exchangers", Tokyo: Proceedings of Absorption Heat Pump Conference, September 30-October 2, pp. 377-382.

Ji, W., Setterwall, F., 1994, "Effects of Surfactants on the Stability of Falling Liquid Films", New Orleans: Proceedings of the International Absorption Heat Pump Conference, ASME AES-Vol. 31, January 19-21, pp. 33-40.

Jung, S., Sgamboti, C., Perez-Blanco, H., 1994, "An Experimental Study of the Effect of Some Additives on Falling Film Absorption", New Orleans: Proceedings of the International Absorption Heat Pump Conference, ASME AES-Vol. 31, January 19-21, pp. 49-56.

Kamoshida, J., Enonoto, E., Isshiki, N., Katayama, K., 1991, "Vapor Absorption/Condensation Heat Transfer with a Bubble Dispersed System", Tokyo: Proceedings of Absorption Heat Pump Conference, September 30-October 2, pp. 189-194.

Kamoshida, J., Isshiki, N., 1994, "Heat Transfer to Water and Water/Lithium Halide Salt Solutions in Nucleate Pool Boiling", New Orleans: Proceedings of the International Absorption Heat Pump Conference, ASME AES-Vol. 31, January 19-21, pp. 501-508.

Kim, K. J., Berman, N. S., Wood, B. D., 1994, "Enhanced Heat/Mass Transfer Mechanisms Using Additives for Vertical Falling Film Absorber", New Orleans: Proceedings of the International Absorption Heat Pump Conference, ASME AES-Vol. 31, January. 19-21, pp. 41-48.

Le Goff, P., Ramadane, A., Louis, G., 1988, "Co-Counter or Cross-Current Gas-Liquid Absorption with Integrated or Separate Heat Exchange", London: Proceedings of Absorption Heat Pumps Workshop, April 12-14, pp. 353-361.

Machielsen, C.H.M., Becker, H., Westra, J.J.W., 1988, "The Use of Compact Heat and Mass Exchangers in Absorption Heat Pumps", London: Proceedings of Absorption Heat Pumps Workshop, April 12-14, pp. 409-418.

Merrill, T., Setoguchi, T., Perez-Blanco, H., 1994, "Compact Bubble Absorber Design and Analysis", New Orleans: Proceedings of the International Absorption Heat Pump Conference, ASME AES-Vol. 31, January 19-21, pp. 217-224.

Miller, W.A., Perez-Blanco, H., 1994, "Vertical-Tube Aqueous LiBr Falling Film Absorption Using Advanced Surfaces", New Orleans: Proceedings of the International Absorption Heat Pump Conference, ASME AES-Vol. 31, January 19-21, pp. 185-202.

Minkowycz, W.J., Sparrow, E.M., 1966, "Condensation Heat Transfer in the Presence of Noncondensables, Interfacial Resistance, Superheating, Variable Properties and Diffusion", Int. J. Heat Mass Transfer, Vol. 9, pp. 1125-1144.

Nomura, T., Nishimura, N., Wei, S., Yamaguchi, S., Kawakami, R., 1994, "Heat/Mass Transfer Mechanism in the Absorber of Water/LiBr Convectional Absorption Refrigerator", New Orleans: Proceedings of the International Absorption Heat Pump Conference, ASME AES-Vol. 31, January 19-21, pp. 203-208.

Ogawa, M., Hoshida, T., Oda, Y., 1991, "Heat Transfer Analysis for the Vertical Condenser/Absorber of an Air-Cooling Absorption Chiller/Heater", Tokyo: Proceedings of Absorption Heat Pump Conference, September 30-October 2, pp. 225-230.

Patnaik, V., Miller, W.A., Perez-Blanco, H., 1994, "Empirical Methodology for the Design of Vertical Tube Absorbers", ASHRAE Trans., Vol. 100, Pt. 2, Paper# 3801.

Patnaik, V., Perez-Blanco, H., 1994, "A Counterflow Heat-Exchanger Analysis for the Design of Falling-Film Absorbers", New Orleans: Proceedings of the International Absorption Heat Pump Conference, ASME AES-Vol. 31, January 19-21, pp. 209-216.

Persson, L.H., Holmberg, P.A., 1994, "Transfer Falling Film Desorption of Concentrated Lithium-Bromide Aqueous Solutions: Octanol Additive", New Orleans: Proceedings of the International Absorption Heat Pump Conference, ASME AES-Vol. 31, January 19-21, pp. 57-64.

Radermacher, R., Alefeld, G., 1982, "Lithiumbromid-Wasser-Losungen als Absorber fur Ammoniak oder Methylamin", Brennstoff, Waerme, Kraft, Vol. 34, Nr. 1, pp. 31-38.

Rignac, J.P., Huor, M.H., Bugarel, R., 1988, "Transfer Heat Pump Absorber: Film Tubular Absorption with Tangential Feed/Turbulence Promoter", London: Proceedings of Absorption Heat Pumps Workshop, April 12-14, pp. 362-371.

Rush, W.F., Wurm, J., Perez-Blanco, H., 1991, "A Brief Review of the Uses and Effect of Additives for Absorption Enhancement", Tokyo: Proceedings of Absorption Heat Pump Conference, September 30-October 2, pp. 183-188.

Ryan, W.A., 1994,"Water Absorption in an Adiabatic Spray of Aqueous Lithium Bromide Solution", New Orleans: Proceedings of the International Absorption Heat Pump Conference, ASME AES-Vol. 31, January 19-21, pp. 155-162.

Serpente, C. P., Kernen, M., Seewald, J. S., Perez-Blanco, H., 1994, "2 kW Lithium-Bromide Absorption Machine with Heat Recovery/Recirculation for Novel Fluid Testing", New Orleans: Proceedings of the International Absorption Heat Pump Conference, ASME AES-Vol. 31, January 19-21, pp. 65-72.

Setoguchi, T., Perez-Blanco, H., 1991, "Effect of Additives in Ammonia Absorption in Water in a Stagnant Pool", Tokyo: Proceedings of Absorption Heat Pump Conference, September 30-October 2, pp. 177-182,

Strigle, R.F., August 1993, "Understand Flow Phenomena in Packed Columns", Chem. Eng. Progress, pp. 79-83.

Smith, I.E., Swallow, F.E., 1988, "Design Development and Performance Evaluation of a Rotating Disc Vapour Absorber", London: Proceedings of Absorption Heat Pumps Workshop, April 12-14, pp. 373-382.

Schwarzer, B. P., Semnani Rahbar, M., Le Goff, P., 1994, "A Spiral Fin Tube: A Novel Type of Falling Film Heat and Mass Exchanger", New Orleans: Proceedings of the International Absorption Heat Pump Conference, ASME AES-Vol. 31, January 19-21, pp. 179-184.

Vliet, G. C., Chen, W., 1994, "Location of Non-Absorbable Gases in a Simplified Absorber Geometry", New Orleans: Proceedings of the International Absorption Heat Pump Conference, ASME AES-Vol. 31, January 19-21, pp. 171-178.

Vliet, G.C., Cosenza, F.B., 1991, "Absorption Phenomenon in Water-Lithium Bromide Films", Tokyo: Proceedings of Absorption Heat Pump Conference, September 30-October 2., pp. 53-61.

Adsorption Systems

Baoqi, H., Yingqiu, L., Qin, W., 1994, "Performance of Adsorption/Heat-Mass Transfer of Zeolite 13X-H_2O: Absorption Cooling System", New Orleans: Proceedings of the International Absorption Heat Pump Conference, ASME AES-Vol. 31, January 19-21, pp. 445-456.

Bjurstrom, H., Lewis, D., Suda, S., 1988, "Simulation of Periodic Heat Pumps Based on Metal Hydrides", London: Proceedings of Absorption Heat Pumps Workshop, April 12-14, pp. 110-120.

Boelman, E., Furuta, Y., Tanaka, T., Kashiwagi, T., 1991, "Adsorption-Type Heat-Pump Chiller Driven By Low-Grade Waste-Heat", Tokyo: Proceedings of Absorption Heat Pump Conference, September 30-October 2, pp. 291-296.

Bouqard, J., Jadot, R., Poulain, V., 1994, "Solid-Gas Reactions Applied to Thermotransformer Design", New Orleans: Proceedings of the International Absorption Heat Pump Conference, ASME AES-Vol. 31, January 19-21, pp. 413-418.

Critoph, R.E., Turner, H.L., Shelton, S.V., Miles, D.J., 1988, "Thermal Wave Heat Pump Cycle", London: Proceedings of Absorption Heat Pumps Workshop, April 12-14, pp.139-144.

de Beijer, H.A., Klein Horsman, J.W., 1994, "S.W.E.A.T. Thermomechanical Heat Pump Storage System", New Orleans: Proceedings of the International Absorption Heat Pump Conference, ASME AES-Vol. 31, January 19-21, pp. 457-462.

de Beijer, H.A., Klein-Horsman, J.W., de Beijer R.T.B., B.V., 1991, "S.W.E.A.T. Thermochemical Heat Pump Storage System", Tokyo: Proceedings of Absorption Heat Pump Conference, September 30-October 2, pp. 309-314.

Guilleminot, J. J., Chalfen, J. B., Choisier, A., 1994, "Heat and Mass Transfer Characteristics of Composites for Adsorption Heat Pumps", New Orleans: Proceedings of the International Absorption Heat Pump Conference, ASME AES-Vol. 31, January 19-21, pp. 401-406.

Hisaki, H., Kobayashi, N., Yonezawa, Y., Morikawa, A., 1994, "Development of Ice-Thermal Storage System Using an Adsorption Chiller", New Orleans: Proceedings of the International Absorption Heat Pump Conference, ASME AES-Vol. 31, January 19-21, pp. 439-444.

Knobbout, J.A., Zegers, P.,1988, "R&D of Absorption and Adsorption Heat Pumps and Heat Transformers in the CEC Programme", London: Proceedings of Absorption Heat Pumps Workshop, April 12-14, pp. 155-159.

Lai, H., Li, C., Zheng, D., Fu, J., 1994, "Modeling of Cycling Process of CaCl/CHOH Solid-Vapor Chemical Heat Pump", New Orleans: Proceedings of the International Absorption Heat Pump Conference, ASME AES-Vol. 31, January 19-21, pp. 419-424.

Mamiya, T., Nikai, I., 1994, "Study on Heat Transfer in Tube Plate Adsorption Reactor", New Orleans: Proceedings of the International Absorption Heat Pump Conference, ASME AES-Vol. 31, January 19-21, pp. 425-432.

Mazet, N., Meyer, P., Neveu, P., Spinner, B., 1994, "Concept/Study of Double Effect Refrigeration Machine Based on Sorption of Solid/Ammonia Gas", New Orleans: Proceedings of the International Absorption Heat Pump Conference, ASME AES-Vol. 31, January 19-21, pp. 407-412.

Meunier, F., Douss, N., 1988, "Adsorptive Heat Pumps Active Carbon-Methanol and Zeolite-Water Pairs", London: Proceedings of Absorption Heat Pumps Workshop, April 12-14, pp. 100-109.

Rockenfeller, U., Kirol, L., 1994, "HVAC and Heat Pump Development Employing Complex Compound Working Media", New Orleans: Proceedings of the International Absorption Heat Pump Conference, ASME AES-Vol. 3, January 19-21, pp. 433-438.

Yanagi, H., Okamoto, N., Komatsu, F., Ino, N., Ogura, M.,1991, "Development of Adsorption Refrigerators Using Silica Gel-Water Pairs", Tokyo: Proceedings of Absorption Heat Pump Conference, September 30-October 2, pp. 115-120.

Zhu, R., Han, B., Yu, Y., 1991, "An Investigation on Zeolite-Water Adsorption Refrigeration", Tokyo: Proceedings of Absorption Heat Pump Conference, September 30-October 2, pp. 303-308.

Chapter 5

Boryta, D.A., 1970, "Solubility of Lithium Bromide in Water Between -50°C and 100°C (45 to 70% Lithium Bromide)", J. Chem. Eng. Data, Vol. 15, No. 1, pp. 142-144.

Ellington, R.T., Kunst, G., Peck, R.E., Reed, J.F., 1957, "The Absorption Cooling Process: A Critical Literature Review", Institute of Gas Technology Research Bulletin 14, Library of Congress Catalog # 54-12110.

Esaki, S., 1991, "Discharge of Hydrogen Using a Palladium Tube in an Absorption Chiller Heater", Tokyo: Proceedings of Absorption Heat Pump Conference, September 30-October 2, pp. 159-164.

Foote Technical Data, Bulletin 145, Foote Mineral Company.

Krueger, R.H., Dockus, K.F., Rush, W.F., February 1964, "Lithium Chromate: Corrosion Inhibitor for Lithium Bromide Absorption Refrigeration Systems", ASHRAE J., Vol. 6, pp. 40-44.

Linke, W.F., Seidell, A., 1941, Solubilities of Inorganic and Metal Organic Compounds; A Compilation of Quantitative Solubility Data from the Periodical Literature, 4th edition, Van Nostrand, New York.

McBride R.B., McKinley, D.L., 1965, "A New Hydrogen Recovery Route", Chemical Engineering Progress, Vol. 61, No. 3, pp. 81-85.

Murray, J.G., 1993, "Purge Systems for Absorption Chillers", ASHRAE Trans., Vol. 99, Pt. 1, pp. 1485-1494.

Ogawa, K., Isshiki, N., Yasuda, K., 1991, "Optimization of Liquid Film Thickness in Absorption Heat Transfer Time Duration", Tokyo: Proceedings of Absorption Heat Pump Conference, September 30-October 2, pp. 383-388.

Reay, D.A., 1988, "The Role of Improved Heat Exchangers in Absorption Cycle Heat Pump Optimization", London: Proceedings of Absorption Heat Pumps Workshop, April 12-14, pp. 428-438.

Setterwall, F., Strenger, U., 1994, "Investigations on Lammellas in an Absorption Heat Pump", New Orleans: Proceedings of the International Absorption Heat Pump Conference, ASME AES-Vol. 31, January 19-21, pp. 149-154.

Setterwall, F., Yao, W., Ji, W., Bjurstrom, H., 1991, "Heat Transfer Additives in Absorption Heat Pumps", Tokyo: Proceedings of Absorption Heat Pump Conference, September 30-October 2, pp. 73-78.

Sparrow, E.M., Minkowycz, W.J., Saddy, M., 1967, Forced Convection Condensation in the Presence of Noncondensables and Interfacial Resistance", Int. J. Heat Mass Transfer. Vol. 10, pp. 1829-11845.

Vliet, G.C., Chen, W., 1994, "Location of Non-Absorbable Gases in a Simplified Absorber Geometry", New Orleans: Proceedings of the International Absorption Heat Pump Conference, ASME AES-Vol. 31, January 19-21, pp. 171-178.

Chapter 6

AGCC, 1995, "Natural Gas Cooling Equipment Guide", 2nd edition, American Gas Cooling Center, Arlington, Va.

Eisa, M.A.R., Devotta, S., Holland, F.A., 1986, "Thermodynamic Design Data for Absorption Heat Pump Systems Operating on Water-Lithium Bromide. I. Cooling", Appl. Energy, Vol. 24, pp. 287-301.

Eisa, M.A.R., Holland, F.A., 1986, "A Study of the Operating Parameters in a Water-Lithium Bromide Absorption Cooler", Energy Res., Vol. 10, pp. 137-144.

Eisa, M.A.R., Devotta, S., Holland, F.A., 1985, "A Study of Economiser Performance in a Water-Lithium Bromide Absorption Cooler", Int. J. Heat Mass Transfer, Vol. 28, No. 12, pp. 2323-2329.

Eisa, M.A.R., Diggory, P.J., Holland, F.A., 1986, "Experimental Studies to Determine the Effect of Absorber Reflux on the Performance of a Water-Lithium-Bromide Absorption Cooler", Energy Res. , Vol. 10, pp. 333-341.

Eriksson, K., Jernqvist, A., January, 1989, "Heat Transformer with Self-Circulation: Design and Preliminary Operational Data", Int. J. Refrig., Vol. 12, pp. 15-20.

Koeppel, E.A., Klein, S.A.,Mitchell, J.W., 1995, "Commercial Absorption Chiller Models for Evaluation of Control Strategies", ASHRAE Trans., Vol. 101, Pt 1, pp. 1175-1184.

Landauro-Paredes, J.M., Watson, F.A., Holland, F.A., November 1983, "Experimental Study of the Operating Characteristics of a Water-Lithium Bromide Absorption Cooler", Chem. Eng. Res. Des., Vol. 61, pp. 362-370.

Leigh, R.W., Isaacs H.S., Kirley, J., Ravve A., 1989, "Capital Cost Reductions in Absorption Chillers", ASHRAE Trans., Vol.95, Pt. 1, CH-89-15-1

Sano, J., Hihara, E., Sato, T., Nagaoka, Y., Nishiyama, N., 1991, "Dynamic Simulation of an Absorption Refrigerating Machine", Tokyo: Proceedings of Absorption Heat Pump Conference, September 30-October 2, pp. 133-138.

Absorption Heat Transformers (Type II)

Ahachad, M., Charia, M., 1994, "Absorption Heat Transformer Applications to Absorption Refrigerating Machine", New Orleans: Proceedings of the International Absorption Heat Pump Conference, ASME AES-Vol. 31, January 19-21, pp. 101-108.

Berntsson, T., Holmberg, P., Liu, Y., Wimby, M., 1988, "Research Activities on Heat Transformers at Chalmers University of Technology Sweden", London: Proceedings of Absorption Heat Pumps Workshop, April 12-14, pp. 322-331.

Bisio, G., Pisoni, C., 1991, "Thermodynamic Analysis of Heat Transformers", Tokyo: Proceedings of Absorption Heat Pump Conference, September 30-October 2, pp. 315-320.

Clark, E.C., Morgan, O., August 1981, "Chemical Heat Pumps for Industry", Atlanta: Proceedings 16th Intersociety Energy Conversion Engineering Conference, pp. 866-870.

Cohen, G., Salvat, J., Rojey, A., August 1979, "A New Absorption-Cycle Process for Upgrading Waste Heat", Proceedings 14th Intersociety Energy Conversion Engineering Conference, Vol. 2, pp. 1720-1724.

Dong, T., Qian, B., Weng, J., Yu, X., 1991, "Lithium Bromide-Water Absorption Heat Transformer Using 78°C Waste Heat", Tokyo: Proceedings of Absorption Heat Pump Conference, September 30-October 2, pp. 357-362.

Eisa, M.A.R., Best, R., Holland, F.A., 1986, "Thermodynamic Design Data for Absorption Heat Transformers. I. Operating on Water-Lithium Bromide", Heat Recovery Syst., Vol. 6, No. 5, pp. 421-432.

Eriksson, K., Jernqvist, A., January 1989, "Heat Transformer with Self-Circulation: Design and Preliminary Operational Data", Int. J. Refrig., Vol. 12, pp. 15-20.

Fournier, P., Kashiwagi, T., 1991, "Simulation of a New Absorption Temperature Amplifier Using a TFE-E181 Pair for Low Level Temperature Upgrading", ASHRAE Trans., Vol. 97, Pt. 1, pp. 156-162.

George, J.M., Srinivasa, S., 1993, "Experiments on a Vapour Absorption Heat Transformer", Int. J. Refrig., Vol. 16, No. 2, pp. 107-119.

Grossman, G, 1991, "Absorption Heat Transformers for Process Heat Generation From Solar Ponds", ASHRAE Trans., Vol. 97, Pt. 1, pp. 420-433.

Grossman, G., 1985, "Multistage Absorption Heat Transformers for Industrial Applications", ASHRAE Trans., Vol. 91, Pt. 2B, pp. 2047-2061.

Herold, K.E., Moran, M.J., 1985, "A Thermodynamic Investigation of an Absorption Temperature Boosting Heat Pump Cycle," Proceedings of Advanced Energy Systems Symposium at ASME Winter Annual Meeting, Miami, Florida, AES-Vol. 1, pp. 81-88.

Holmberg, P., 1988, "System Studies and Optimization of the Single-Stage Absorption Heat Transformer Cycle", Chalmers University of Technology, Sweden, pp. 1-94.

Holmberg, P., Berntsson, T., 1990, "Alternative Working Fluids in Heat Transformers", ASHRAE Trans., Vol. 96, Pt. 1, pp. 1582-1589.

Iyoki, S., Uemura, T., May 1990, "Performance Characteristics of the Water-Lithium Bromide-Zinc Chloride-Calcium Bromide, Absorption Refrigerating Machine, Absorption Heat Pump and Absorption Heat Transformer", Int. J. Refrig., Vol. 13, pp. 191-196.

Kamoshida, J., Hirata, Y., Isshiki, N., Katayama, K., 1991, "Absorption Cycle with Many Kinds of Solutions for Storing Power/Upgrading Temperature", Tokyo: Proceedings of Absorption Heat Pump Conference, September 30-October 2, pp. 279-284.

Knobbout, J.A., Zegers, P., 1988, "R&D of Absorption and Adsorption Heat Pumps and Heat Transformers in the CEC Programme", London: Proceedings of Absorption Heat Pumps Workshop, April 12-14, pp. 155-159.

Kouremenos, D.A., 1985, "A Tutorial on Reversed NH_3/H_2O Absorption Cycles for Solar Applications", Solar Energy, Vol. 34, No. 1, pp. 101-115.

Krom, R.M., van Buren, J.E., 1988, "Technical-Economical Aspects of an Absorption Heat Pump Type II at Hoogvens Ijmuiden", London: Proceedings of Absorption Heat Pumps Workshop, April 12-14, pp. 76-84.

Mohanty, B., Paloso, G., 1991, "Boosting Low Temperature Geothermal Heat Source by Absorption Heat Transformer for Power Generation", Tokyo: Proceedings of Absorption Heat Pump Conference, September 30-October 2, pp. 351-356.

Stephan, K., Hengerer, R., 1993, "Heat Transformation with the Ternary Working Fluid TFE-H2O-E181", Int. J. Refrig., Vol. 16, Pt 2, pp. 120-128.

Zhuo, C.Z., Machielsen, C.H.M., 1994, "Experimental Measurements of an Absorption Heat Transformer with the Working Pair TFE-PYR", New Orleans: Proceedings of the International Absorption Heat Pump Conference, ASME AES-Vol. 31, January 19-21, pp. 321-326.

Chapter 7

AGCC, 1995, "Natural Gas Cooling Equipment Guide", 2nd edition, American Gas Cooling Center, Arlington, Va.

Alefeld, G., Radermacher, R., 1994, Heat Conversion Systems, CRC Press, Boca Raton, FL.

Alefeld, G., Ziegler, F., 1985, "Advanced Heat Pump and Air Conditioning Cycles for the Working Pair H_2O/LiBr: Domestic and Commercial Applications", ASHRAE Trans., Vol. 91, 2B, pp. 2062-2071.

DeVuono, A.C., Christensen, R.N., Landstrom, D.K., Wilkinson, W.H., Ryan, W.A., 1990, "Development of a Residential Gas-Fired Double-Effect Air Conditioner-Heater Using Water and Lithium Bromide", <u>ASHRAE Trans.</u>, Vol. 96, Pt. 1, pp. 1494-1498.

Fallek, M., 1985, "Parallel Flow Chiller Heater", <u>ASHRAE Trans.</u>, Vol. 91, Pt. 2B, pp. 2095-2102.

Kurosawa, S., Fufimaki, S., 1989, "Development of Air Cooled Double Effect Gas Fired Absorption Water Chiller Heater", <u>ASHRAE Trans.</u>, Vol. 95, Pt. 1, pp. 318-325.

Kurosawa, S., Yoshikawa, M., 1982, "The Highest Efficiency Gas Direct-Fired Absorption Water Heater-Chiller", <u>ASHRAE Trans.</u>, Vol. 88, Pt. 1, pp. 401-415.

Petty, S.E., Meacham, H.C., Cook, F.B., 1990, "Status of the Double-effect Absorption Heat Pump", <u>ASHRAE Trans.</u>, Vol. 96, Pt. 1, AT-90-27-2

Wilkinson, W.H., 1987, "What are the Performance Limits for Double Effect Absorption Cycles?", <u>ASHRAE Trans.</u>, Vol. 93, Pt. 2, pp. 2429-2441.

Chapter 8

Alefeld, G., Demmel, S., Kern, W., Scharfe, J., Riesch, P., 1991, "Advanced Absorption Cycles and Systems for Environmental Protection", Tokyo: Proceedings of Absorption Heat Pump Conference , September 30-October 2, pp. 9-18.

Alefeld, G., Radermacher, R., 1994, <u>Heat Conversion Systems</u>, CRC Press, Boca Raton, FL.

Alefeld, G., Scharfe, J., 1991, "An Absorption Heat Pump Transformer (Heat Pump Type III) For Distillation Plants", Tokyo: Proceedings of Absorption Heat Pump Conference, September 30-October 2, pp. 321-327.

Alefeld, G., Ziegler, F., 1985, "Advanced Heat Pump and Air-Conditioning Cycles for the Working Pair H_2O/LiBr: Industrial Applications", <u>ASHRAE Trans.</u>, Vol. 91, Pt. 2B, pp. 2072-2080.

Alefeld, G., Feuerecker, G., Greiter, I., Kern, W., Riesch, P., Scharfe, J., Ziegler, F., 1992, "Untersuchung fortgeschrittener Absorptionswärmepumpen", <u>IZW-Berichte</u>, Informations-Zentrum Wärmepumpen + Kältetechnik, Karlsruhe.

Arh, S., 1994, "Absorption Heat Pump Transformer Cycle for Simultaneous Heating and Cooling", New Orleans: Proceedings of the International Absorption Heat Pump Conference, ASME AES-Vol. 31, January 19-21, pp. 79-84.

Boer, D., Huor, M. H., Prevest, M., Coronas, A., 1994, "Mixed Vapor Compression-Double Effect Absorption Cycle for A/C: A New High Performance Cycle", New Orleans: Proceedings of the International Absorption Heat Pump Conference, ASME AES-Vol. 31, January 19-21, pp. 483-486.

Bokelmann, H., 1988, "Advanced Absorption Systems", London: Proceedings of Absorption Heat Pumps Workshop, April 12-14, pp. 332-341.

CAC, 1985, Compound Absorption Chiller project performed for DOE by Battelle Memorial Institute.

Erickson, D.C., 1995, "Waste-Heat-Powered Icemaker for Isolated Fishing Villages", ASHRAE Trans., Vol. 101, Pt. 1.

Gommed, K., Grossman, G., 1990, "Performance Analysis of Staged Absorption Heat Pumps: Water Lithium Bromide Systems", ASHRAE Trans., Vol. 96, Pt. 1, pp. 1590-1598.

Greiter, I., Kern, W., Alefeld, G., 1991, "A 500 kW Heat Pump for Heating at Two Temperature Levels and for Air -Conditioning", Tokyo: Proceedings of Absorption Heat Pump Conference, September 30-October 2, pp. 285-290.

Grossman, G., 1994, "Modular and Flexible Simulation of Advanced Absorption Systems", New Orleans: Proceedings of the International Absorption Heat Pump Conference, ASME AES-Vol. 31, January 19-21, pp. 345-352.

Grossman, G., Zaltash, A., DeVault, R.C., 1995, "Simulation and Performance Analysis of a Four-Effect Lithium Bromide-Water Absorption Chiller", ASHRAE Trans., Vol. 101, Pt. I, pp. 1302-1312.

Herold, K.E., Howe, L.A., Radermacher, R., September 1991, "Analysis of a Hybrid Compression-Absorption Cycle Using Lithium Bromide and Water as the Working Fluid", Int. J. Refrig., Vol. 14, pp. 264-272.

Howe, L.A., Erickson, D.C., 1991, "High-Temperature Industrial Absorption Heat Pumping", Tokyo: Proceedings of Absorption Heat Pump Conference, September 30-October 2, pp. 165-170.

Isshiki, N., 1991 , "Studies and Perspectives on Absorption Technology", Tokyo: Proceedings of Absorption Heat Pump Conference, September 30-October 22, pp. 3-8.

Kahn, R., Scharfe, J., Haberle, A., Gunzbourg, J., Larger, D., 1991, "An Absorption Heat Pump for Water Desalination", Tokyo: Proceedings of Absorption Heat Pump Conference, September 30-October 2, pp. 231-236.

Labidi, J., Schwarzer, B. P., Le Goff, P., 1994, "Absorption Heat Pump Composed of Multiple Stages: Independent or Belonging to a Unique Column", New Orleans: Proceedings of the International Absorption Heat Pump Conference, ASME AES-Vol. 31, January 19-21, pp. 251-256.

McNeely, L.A., 1979, "Thermodynamic Properties of Aqueous Solutions of Lithium Bromide", ASHRAE Trans., Vol. 85, Pt. I, pp. 413-434.

Mohdal R.J., Hayes, F.C., 1992, "Development and Proof-Testing of Advanced Absorption Refrigeration Cycle Concepts, Report on Phases I and IA", ORNL Report ORNL/Sub/86-17498/1.

Ohuchi, T., Kunugi, Y., Aizawa, M., Kawakami, R., 1991, "Development of Absorption Air-Conditioners", Tokyo: Proceedings of Absorption Heat Pump Conference, September 30-October 2, pp. 219-224.

Ouimette, M.S., Herold, K.E., 1994, "Performance Modelling of a Triple-Effect Absorption Cycle", New Orleans: Proceedings International Absorption Heat Pump Conference, January 19-21, pp. 233-242.

Painchaud, G., 1994, "Evaluation of Advanced Absorption Technologies for Canadian Applications", New Orleans: Proceedings of the International Absorption Heat Pump Conference, ASME AES-Vol. 31, January 19-21, pp. 353-360.

Perez-Blanco, H., 1993, "Conceptual Design of a High-Efficiency Absorption Cooling Cycle", Rev. Int. Froid, Vol. 16, No. 6, pp. 429-433.

Rose, D. T., Zuritz, C. A., Perez-Blanco, H., 1994, "Thermodynamic Analysis/Pilot Plant Design for a Solar Assisted Double-Effect Refrigeration Cycle", New Orleans: Proceedings of the International Absorption Heat Pump Conference, ASME AES-Vol. 31, January 19-21, pp. 109-116.

Scharfe, J., 1988, "The Heat-Pump-Transformer", London: Proceedings of Absorption Heat Pumps Workshop, April 12-14, pp. 342-344.

Ziegler, F., Alefeld, G., 1994, "Comparison of Multi-Effect Absorption Cycles", New Orleans: Proceedings of the International Absorption Heat Pump Conference, ASME AES-Vol. 31, January 19-21, pp. 257-264.

Ziegler, F., Brandl, F., Völkl, J., Alefeld, G., 1985, "A Cascading Two-Stage Sorption Chiller System Consisting of a Water-Zeolite High Temperature Stage and a Water-LiBr Low-Temperature Stage", Paris: Absorption Heat Pumps Congress, March 20-22.

Air-Cooled Systems

Mardorf, L., 1994, "Controllable Cycle Investigation of Direct Fired Absorption Heat Pump for Residential Heat Systems", New Orleans: Proceedings of the International Absorption Heat Pump Conference, ASME AES-Vol. 31, January 19-21, pp. 339-344.

Ogawa, M., Hoshida, T., Oda, Y., 1991, "Heat Transfer Analysis for the Vertical Condenser/Absorber of an Air-Cooling Absorption Chiller/Heater", Tokyo: Proceedings of Absorption Heat Pump Conference, September 30-October 2, pp. 225-230.

Oh, M.D., Kim, S.C., Kim, Y.L., Kim, Y.I., 1994, "Cycle Analysis of Air-Cooled Double-Effect Absorption Heat Pump with Parallel Flow Type", New Orleans: Proceedings of the International Absorption Heat Pump Conference, ASME AES-Vol. 31, January 19-21, pp. 117-124.

Ohmori, S., Furukawa, T., 1991, "The Characteristics of Air-Cooled Absorption Refrigeration Machine for A/C", Tokyo: Proceedings of Absorption Heat Pump Conference, September 30-October 2, pp. 207-212.

Okano, T., Asawa, Y., Fujimoto, M., Nishiyama, N., Sanai, Y., 1994, "Development of an Air-Cooled Absorption Refrigerating Machine Using a New Working Fluid", New Orleans: Proceedings of the International Absorption Heat Pump Conference, ASME AES-Vol. 31, January 19-21, p. 311.

Sawada, N., Tanaka, T., Ikumi, Y., Kobayashi, T., Abe, K., 1991, "Study on Air-Cooled Absorption System for Light Commercial Use", Tokyo: Proceedings of Absorption Heat Pump Conference, September 30-October 2, pp. 213-218.

Tongu, S., Makino, Y., Ohnishi, K., Nakatsugawa, S., 1994, "Practical Operating of Air-Cooled Double-Effect Absorption Chiller-Heater by Lithium-Bromide/Aqueous", New Orleans: Proceedings of the International Absorption Heat Pump Conference, ASME AES-Vol. 31, January 19-21, pp. 125-132.

Applications

Abrahamsson, K., Aly, G., Jernqvist, A., Stenstrom, S., 1994, "Applications of Absorption Heat Cycles in the Pulp and Paper Industry", New Orleans: Proceedings of the International Absorption Heat Pump Conference, ASME AES-Vol. 31, January 19-21, pp. 295-302.

Ataer, A.E., Kilkis, B., 1994, "An Analysis of the Solar Absorption Cycle When Coupled with In-Slab Radiant Cooling Panels", New Orleans: Proceedings of the International Absorption Heat Pump Conference, ASME AES-Vol. 31, January 19-21, pp. 385-392.

Bouma, J.W.J., 1991, "International Users Club of Sorption Systems", Tokyo: Proceedings of Absorption Heat Pump Conference, September 30-October 2, p. 261.

Grossman, G., Johannsen, A., 1981, "Solar Cooling and Air Conditioning", Prog. Energy Combust. Sci., Vol. 7, pp. 185-228.

Lu, Z., Yu-Chi, B., Fan, L., Tang, J., Wang, C., 1991, "Absorption Heat Pumps Powered by Industrial Waste Heat", Tokyo: Proceedings of Absorption Heat Pump Conference, September 30-October 2, pp. 369-376.

Perez-Blanco, H., Grossman, G., 1981, "Cycle and Performance Analysis of Absorption Heat Pump for Waste Heat Utilization", ORNL Report, ORNL/TM-7852.

Zhou, Z., 1991, "Recent Development of Absorption Refrigerating Machine and Heat Pump in China", Tokyo: Proceedings of Absorption Heat Pump Conference, September 30-October 2, pp. 363-368.

Ziegler, F., Kahn, R., Summerer, F., Alefeld, G., 1993, "Multi-Effect Absorption Chillers", Int. J. Refrig., Vol. 16, No. 5.

Ziegler, F., Feuerecker, G., Alefeld, G., 1993, "Evaluation of Complex Energy Conversion Systems by Advanced Thermodynamic Analysis", <u>Energy Systems and Ecology, Proc. of the Int. Conf.</u>, Cracow, Poland, July 5-9, 1993, Vol. 1, pp. 449-456.

Integration of Absorption Systems in Cogeneration Facilities

Homma, R., Nishiyama, N., Wakimizu, H., 1994, "Simulation and Research of Single-Effect Absorption Refrigerators Driven by Waste Hot Water", New Orleans: Proceedings of the International Absorption Heat Pump Conference, ASME AES-Vol. 31, January 19-21, pp. 273-278.

Kashiwagi, T., 1991, "Importance of Thermally Activated Heat Pumps for Solving Global Environmental Problems", Tokyo: Proceedings of Absorption Heat Pump Conference, September 30-October 2, pp. 63-72.

Spinner, B., Mauran, S., 1988, "Industrial Heat Recovery: Heat Upgrading or Cold Product Using Solid-Gas Chemical Heat Pump", London: Proceedings of Absorption Heat Pumps Workshop, April 12-14, pp. 91-99.

Other Cycles with Absorption Related Technology

Amrane, K., Rane M.V. , Radermacher R., 1991."Performance Curves for Single-Stage Vapor Compression Cycles with Solution Circuit", <u>ASME J. Eng. Gas Turbines and Power</u>, Vol. 113, No. 2., pp. 221-227.

Cerepnalkovsk, I., 1991, "Improved Absorption-Compression Heat Pump: Geothermal Water as a Heat Source", Tokyo: Proceedings of Absorption Heat Pump Conference, September 30-October 2, pp. 109-114.

Cheron, J., 1988, "Absorption Cycles with Auxiliary Fluid", London: Proceedings of Absorption Heat Pumps Workshop, April 12-14, pp. 317-321.

Daltrophe, N.C., Jelinek, M., Borde, I., 1994, "Heat and Mass Transfer in a Jet Ejector for Absorption Systems", New Orleans: Proceedings of the International Absorption Heat Pump Conference, ASME AES-Vol. 31, January 19-21, pp. 327-332.

Groll, E.A., Radermacher, R., 1994, "Vapor Compression Heat Pump with Solution Circuit and Desorber/Absorber Heat Exchange", New Orleans: Proceedings of the International Absorption Heat Pump Conference, ASME AES-Vol. 31, January 19-21, pp. 463-470.

Herold K.H., Howe L.A. , Radermacher R., 1991, "Analysis of a Hybrid Compression/Absorption Cycle Using Lithium Bromide and Water as the Working Fluid", <u>Int. J. of Refrig.</u>, Vol. 14, pp. 264-272.

Hodgett, D.L., Aahlby, L., 1988, "Compression-Absorption Heat Pumps.", London: Proceedings of Absorption Heat Pumps Workshop, April 12-14, pp. 204-215.

Isshiki, N., 1991 , "Studies and Perspectives on Absorption Technology", Tokyo: Proceedings of Absorption Heat Pump Conference, September 30-October 2, pp. 423-438.

Itard, L. C., Machielsen, C. H. M., 1994, "Study for the Optimization of a Compression Hybrid Cycle for the Working Pair NH_3/H_2O", New Orleans: Proceedings of the International Absorption Heat Pump Conference, ASME AES-Vol. 31, January 19-21, pp. 17-24.

Kawada, A., Otake, M., Toyofuku, M., 1991, "Absorption Compression Heat Pump Using TFE/EL8L", Tokyo: Proceedings of Absorption Heat Pump Conference, September 30-October 2, pp. 121-126.

Kawada, A., Furutera, M., Hoshida, T., Yumikura, T., Izawa, H., 1994, "NH_3/H_2O System Compression Hybrid Pump Evaluation of Temperature Amplifier Type Heat Pump", New Orleans: Proceedings of the International Absorption Heat Pump Conference, ASME AES-Vol. 31, January 19-21, pp. 477-482.

Pourreza-Djourshari S. , Radermacher R. , 1986, "Calculation of the Performance of Vapor Compression Heat pumps with a Two-Stage Solution Circuit", Int. J. of Refrig., Vol. 9, pp. 245-250.

Radermacher R. 1988, "An Example for the Manipulation of Effective Vapor Pressure Curves by Thermodynamics Cycles" ASME J. Eng. Gas Turbines and Power, Vol. 110, pp. 647-651.

Radermacher, R., Herold, K.E., Howe, L.A., 1988, "Combined Vapor Compression/Absorption Cycles", London: Proceedings of Absorption Heat Pumps Workshop, April 12-14, pp. 225-234.

Radermacher R., Howe L.A., 1988, "Temperature Transformation for High Temperature Heat Pumps" ASME J. Eng. Gas Turbines and Power, Vol. 110, pp. 652-657.

Radermacher, R., Zheng, J., Herold, K.E., 1988, "Vapor Compression Heat Pump with Two-Stage Solution Circuit: Proof of Concept Unit", London: Proceedings of Absorption Heat Pumps Workshop, April 12-14, pp. 216-224.

Rahman, M.M., Gui, F., Scaringe, R.P., 1994, "Design/Experimental Performance Evaluation of a Hybrid Chemical/Mechanical Heat Pump", New Orleans: Proceedings of the International Absorption Heat Pump Conference, ASME AES-Vol. 31, January 19-21, pp. 487-492.

Ramshaw, C., Winnington, T., 1988, "An Intensified Absorption Heat Pump (ROTEX)", London: Proceedings of Absorption Heat Pumps Workshop, April 12-14, pp. 258-268.

Rane M.V., Amrane K., Radermacher R., "Performance Enhancement of a Two-stage Vapor Compression Heat Pump with Solution Circuits by Eliminating the Rectifier", Int. J. of Refrig., Vol. 16, No. 4, 1993, pp. 247-257.

Rane, M.V., Radermacher, R., 1991, "Two-Stage Vapor Compression Heat Pump with Solutions Circuits: Performance Enhancement with a Bleed Line", Tokyo: Proceedings of Absorption Heat Pump Conference, September 30-October 2, pp. 97-102.

Rane, M.V., Radermacher, R., 1993, "Feasibility of a Two-Stage Vapor Compression Heat Pump with Ammonia-Water Solution Circuits: Experimental Results", Int. J. Refrig., Vol. 16, No. 4, pp. 258-264.

Sawada, N., Minato, K., Kunugi, Y., Mochizuki, T., Kashiwagi, T., 1994, "Cycle Simulation/COP Evaluation of Compression Hybrid Heat Pump: Heat Amplifier Type", New Orleans: Proceedings of the International Absorption Heat Pump Conference, ASME AES-Vol. 31, January 19-21, pp. 471-476.

Smith, I.E., Swallow, F.E., 1988, "Design Development and Performance Evaluation of a Rotating Disc Vapour Absorber", London: Proceedings of Absorption Heat Pumps Workshop, April 12-14, pp. 373-382.

Torstensson, H., Nowacki, J.E., 1991, "A Sorption/Compression Heat Pump Using Exhaust Air as Heat Source", Tokyo: Proceedings of Absorption Heat Pump Conference, September 30-October 2, pp. 103-108.

Trepp, Ch., Matthys, H., Niederhauser, Th., 1988, "Absorption Heat Pumps and Compression Heat Pumps with Solution Circuit at the ETH Zurich", London: Proceedings of Absorption Heat Pumps Workshop, April 12-14, pp. 173-182.

Sunye, R., Prevost, M., Bugarel, R., 1988, "High Temperature Sorption Cycle for Heat Pumping: Compressor Aided Heat Transformer", London: Proceedings of Absorption Heat Pumps Workshop, April 12-14, pp. 197-203.

Wilkinson W.H., 1990, "Alternative Dublesorb Concepts", ASHRAE Trans., Vol. 96, Pt. 1, AT-90-19-4.

Ziegler, F., Alefeld, G., 1989, "Compression Absorption Heat Pumps for High Temperature Applications", Proc. of the IEA Heat Pump Center Workshop "High Temperature Heat Pumps", November 1989, Hannover, Report No. PC-WR-5, IEA Heat Pump Center, Karlsruhe, pp. 131-145.

Chapter 9

Alefeld, G., Radermacher, R., 1994, Heat Conversion Systems, CRC Press, Boca Raton, FL.

Bassols-Rheinfel, J., 1988, "Experimental/Numeric Optimization of the Heat Exchange Surfaces of a Sorption Heat Pump", London: Proceedings of Absorption Heat Pumps Workshop, April 12-14, pp. 419-427.

Berghmans, J., 1991, "Safety Aspects of Heat Pump Working Fluids", Tokyo: Proceedings of Absorption Heat Pump Conference, September 30-October 2, pp. 31.

Bogart, M.J.P., 1981, Ammonia Absorption Refrigeration in Industrial Processes, Gulf Publishing Co., Houston.

Bogart, M.J.P., July 1982, "Ammonia Absorption Refrigeration", Plant/Operations Progr., Vol. 1, pp. 147-151.

Bojanowski, W., January 1985, "Some Serious Safety Problems Using Ammonia-Operational Experience and Recommendations", Int. J. Refrig., Vol. 8, No. 1, pp. 56-57.

Christensen, R.N., Zhang, C., Garimella, S., 1993, "Parametric Analyses and Optimization of the Performance of an Ammonia/Water Evaporator Dryer System", ASHRAE Trans., Vol. 99, Pt. 1, pp. 1503-1510.

Didion, D., Radermacher, R., November 1984, "Part-Load Performance Characteristics of Residential Absorption Chillers and Heat Pumps", Int. J. Refrig., Vol. 7, pp. 393-398.

Domanski, P.A., Didion, D.A., Doyle, J.P., 1994, "Evaluation of Suction-Line/Liquid-Line Heat Exchange in the Refrigeration Cycle", Int. J. Refrig., Vol. 17, No. 7, pp. 487-493.

Erickson., D.C., 1991, "Isaac Solar Refrigerator", Tokyo: Proceedings of Absorption Heat Pump Conference, September 30-October 2, pp. 237-242.

Fujimaki, S., Kawakami, R., Takei, T., Fujiwara, I., Nakashima, Y., 1994, "Analysis of Technical Tasks for Improving the Safety of Ammonia Absorption Heat Pumps", New Orleans: Proceedings of the International Absorption Heat Pump Conference, ASME AES-Vol. 31, January 19-21, pp. 279-286.

Garimella, S., Christensen, R. N., 1994, "Gas-Fired Absorption Systems for Space Conditioning in Recreational Vehicles", New Orleans: Proceedings of the International Absorption Heat Pump Conference, ASME AES-Vol. 31, January 19-21, pp. 287-294.

Herbine, G.S., Perez-Blanco, H., 1995, "Model of an Ammonia-Water Bubble Absorber", ASHRAE Trans., Vol. 101, Pt. 1.

Howe L.A. , Radermacher R. , Herold K.E. , 1989, "Combined Cycles for Engine-Driven Heat Pumps", Int. J. Refrig., Vol 12, , pp. 21-28.

Inoue, N., Iizuka, H., Ninomiya, Y., Watanabe, K., 1994, "COP Evaluation for Advanced Ammonia-Based Absorption Cycles", New Orleans: Proceedings of the International Absorption Heat Pump Conference, ASME AES-Vol. 31, January 19-21, pp. 1-6.

Jeong, S., Kang, B.H., Lee, C.S., Karng, S.W., 1994, "Computer Simulation on Dynamic Behavior of a Hot Water Driven Absorption Chiller", New Orleans: Proceedings of the International Absorption Heat Pump Conference, ASME AES-Vol. 31, January 19-21, pp. 333-338.

Johnston, A.M., 1980, "Ammonia/Water Absorption Cycles with Relatively High Generator Temperatures", Solar Energy, Vol. 25, pp. 243-254.

Kahn, R., Alefeld, G., Hammerer, S., Pfeifer, R., Tomasek, M., 1994, "An Ammonia-Water Absorption Cycle with High Temperature Lift", New Orleans: Proceedings of the International Absorption Heat Pump Conference, ASME AES-Vol. 31, January 19-21, pp. 93-100.

Lazzarin, R.M., March, 1988, "Commercially Available Absorption Heat Pumps: Some Experimental Tests", Int. J. Refrig., Vol. 11, pp. 96-99.

Machielsen, C.H.M., 1990, "Research Activities on Absorption Systems for heating, Cooling and Industrial Use", ASHRAE Trans., Vol. 96, Pt. 1, AT-90-30-1.

Mardorf, L., 1994, "Controllable Cycle Investigation of Direct Fired Absorption Heat Pump for Residential Heat Systems", New Orleans: Proceedings of the International Absorption Heat Pump Conference, ASME AES-Vol. 31, January 19-21, pp. 339-344.

McLinden, M. O. and Klein, S. A., 1985,"Steady-State Modeling of Absorption Heat Pumps With a Comparison to Experiments," ASHRAE Trans., Vol. 91,Pt. 2B, pp. 1793-1807.

McLinden, M. O. and Klein, S. A., 1983,"Simulation of an Absorption Heat Pump Solar Heating and Cooling System," Solar Energy, Vol. 31, No. 5, pp. 473-482.

McLinden M.O., Radermacher R., 1985, "An Experimental Comparison of Ammonia/Water and Ammonia/Water/Lithium-bromide in an Absorption Heat Pump", ASHRAE Trans., Vol. 91, Pat 2 #HI-85-36 No. 4.

Merrill, T., Setoguchi, T., Perez-Blanco, H., 1994, "Compact Bubble Absorber Design and Analysis", New Orleans: Proceedings of the International Absorption Heat Pump Conference, ASME AES-Vol. 31, January 19-21, pp. 217-224.

Minato, K., Takagi, S., Matsuki, K., 1991, "Anticorrosive Method for Ammonia Absorption Refrigerating Cycle", Tokyo: Proceedings of Absorption Heat Pump Conference, September 30-October 2. pp. 249-254.

Pentz, N.E., Metzger, J.F., Bonar, H., II, Molsbee, L., Polhemus, J., 1983, "A Guide to Good Practices for the Operation of an Ammonia Refrigeration System", International Institute of Ammonia Refrigeration, pp. 1-19.

Perez-Blanco, H., 1988, "A Model of an Ammonia Water Falling Film Absorber", ASHRAE Trans., Vol. 94, Pt. 1, pp. 467-482.

Radermacher, R., Klein, S. A. and D. A. Didion, 1983,"Investigation of the Part-Load Performance of an Absorption Chiller," ASHRAE Trans., Vol. 89, Pt. 1.

Setoguchi, T., Perez-Blanco, H., 1991, "Effect of Additives in Ammonia Absorption in Water in a Stagnant Pool", Tokyo: Proceedings of Absorption Heat Pump Conference, September 30-October 2, pp. 177-182.

Chapter 10

Alefeld, G., 1983, "Double-Effect, Triple-Effect and Quadruple-Effect Absorption Machines", Paris: Proceedings 16th International Congress of Refrigeration, Vol. 2, pp. 951-956.

Alefeld, G., Radermacher, R., 1994, Heat Conversion Systems, CRC Press, Boca Raton, FL.

Cheung, K. , Hwang, Y. , Judge, J.F. , Kolos, K., Singh, A. , Radermacher, R., 1995, "Performance Assessment of Multi-Stage Absorption Cycles", submitted for publication to <u>Int. J. Refrig.</u>

DeVault, R.C., Marsala J., 1990, "Ammonia-Water Triple-Effect Absorption Cycle", <u>ASHRAE Trans.</u>, Vol. 96, Pt. 1, AT-90-27-1.

Ivester, D.N., Shelton, S.V., 1994 "Varying Heat Exchanger Parameters in a Triple-Effect Absorption Cycle", New Orleans: Proceedings International Absorption Heat Pump Conference, January 19-21, pp. 243-250.

Ouimette, M.S., Herold, K.E., 1994, "Performance Modelling of a Triple-Effect Absorption Cycle", New Orleans: Proceedings International Absorption Heat Pump Conference, January 19-21, pp. 233-242.

Pride, R.D., Randall, J.D., Miriam, J.M., 1988, "Thermodynamic Considerations in the Design of High Lift Absorption Heat Pumps", London: Proceedings of Absorption Heat Pumps Workshop, April 12-14, pp. 23-31.

Richter, K.H., 1962, "Multi-Stage Absorption Refrigeration Systems", <u>J. of Refrig.</u>, 1962, September /October pp 105-111.

Chapter 11

Altenkirch, E., 1913, "Reversible Absorptionsmaschinen", <u>Zeitschrift fuer die gesamte Kaelteindustrie</u>, Vol. 20, Januar, pp. 1-8 and 115-119 and 150-161.

Altenkirch, E., Tenckhoff, B., 1914, "Absorptionskaeltemaschine zur kontinuierlichen Erzeugung von Kaelte und Waerme oder acuh von Arbeit.", September 22, German Patent 278,076.

Bassols, J., Schneider, R., Veelken, H., Kuckelkorn, B., Ohrt, D., 1994, "1st Operation Results of a Gas-Fired 250 kW Absorption Heat Pump with Plate-Fin Heat Exchangers", New Orleans: Proceedings of the International Absorption Heat Pump Conference, ASME AES-Vol. 31, January 19-21, pp. 73-78.

Bassols, J., 1988, "Experimental/Numeric Optimization of the Heat Exchange Surfaces of a Sorption Heat Pump", London: Proceedings of Absorption Heat Pumps Workshop, April 12-14, pp. 419-427.

Erickson, D.C., Rane, M.V., 1992,"The GAX Family of Absorption Cycles." <u>IEA Heat Pump Newsletter</u>. Volume 10. No. 4.

Groll, E.A., Radermacher, R., 1994, "Vapor Compression Cycle with Solution Circuit and Desorber/Absorber Heat Exchange", <u>ASHRAE Trans.</u>, Vol. 100, Pt. 1, pp. 73-83.

Murphy, K.P., Phillips, B.A., January 1984, "Development of a Residential Gas Absorption Heat Pump", Int. J. Refrig., Vol. 7, pp. 56-58.

Hanna, W.T., Wilkinson, W.H., Saunders, J.H., Philips, D.B., 1995, "Pinch-Point Analysis: An Aid to Understanding the GAX Absorption Cycle", ASHRAE Trans., Vol. 101, Pt. 1., pp. 1189-1198.

Herold, K.E., He, X., Erickson, D.C., Rane, M.V., 1991, "The Branched GAX Absorption Heat Pump Cycle", Tokyo: Proceedings of Absorption Heat Pump Conference, September 30-October 2, pp. 127-132.

Phillips, B.A., November 1986, "A New Future for Absorption?", ASHRAE J. Vol. 28, pp. 38-42.

Phillips, B.A., 1985, "High Efficiency Absorption Cycles for Residential Heating and Cooling", Society of Automotive Engineers Inc. 859331, 2.229-2.234.

Phillips, B.A., 1990, "Development of a High-Efficiency Gas-Fired Absorption Heat Pump for Residential and Small Commercial Applications, Phase 1 Final Report", ORNL Report, ORNL/Sub/86-24610/1.

Rane, M.V., Erickson, D.C., 1994, "Advanced Absorption Cycle: Vapor Exchange GAX ", New Orleans: Proceedings of the International Absorption Heat Pump Conference, ASME AES-Vol. 31, January 19-21, pp. 25-32.

Scharfe, J., Ziegler, F., Radermacher, R., November 1986, "Analysis of Advantages and Limitations of Absorber-Generator Heat", Int. J. Refrig., Vol. 9, pp. 326-333.

Tae Kang, Y., Christensen, R. N., 1994, "Development of a Counter-Current Model for a Vertical Fluted Tube GAX Absorber", New Orleans: Proceedings of the International Absorption Heat Pump Conference, ASME AES-Vol. 31, January 19-21, pp. 7-16.

Chapter 12

ASHRAE Handbook of Fundamentals, 1989, ASHRAE, New York.

Bird, B.R., Stewart, W.E., Lightfoot, E.N., 1976, Transport Phenomena, John Wiley & Sons, New York.

Bourseau, P., Bugarel, R., July 1986, "Absorption-Diffusion Machines: Comparison of the Performances of NH_3-H_2O and NH_3-NaSCN", Int. J. of Refrig., Vol. 9, pp. 206-214.

Bourseau, P., Mora, J.C., Bugarel R., July 1987, "Coupling of an Absorption-Diffusion Refrigeration Machine and Solar Flat-Plate Collector", Int. J. Refrig., Vol. 10, pp. 209-216.

Buche, W., June 1935, "Influences of Heating on the Liquid Circulation with Thermo-Siphon", Z. Gesamte Kalte-Ind., No. 6, pp. 101-107.

Chen, J., 1995, Further Development of the Diffusion Absorption Refrigerator, Ph.D. Dissertation, University of Maryland.

Chen, J. and K.E. Herold, "Buoyancy Effects on the Mass Transfer in Absorption with a Non-Absorbable Gas", submitted to Int. J. Heat Fluid Flow, January 1995.

Chen, J., K.J. Kim, and K.E. Herold, "Performance Enhancement of a Diffusion-Absorption Refrigerator", submitted to Int. J. Refrig., March 1995.

Eber, N., 1967, "A New Analysis of Rectification in Absorption Refrigeration", Madrid: XIIth International Congress of Refrigeration, Vol. 2, pp. 1339-1351.

Eber, N., 1975, "New Compact Heat Exchanger for Absorption Cooling Units", Moscow: XIVth International Congress of Refrigeration, Vol. 2, No. B2.46, pp. 886-892.

Geppert, H., November 1900, "Progress of Producing Cold", U.S. Patent 662,690.

Gajczak, S., 1959, "Study of a Refrigerating Machine of the Platen-Munters Type", Int. J. Refrig., Vol. 5, No. 3, pp. 115-119.

Hassoon, H.M., Fitt, P.W., 1991, "Mathematical Modeling of a Steam Injected Bubble Pump", ASHRAE Trans., Vol. 97, Pt. 1, pp. 163-171.

Hirschfelder, J.O., Bird, R.B., 1949, "Viscosity and Other Physical Properties of Gases and Gas Mixtures", Trans. ASME, Vol. 71, pp. 921-937.

Kim, J.K., Shi, Z., Chen, J., Herold, K.E., 1995, "Hotel Room Air Conditioner Design Based on the Diffusion-Absorption Cycle", ASHRAE Trans., Vol. 101, Pt. 1, pp. 1290-1301.

Kouremenous, D.A., Stegou-Sagia, A., 1988, "Measuring the Evaporation of NH_3 in Triple-Fluid Gas Absorption Units", Int. J. Refrig., May, Vol. 11, pp. 153-158.

Kouremenos, D.A., Stegou-Sagia, A., September 1988, "Use of Helium Instead of Hydrogen in Inert Gas Absorption Refrigeration", Int. J. Refrig., Vol. 11, pp. 336-341.

Kouremenos, D.A., 1973, "The Thermodynamic Properties of Ammonia-Hydrogen Gas Mixtures; A Temperature Mass Fraction Diagram for Evaporation Purposes", Proc. VIIIth Int. Congr. of Refrig., pp. 437-443.

Kouremenos, D.A., Stegou-Sagia A., May 1987, "Measuring the Evaporation of NH_3 in Triple-Fluid Gas Absorption Units", XVIIth Int. Congr. of Refrig., Vol. 11, pp. 153-158.

Kouremenos, D.A., Stegou-Sagia, A., 1986, "Evaporation of NH_3 in NH_3/H_2 Atmosphere by 25 bar for Neutral Gas Absorption Refrigeration Units", Forsch. Ingenieurwes., Vol. 5, pp. 153-161.

Kouremenos, D.A., Stegou-Sagia, A., September 1988, "Use of Helium Instead of Hydrogen in Inert Gas Absorption Refrigeration", Int. J. of Refrig., Vol. 11, pp. 336-341.

Kouremenos, D.A., Stegou-Sagia, A., Antonopoulos, A.K., 1990, "The Irreversible Evaporation of Ammonia in Vertical Annular Evaporators Using Helium as Inert Gas", ASME AES-Vol. 18, pp. 15-24. `

Kouremenos, D.A., Antonopoulos, K.A., Stegou-Sagia, A., June 1991, "Absorption of Ammonia in Refrigerators with Helium as Inert Gas", Athens, Greece: Analysis of Thermal and Energy System, Proceedings of the International Conference, pp. 701-717.

Landolt, Börnstein, 1968, <u>Zahlenwerte und Funktionen</u>, Vol. 2, Springer-Verlag, Berlin, pp. 54-59.

Leibundgut, H.J., Schamaun, J., Schuepbach, R., Stierlin, H, 1983, "Absorption Type Heat Pump with Inert Gas for Domestic Heating", Paris: XIVth International Congress of Refrigeration, pp. 586-595.

Lucas, P., 1967, "A Boiler for Absorption Units of the Inert Gas Type Having a Wide Range of Input Power", <u>XIIth Int. Congr. of Refrig.</u>, Vol. 2, No. 3.50, pp. 1353-1359.

Mason, E.A., Monchick, L., 1962, "Transport Properties of Polar-Gas Mixtures", <u>J. Chem. Phys.</u>, Vol., 36, pp. 2746-2755.

Miller, C.G., October 1972, "A Program for Calculating Expansion-Tube Flow Quantities for Real Gas Mixtures and Comparison with Experimental Results", NASA Technical Report, NASA, TN-D-6830.

Narayankhedkar, K.G., Maiya, M.P., November 1985, "Investigations on Triple Fluid Vapor Absorption Refrigerator." <u>Int. J. Refrig.</u>, Vol. 8, pp. 335-342.

Niebergall, W., 1981, <u>Sorptions-kältemaschinen</u>, Springer-Verlag, Berlin, pp. 151-152.

Osipov, Y.V., Yakov, N.P., Nekrasov, N.N., July/August 1972, "Heat and Mass Transfer During Absorption of Ammonia by an Aqueous Solution of Ammonia Hydrogen-Ammonia Mixture", <u>Heat Transfer-Sov. Res.</u>, Vol. 4, No. 4, pp. 1-8.

Otto, D., 1967, "Theoretical and Experimental Investigation of Absorption Refrigerating Circuits with Hydrogen as the Conveying Medium", Madrid: XIIth International Congress of Refrigeration, Vol. 2, No. 3.18, pp. 1361-1367.

Von Platen, B.C., Munters, C.G., 1928, "Refrigerator", U.S. Patent 1,685,764.

Rakshit, A.B., Roy, C.S., 1974, "Viscosity and Polar-Nonpolar Interactions in Mixtures of Inert Gases with Ammonia", <u>Physica</u>, 78, pp. 153-164.

Reid, R.C., Prausnitz, J.M., Poling, B.E., 1987, <u>The Properties of Gases and Liquids</u>, 4th Edition, McGraw-Hill, New York.

Sellerio, U., 1951, "Device to Speed the Circulation of Water Solutions in Household Absorption Refrigerators", London: VIIIth International Congress of Refrigeration, Vol. 8, pp. 453-459.

Srivastava, I.B., 1962, "Mutual Diffusion of Binary Mixtures of Ammonia with He, Ne and Xe", Indian J. Phys., Vol., 36, pp. 193-199.

Stierlin, H., Ferguson, J.R., December 1988, "Diffusion Absorption Heat Pump (DAR)", London: Proceedings of Workshop on Absorption Heat Pumps, pp. 247-257.

Stierlin, H., 1967, "Latest Developments in Domestic Absorption Refrigerators and the Future Outlook", Madrid: XIIth International Congress of Refrigeration, Vol. 2, pp. 1323-1337.

Stierlin, H., 1971, "Multiple Hydrogen Circuits in Absorption Deep Freezers", Washington: XIIIth International Congress of Refrigeration, Vol. 2, pp. 526-533.

Stierlin, H, Ferguson, J.R., 1990, "Diffusion Absorption Heat Pump (DAR)", ASHRAE Trans., Vol. 96, Pt. 1, pp. 3319-3328.

Wang, L., 1992, "Simulation of a Diffusion-Absorption Heat Pump Using Helium as Auxiliary Gas", M.S. Thesis, University of Maryland.

Watts, F.G., Gulland, C.K., July/August 1958, "Triple-Fluid Vapor Absorption Refrigerators", J. Refrig., pp. 107-115.

Ziegler, B., Trepp, C., March 1984, "Equation of State for Ammonia-Water Mixtures", Int. J. Refrig., Vol. 7, No. 2, pp. 101-106.

Appendix A

Battelle Memorial Institute, March, 1982 ,"Thermophysical Properties of Aqueous Lithium Bromide Solutions", Sun Oil Program Final Report.

Beattie, J.A., 1928, International Critical Tables, LiBr Data.

Bichowsky, F.R., Rossini, F.D.,1936, "The Thermochemistry of the Chemical Substances LiBr Data Section", Reinhold, New York, pp. 1.

Bogatykh, S.A., Evonvich, I.D.,1963, "A Study of the Viscosities of Aqueous Solutions of LiCl, LiBr and $CaCl_2$ Applicable to the Normal Drying of Gases", Zh. Prikl. Khim., Vol. 36, No. 8, pp. 1867-1868.

Bogatykh, S.A., Evonvich, I.D., 1965, "Investigation of Density of Aqueous LiBr, LiCl and $CaCl_2$ Solutions in Relation to Conditions of Gas Drying", Zh. Prikl. Khim..Vol. 38, No. 4, pp. 945-946.

Boryta, D.A., 1970, "Solubility of Lithium Bromide in Water between -50 and +100°C (40 to 70% LiBr)", J. Chem. Eng. Data, Vol. 15, No. 1, pp. 142-144.

Boryta, D.A., Maas, A.J., Grant, C.B., 1975, "Vapor Pressure-Temperature-Concentration Relationship for System Lithium Bromide and Water (40-70% LiBr)", J. Chem. Eng. Data , Vol. 20, No. 3, pp. 316-319.

Di Guillo, R.M., Lee, R.J., Jeter, S.M., Teja, A.S., 1990, "Properties of Lithium Bromide-Water Solutions at High Temperatures and Concentrations. I. Thermal Conductivity", ASHRAE Trans.., Vol. 96, Pt. 1, pp. 702-708.

Eigen, M., Wicke, E., 1951, "Ionenhydration und spezifische Warme waBriger Elekrolytlosungen", Z. Electrochemie, Vol. 55, No. 5, pp. 354-363.

Ellington, R.T., Kunst, G., Peck, R.E., Reed, J.F., 1957, "The Absorption Cooling Process", IGT Res. Bull., Vol. 14.

Fedorov, M.K., Antonov, N.A., L'vov, S.N., 1976, "Vapor Pressure of Saturated Aqueous Solutions of LiCl LiBr and LiI and Thermodynamic Characteristics of the Solvent in These Systems at Temperatures of 150-350°C and Pressure Up to 1500 Bar", Zh. Prikl. Khim., Vol. 49, No. 6, pp. 1226-1232.

Feuerecker, G., Scharfe, J., Greiter, I., Frank, C., Alefeld, G., 1994, "Measurement of Thermophysical Properties of Aqueous LiBr-Solutions at High Temperature Concentrations", New Orleans: Proceedings of the International Absorption Heat Pump Conference, ASME AES-Vol. 31, January 19-21, pp. 493-500.

Fried, I., Segal, M., 1983, "Electrical Conductivity of Concentrated Lithium Bromide Aqueous Solutions", J. Chem. Eng. Data., Vol. 28, pp. 127-130.

Furukawa, M., Enomoto, E., Sekoguchi, K., 1994, "Boiling Heat Transfer in High Temperature Generator of Absorption Chiller/Heater", New Orleans: Proceedings of the International Absorption Heat Pump Conference, ASME AES-Vol. 31, January 19-21, pp. 517-524.

Haltenberger, W., June 1939, "Enthalpy Concentration Charts from Vapor Pressure Data", Ind. Eng. Chem., pp. 783-786.

ICT, 1928, International Critical Tables-National Research Council, Vol. 3.

Herold, K.E., Moran, M.J., 1987, "A Gibbs Free Energy Expression for Calculating Thermodynamic Properties of Lithium Bromide/Water Solutions," ASHRAE Trans., Vol. 93, Pt. 1, pp. 35-48.

Jauch, K., 1921, "Die Spezifische Warme Wasseriger Salzloungen", Z. Physik., Vol. 4, pp. 441-447 (German).

Jeter, J.M., Moran, J.P., Teja, A.S., 1992, "Properties of Lithium Bromide-Water Solutions at High Temperatures and Concentrations. III. Specific Heat", ASHRAE Trans. Vol. 98, Pt. 1, pp. 137-149.

Kamoshida, J., Isshiki, N., 1994, "Heat Transfer to Water and Water/Lithium Halide Salt Solutions in Nucleate Pool Boiling", New Orleans: Proceedings of the International Absorption Heat Pump Conference, ASME AES-Vol. 31, January 19-21, pp. 501-508.

Lange, E., Schwartz, E., 1928, "Losung-und Verdunnungswamen von Salzen von der aussersten Vednnnung bis zur Sattignng IV. Lithiumbromid", Z. Phy. Chem., Vol. 133, pp. 129-150.

Lee, R.J., DiGuillio, R.M., Jeter, S.M., Teja, A.S., 1990, "Properties of Lithium Bromide-Water Solutions at High Temperatures and Concentrations. II. Density and Viscosity", ASHRAE Trans., Vol. 96, Pt. 1, pp. 709-728.

Lenard, J.L.Y., Jeter, S.M., Teja, A.S., 1992, "Properties of Lithium Bromide-Water Solutions at High Temperatures and Concentrations. IV. Vapor Pressure", ASHRAE Trans., Vol. 98, Pt. 1.

Löwer, H., 1961, Kaltetechnik, Vol. 13, No. 5, pp. 436.

Löwer, H., 1960, Thermodynamishe und physikalische Eigenshaften der wassngen Lithiumbromid-Losung, Ph.D. Dissertation, University of Karlshrue.

Löwer, H., May 1961, "Thermodynamische Eigenschaften Und Warmediagramme DesBinaren Systems Lithiumbromid/Wassen", Kaltetechnik, Vol. 13, Janrganng Heft., pp. 178-184.

Marcus, Y., 1975, "Thermodynamics of Liquid-Liquid Distribution Reactions II. Lithium Bromide-Water-2-Ethyl-1-hexanol System", J. Chem. Eng. Data., Vol. 2, pp. 141-144.

Mashovets, V.P., Baron, N.M., Shcherba, M.U., 1971, "Viscosity and Density of Aqueous LiI LiBr and LiI Solutions at Moderate and Low Temperatures", Zh. Phrikl. Khim, Vol. 44, No. 9, pp. 1981-1986.

McNeely, L.A., 1979, "Thermodynamic Properties of Aqueous Solutions of Lithium Bromide", ASHRAE Trans., Vol. 85, Pt. I, pp. 413-434.

Murakami, K., Sato, H., Watanabe, K., 1994, "Measurement of Bubble-Point Pressures for Binary LiBr/H$_2$O Pair", New Orleans: Proceedings of the International Absorption Heat Pump Conference, ASME AES-Vol. 31, January 19-21, pp. 509-516.

Othmer, D.F., 1940, "Correlating Vapor Pressure and Latent Heat Data - A New Plot", Ind. Eng. Chem., Vol. 32, No. 6, pp. 841-856.

Patterson, M.R., Perez-Blanco, H., 1988, "Numerical Fits of the Properties of Lithium-Bromide Water Solutions", ASHRAE Trans., Vol. 94, Pt. 2, pp. 2059-2077.

Pennington, W., May 1955, "How to Find Accurate Vapor Pressures of LiBr Water Solutions", Refrig. Eng., pp. 57-61.

Patterson, M.R., Crosswhite, R.N., Perez-Blanco, H., January 1990, "A Menu-Driven Program for Determining Properties of Aqueous Lithium Bromide Solutions", ORNL ORNL/TM-1131.

Ramabrahamam, K., Suryanarayana, M., 1973, "Verification of Bachem's Relation at Various Temperatures in Aqueous and Non- Aqueous Solutions of Some Lithium Salts", Ind. J. Pure Appl. Phys., Vol. 11, No. 1, pp. 71-73.

Ramabrahmam, K., Suryanavayana, M., 1973, "Temperature Variation of Ultrasonic Velocity Adib. Comp. and Molar Sound Velocity in Aqueous and Non-Aqueous Solutions of Some Lithium Salts", Ind. J. Pure Appl. Phys., Vol. 11, pp. 99-105.

Robinson, R.A., McCoach, H.J., 1947, "Osmotic and Activity Coefficients of Lithium Bromide and Calcium Bromide Solutions", J. Am. Chem. Soc., Vol. 69, pp. 2244.

Stokes, R. H., 1950, J. Amer. Chem. Soc., Vol. 72, pp. 2243.

Uemura, T., Hasaba, S., December, 1964, "Studies on the Lithium Bromide-Water Absorption Refrigerating Machine", Technical Report of Kansai University, Vol. 6, pp. 31-55.

Uemura, T., 1972, Refrigeration, Vol. 47, pp. 105-108.

Appendix B

Anand, G., Christensen, R.N., Richards, D.E., 1988, "Forced Convection Condensation of Ammonia inside Coiled Fluted Tubes", ASHRAE Trans., Vol. 94, Pt. 2, pp. 1132-1158.

ASHRAE Handbook 1977 Fundamentals, ASHRAE, New York, New York.

Bogart, M.J.P., July 1982, "Ammonia Absorption Refrigeration", Plant/Operations Progress, Vol. 1, Pt 3, pp. 147-151.

Bogart, M., 1980, Ammonia Absorption Refrigeration in Industrial Processes, Gulf Publishing Company.

Bojanowski, W., January 1985, "Some Serious Safety Problems Using Ammonia Operational Experience and Recommendations", Int. J. Refrig., Vol. 9, pp. 56-57.

Bulkley, W.L., Swartz, R.H., July 1951, "Temperature-Pressure-Concentration Chart for Ammonia-Water Solutions", Refrig Eng., Vol. 59, pp. 660-661.

Clifford, I.L., and Hunter, E., 1933,"The Ammonia-Water System at Temperatures up to 150°C and at Pressures up to Twenty Atmospheres", J. Phys. Chem., Vol. 37, pp. 101-118.

Clifford, I.L., Hunter, E., 1933, "The System Ammonia-Water at Temperatures up to 150°C and at Pressures up to 20 Atm.", J. Phys. Chem., Vol. 37, pp. 101-118.

Döring, R., "Thermodynamische Eigenschaften von Ammoniak (R-717)", [(Klima-und Kglte-Ingenieur) Klima- +Ktilte-Ingenieur: Extra; 51 ISBN 3-7880-7093-5.

Dvorak, K., Boublik, T., 1963, "Liquid-Vapour Equilibria. XXIX.   Measurement of Equilibrium Data in Systems with High Equilibrium Ratio of Components", Col. Szech. Chem. Commun., Vol. 28, pp. 1249-1255.

Edwards, T.J., Newman, J., Prausnitz, J.M., March 1975, "Thermodynamics of Aqueous Solutions Containing Volatile Weak Electrolytes", AIChE J., Vol. 21, No. 2, pp. 245-259.

Edwards, T.J., Newman, J., Prausnitz, J.M., 1978, "Thermodynamics of Vapor-Liquid Equilibria for the Ammonia-Water System", Ind. Eng. Chem. Fund., Vol. 17, No. 4, pp. 264-269.

Edwards, T.J., 1977, "Vapor-Liquid Equilibria in Multicomponent Aqueous Solutions of Volatile Electrolytes", Ph.D. Dissertation, University of California, Berkeley, pp. 157.

Edwards, T.J., Maurer, G., Newman, J., Prausnitz, J.M., November 1978, "Vapor-Liquid Equilibria in Multicomponent Aqueous Solutions of Volatile Weak Electrolytes", AIChE J., Vol. 24, No. 6, pp. 966-976.

El-Sayed, Y.M., and Tribus, M., 1985, "Thermodynamic Properties of Water-Ammonia Mixtures: Theoretical Implementation for Use in Power Cycles Analysis", paper presented at 1985 ASME Winter Meeting, Miami Beach, Florida, November 17-22.

Fenton, D.L., Noeth, A.F., Gorton, R.L., 1991, "Absorption of Ammonia into Water", ASHRAE Trans., Vol. 97, Pt 1, pp. 204-213.

Foote, H.W., 1921, "Equilibrium in the System Ammonia: Water: Ammonium Thiocyanate", J. Amer. Chem. Soc., Vol. 43, No. 1, pp. 1031-1038.

Gillespie, P.C., Wilding, W.V., and Wilson, G.M., October 1985, "Vapor Liquid Equilibrium Measurements on the Ammonia Water System from 313 K to 589 K", Research Report RR-90 Gas Producers Association: Wiltec Research Company Inc., Provo, Utah.

Gillespie, P.C., Wilding, W.V., and Wilson, G.M., January 1987, "Vapor Liquid Equilibrium Measurements on the Ammonia Water System from 313 K to 589 K", Research Report RR-90 Gas Producers Association: Wiltec Research Company Inc., Provo, Utah.

Guillevic, J.L., 1983, Doctorate Thesis at Ecole Nationale Supdrieure des Mines de Paris, "Determination de Donn'eesd'Equilibre Complètes du Système Eau-Ammoniac par une Nouvelle Méthode Statique" (French text).

Gupta, C.P., Sharma, C.P., 1975, "Entropy and Availability Function Values of Saturated Ammonia-Water Mixtures", A.S.M.E. Reprint 75-WA/PID.

Haar, L., Gallagher, J.S., 1978, " Thermodynamic Properties of Ammonia", J. Phys. Chem. Ref. Data, Vol. 7, No. 3, pp. 635-792.

Herold, K.E., Han, K., Moran, M.J., 1988, "AMMWAT: A Computer Program for Calculating the Thermodynamic Properties of Ammonia and Water Mixture Using a Gibbs Free Energy Formulation", Proc. ASME Winter Annu. Meet., AES-Vol. 4, pp. 65-75.

Ibrahim, O.M., Klein, S.A., 1993, "Thermodynamic Properties of Ammonia Water Mixtures", ASHRAE Trans., Vol. 99, Pt 1, pp. 1495-1502.

IIF-IIR, 1981, "Tables et diagrammes pour l'industrie du froid - Tables and diagrams for the Refrigeration Industry," BRO R717, 31 pages.

IIR, 1994, "Thermodynamic and Physical Properties of NH_3 - H_2O", International Institute of Refrigeration, 177 Boulevard Malesherbes, 75017 Paris, France, Tel: 42 27 32 35, FAX: 47 63 17 98, ISBN: 2-903633-66-5.

Jain, P.C., Gable, G.K., 1971, "Equilibrium Property Data Equations for Aqua-Ammonia Mixtures", ASHRAE Trans., Vol. 77, Pt. 1, pp. 149-151.

Jennings, B.H., 1965, "Ammonia-Water Properties (Experimentally-Determined P, V, T, x Liquid Phase Data)", Paper presented at ASHRAE Semi-annual Meeting in Chicago, January 25-28.

Kuzman, Raznjevic, Handbook of Thermodynamic Tables and Charts, McGraw Hill.

Macriss, R.A., Eakin, B.E., Ellington, R.T., and Huebler, J., 1964, "Physical and Thermodynamic Properties of Ammonia-Water Mixtures", Inst. Gas Tech., Chicago, Res. Bull., No. 34.

Jennings, B.H., 1965, "Ammonia-Water Properties (Experimentally determined PVTX Liquid-Phase Data)", ASHRAE Trans., Vol. 71, Pt. 1, pp. 21-29.

Jennings, B.H., Shannon, F.P., May 1938, "The Thermodynamics of Absorption Refrigeration", J. A.S.R.E. (Refrig. Eng.), pp. 333-336.

Johnson, E.F., Jr., Molstad, M.C., 1951, "Thermodynamic Properties of Aqueous Lithium Chloride Solutions", J. Phys. Colloidal Chem., Vol. 55, pp. 257-281.

Jones, M.E., 1963, "Ammonia Equilibrium Between Vapor and Liquid Aqueous Phases at Elevated Temperatures", J. Phys. Chem., Vol. 67, pp. 1113-1115.

Kouremenos, D.A., Stegou-Sagia, A., Rogdakis, E., 1987, "The Excess Enthalpy and Volume of the NH_3/H_2O Vapor Mixture", Vienna, Austria: XVIIth International Congress of Refrigeration, Vol. B, pp. 203-210.

Kracek, F.C., 1930, "Vapor Pressures of Solutions and the Ramsay-Young Rule. Application to the Complete System Water-Ammonia", J. Phys. Chem., Vol. 34, pp. 499-521.

Kumagai, A., Date, K., Iwasaki, H., 1976 "Tait Equation for Liquid Ammonia", J. Chem. Eng. Data, Vol. 21, No. 2, pp. 226-227.

Macriss, R.A., Eakin, B.E., 1964, "Thermodynamic Properties of Ammonia-Water Solutions Extended to Higher Temperature and Pressure", ASHRAE Trans., Vol. 70, pp. 319-327.

Macriss, R.A., Eakin, B.E., Ellington, R.T., Huebler, J., September 1964, "Physical and Thermodynamic Properties of Ammonia-Water Mixtures", AGA & IGT Res. Bull., No. 34.

Mittach, A., Kuss, E. Schlueter, H., 1926, "Dichten und Dampfdrucke von waBrigen Ammoniaklosugen und von flussigem Stickstofftetroxyd fur das Temperaturgebiet Obisbo", Z. Anorg. Allg. Chem., Vol. 159, pp. 1-36.

Morgan, O.M., Maass, O., 1931, "An Investigation of the Equilibria Existing in Gas-Water Systems Forming Electrolytes", Can. J. Res., Vol. 5, pp. 162-199.

Neuhausen, B.S., Patrick, W.A., 1921, "A Study of the System Ammonia-Water as a Basis for a Theory of the Solution of Gases in Liquids", J. Phys. Chem., Vol. 25, pp. 693-720.

Park, Y.M., Sonntag, R.E., 1990, "Thermodynamic Properties of Ammonia Water Mixtures: A Generalized Equation of State Approach", ASHRAE Trans., Vol. 96, Pt. 1, pp. 150-159.

Perman, E. P., 1901, "Vapor Pressure of Aqueous Ammonia Solution. 1., J. Chem. Soc. Vol. 79, pp. 718-725.

Perman, E.P., 1903, "Vapour Pressure of Aqueous Ammonia Solution. II.", J. Chem. Soc., Vol. 83, pp. 1168-1183.

Pierre, B., August 1959, "Total Vapour Pressure in Bars Over Ammonia-Water Solutions", Kyltek, Tidsk, Sheet 14 (English text).

Polak, J., and Lu, B.C.Y., 1975, "Vapor-Liquid Equilibrium in System Ammonia-Water at 14.69 and 65 psia", J. Chem. Eng. Data, Vol. 20, No. 2.

Redlich, 0., and Kwong, J.N.S., 1949, "An Equation of State: Fugacities of Gaseous Solutions", Chem. Rev., Vol. 44, pp. 233.

Redlich, 0., and Kister, A.T., February 1948, "Algebraic Representation of Thermodynamic Properties and the Classification of Solutions", Ind. Eng. Chem., Vol. 40, No. 2, pp. 345-348.

Rizvi S.S.H., 1985, Ammonia-Water Equilibrium, Ph.D. Thesis, University of Calgary, Calgary, Alberta.

Scatchard, G., Epstein, L.F., Warburton, J., Cody, P.J., May 1947, "Thermodynamic Properties of Saturated Liquid and Vapor of Ammonia Water Mixtures ", J. A.S.R.E. (Refrig. Eng.), pp. 413-452.

Schulz, S.C.G., 1973, "Equations of State for the System Ammonia-Water for Use with Computers", Proc. of 13th Int. Congr. Refrig.-1971, pp. 431-436.

Sun, S.B.K., Storvick, T.S., 1979, "Viscosity of Ammonia at High Temperature and Pressure", J. Chem. Eng. Data, 24, Vol. 2, pp. 88-91.

Tsiklis D.S., Linshits L.R. and Goryunova N.P., "Phase Equilibria in the Ammonia-Water System", Russian J. Phys. Chem. USSR, Vol. 39, pp. 1590.

Wilson, T.A., February 1925, "The Total and Partial Vapor Pressures of Aqueous Ammonia Solutions", Bulletin No. 146 of the University of Illinois Engineering Experimental Station, Urbana, Illinois, University of Illinois.

Won, K.W., Selled, F.T., and Walker, C.K., 1980, "Vapor-Liquid Equilibria of the Ammonia-Water System", Paper presented at the 11 International Symposium on Phase Equilibrium and Fluid Properties in the Chemical Industry, Berlin, March 17-21.

Wucherer, J., 1932, "Measurements of Pressure, Temperature and Composition of Liquids and Vaporous Phase of Ammonia-Water Mixtures at the Saturation Point", 2. Gesamt. Kalte-ind. Vol. 39, pp. 97-104, 136-140, (German text).

Zander, M., Thomas, W., 1979, " Some Thermodynamic Properties of Liquid NH_3: PVT Data Vapor Pressure and Critical Temperature", J. Chem. Eng. Data , Vol. 24, No. 1, pp. 1-2.

Ziegler, B., Trepp, Ch., March 1984, "Equation of State for Ammonia-Water Mixtures", Int. J. Refrig. Vol. 7, No. 2, pp. 101-106.

Appendix C

Abrahamsson, K., Jernqvist, A., 1994, "Modelling and Simulation of Absorption Heat Cycles", New Orleans: Proceedings of the International Absorption Heat Pump Conference, ASME AES-Vol. 31, January 19-21, pp. 361-368.

Bedard, G.S., 1994, "Varied Flow Rate Applications for Commercial Absorption Liquid Chillers", New Orleans: Proceedings of the International Absorption Heat Pump Conference, ASME AES-Vol. 31, January 19-21, pp. 133-140.

Homma, R., Nishiyama, N., Wakimizu, H., 1994, "Simulation and Research of Single-Effect Absorption Refrigerators Driven by Waste Hot Water", New Orleans: Proceedings of the International Absorption Heat Pump Conference, ASME AES-Vol. 31, January 19-21, pp. 273-278.

Jeong, S., Kang, B.H., Lee, C.S., Karng, S.W., 1994, "Computer Simulation on Dynamic Behavior of a Hot Water Driven Absorption Chiller", New Orleans: Proceedings of the International Absorption Heat Pump Conference, ASME AES-Vol. 31, January 19-21, pp. 333-338.

Oh, M.D., Kim, S.C., Kim, Y.L., Kim, Y.I., 1994, "Cycle Analysis of Air-Cooled Double-Effect Absorption Heat Pump with Parallel Flow Type", New Orleans: Proceedings of the International Absorption Heat Pump Conference, ASME AES-Vol. 31, January 19-21, pp. 117-124.

Patnaik, V., Perez-Blanco, H., 1994, "A Counterflow Heat-Exchanger Analysis for the Design of Falling-Film Absorbers", New Orleans: Proceedings of the International Absorption Heat Pump Conference, ASME AES-Vol. 31, January 19-21, pp. 209-216.

Pierucci, S., Sogaro, A., Galli, S., Ciancia, A., 1988, "Advanced Simulation of Heat Pump Schemes", London: Proceedings of Absorption Heat Pumps Workshop, April 12-14, pp. 43-54.

Rose, D.T., Zuritz, C.A., Perez-Blanco, H., 1994, "Thermodynamic Analysis/Pilot Plant Design for a Solar Assisted Double-Effect Refrigeration Cycle", New Orleans: Proceedings of the International Absorption Heat Pump Conference, ASME AES-Vol. 31, January 19-21, pp. 109-116.

Zhu, R., Yu, Y., 1991, "Cycle Simulation of a Liquid Absorption System for Air Conditioning", Tokyo: Proceedings of Absorption Heat Pump Conference, September 30-October 2, pp. 139-144.

Appendix D

Becker, R.S., Wentworth, W.E., 1972, General Chemistry, Houghton Mifflin, pp. 737.

Skoog, D.A., West, D.M., 1979, Analytical Chemistry, 3rd edition, Holt, Rinehart & Winston, pp. 190-193.

Appendix F

Grossman, G., Wilk, M., 1994, "Advanced Modular Simulation of Absorption Systems", Int. J. Refrig., Vol. 17, pp. 231-244.

Grossman, G., 1994, "Modular and Flexible Simulation of Advanced Absorption Systems", New Orleans: Proceedings of the International Absorption Heat Pump Conference, ASME AES-Vol. 31, January 19-21, pp. 345-352.

Grossman, G., DeVault, R.C., Creswick, F.A., 1995, "Simulation and Performance Analysis of an Ammonia-Water Absorption Heat Pump Based on the Generator-Absorber Heat Exchange (GAX) Cycle", ASHRAE Trans., Vol. 101, Pt. 1., pp. 1313-1323.

McGahey, K.R., Garimella, S., Cook, F.B., Christensen, R.N., 1994, "Enhancement of the ORNL Absorption Model/Simulation of Double-Effect Absorption Heat Pump", New Orleans: Proceedings of the International Absorption Heat Pump Conference, ASME AES-Vol. 31, January 19-21, pp. 141-148.

INDEX

Lightning Source UK Ltd.
Milton Keynes UK
09 February 2011

167230UK00009B/2/A